# Traffic

## Why We Drive the Way We Do
## (and What It Says About Us)

Tom Vanderbilt

PENGUIN BOOKS

PENGUIN BOOKS

Published by the Penguin Group
Penguin Books Ltd, 80 Strand, London WC2R 0RL, England
Penguin Group (USA) Inc., 375 Hudson Street, New York, New York 10014, USA
Penguin Group (Canada), 90 Eglinton Avenue East, Suite 700, Toronto, Ontario, Canada M4P 2Y3
(a division of Pearson Penguin Canada Inc.)
Penguin Ireland, 25 St Stephen's Green, Dublin 2, Ireland (a division of Penguin Books Ltd)
Penguin Group (Australia), 250 Camberwell Road, Camberwell, Victoria 3124, Australia
(a division of Pearson Australia Group Pty Ltd)
Penguin Books India Pvt Ltd, 11 Community Centre, Panchsheel Park, New Delhi – 110 017, India
Penguin Group (NZ), 67 Apollo Drive, Rosedale, North Shore 0632, New Zealand
(a division of Pearson New Zealand Ltd)
Penguin Books (South Africa) (Pty) Ltd, 24 Sturdee Avenue, Rosebank,
Johannesburg 2196, South Africa

Penguin Books Ltd, Registered Offices: 80 Strand, London WC2R 0RL, England

www.penguin.com

First published in the United States of America by Alfred A. Knopf 2008
First published in Great Britain by Allen Lane 2008
Published in Penguin Books 2009





Printed in England by Clays Ltd, St [illegible]



978–0–141–02739–5

www.greenpenguin.co.uk

Penguin Books is committed to a sustainable future
for our business, our readers and our planet.
The book in your hands is made from paper
certified by the Forest Stewardship Council.

PENGUIN BOOKS

# TRAFFIC

Tom Vanderbilt writes on design, technology, science and culture for many publications, including *Wired*, *Slate*, the *London Review of Books*, *Gourmet*, the *Wall Street Journal*, *Artforum*, *Travel and Leisure*, *Rolling Stone*, *The New York Times Magazine*, *Cabinet*, *Metropolis* and *Popular Science*. He is contributing editor to the award-winning design magazines *I.D.* and *Print*, contributing editor to *Business Week Online*, and contributing writer of the popular blog *Design Observer*. He is the author of two previous books: *Survival City: Adventures Among the Ruins of Atomic America* and *The Sneaker Book*.

# Contents

## Chapter Five

## Why Women Cause More Congestion Than Men (and Other Secrets of Traffic)



## Chapter Six

## Why More Roads Lead to More Traffic (and What to Do About It)



## Chapter Seven

## When Dangerous Roads Are Safer

# Traffic

## Prologue

# Why I Became a Late Merger (and Why You Should Too)

Why does the other lane always seem to be moving faster?

It is a question you have no doubt asked yourself while crawling down some choked highway, watching with mounting frustration as the adjacent cars glide ahead. You drum the wheel with your fingers. You change the radio station. You fixate on one car as a benchmark of your own lack of progress. You try to figure out what that weird button next to the rear-window defroster actually does.

I used to think this was just part of the natural randomness of the highway. Sometimes fate would steer me into the faster lane, sometimes it would relinquish me to the slow lane.

That was until recently, when I had an experience that made me rethink my traditionally passive outlook on the road, and upset the careful set of assumptions that had always guided my behavior in traffic.

I made a major lifestyle change. I became a *late merger.*

Chances are, at some point you have found yourself driving along the highway when a sign announces that the left lane, in which you are traveling, will close one mile ahead, and that you must merge right.

You notice an opening in the right lane and quickly move over. You breathe a sigh, happy to be safely ensconced in the Lane That Will Not End. Then, as the lane creeps to a slow halt, you notice with rising indignation that cars in the lane you have vacated are continuing to speed ahead, out of sight. You quietly seethe and contemplate returning to the much faster left lane—if only you could work an opening. You grimly accept your condition.

One day, not long ago, I had an epiphany on a New Jersey highway. I was having a typical white-knuckle drive among the scenic oil-storage depots and chemical-processing plants of northern Jersey when suddenly, on the approach to the Pulaski Skyway, the sign loomed: LANE ENDS ONE MILE. MERGE RIGHT.

Seized by some rash impulse, I avoided the instinctual tickle at the back of my brain telling me to get in the already crowded right lane. *Just do what the sign says,* that voice usually counsels. Instead, I listened to another, more insistent voice: *Don't be a sucker. You can do better.* I plowed purposefully ahead, oblivious to the hostile stares of other drivers. From the corner of my eye I could see my wife cringing. After passing dozens of cars, I made it to the bottleneck point, where, filled with newfound swagger, I took my rightful turn in the small alternating "zipper" merge that had formed. I merged, and it was clear asphalt ahead. My heart was beating faster. My wife covered her face with her hands.

In the days after, a creeping guilt and confusion took hold. Was I wrong to have done this? Or had I been doing it wrong all my life? Looking for an answer, I posted an anonymous inquiry on Ask MetaFilter, a Web site one can visit to ask random questions and tap into the "hive mind" of an anonymous audience of overeducated and overopinionated geeks. Why should one lane move faster than the other, I wanted to know, and why are people rewarded for merging at the last possible moment? And was my new lifestyle, that of the late merger, somehow deviant?

I was startled by the torrent of responses, and how quickly they came. What struck me most was the passion and conviction with which people argued their various cases—and the fact that while many people seemed to think I was wrong, almost as many seemed to think I was right. Rather than easy consensus, I had stumbled into a gaping divide of irreconcilable belief.

The first camp—let us name it after the bumper sticker that says PRACTICE RANDOM ACTS OF KINDNESS—viewed early mergers as virtuous souls doing the right thing and late mergers as arrogant louts. "Unfortunately, people suck," wrote one Random Acts poster. "They'll try whatever they can to pass you, to better enjoy the traffic jam from a few car lengths ahead of you. . . . People who feel that they have more pressing concerns and are generally more important than you will keep going, and some weak-spined schmuck will let them in further down, slowing your progress even more. This sucks; I'm afraid it's the way of the world."

Another camp, the minority camp—let's call them Live Free or Die, after the license-plate motto of the state of New Hampshire—argued that the late mergers were quite rationally utilizing the highway's maximum capacity, thus making life better for everyone. In their view, the other group's attempts toward politeness and fairness were actually detrimental to all.

It got more complicated. Some argued that late merges caused more accidents. Some said the system worked much better in Germany, and hinted that my dilemma perhaps revealed some national failing in the American character. Some said they were afraid of not being "let in" at the last moment; some said they would actively try to block someone from merging, the way truckers often do.

So what was going on here? Are we not all driving the same road, did we not all pass the same driving tests? What was puzzling was not just the variety of responses but the sense of moral righteousness each person attributed to his or her highway behavior, and the vitriol each person reserved for those holding the opposite view. For the most part, people were not citing traffic laws or actual evidence but their own personal sense of what was right.

I even found someone claiming to have had a conversion experience exactly the opposite of mine. "Until very recently, I was a 'late merger,' " wrote the author, an executive with a software company, in a business magazine. Why had he become a born-again early merger? "Because I came to realize that traffic flowed faster the sooner people merged." He used this as a metaphor for successful team building in corporate America, in which "late mergers" were those who consistently put their own opinions and motives above the greater company. "Early mergers," he wrote, could help push companies to their "maximum communal speed."

But did traffic flow faster when people merged sooner? Or did it just seem more noble to think that it did?

. . .

You may suspect that getting people to merge in a timely fashion, and without killing one another, is less of a traffic problem and more of a *human* problem. The road, more than simply a system of regulations and designs, is a place where many millions of us, with only loose parameters for how to behave, are thrown together daily in a kind of massive petri dish in which all kinds of uncharted, little-understood dynamics are at work. There is no other place where so many people from different walks of life—different ages, races, classes, religions, genders, political preferences, lifestyle choices, levels of psychological stability—mingle so freely.

What do we really know about how it all works? Why do we act the way we do on the road, and what might that say about us? Are certain people predisposed to drive certain ways? Do women behave differently than men? And if, as conventional wisdom has it, drivers have become progressively less civil over the past several decades, why is that so? Is the road a microcosm of society, or its own place with its own set of rules? I have a friend, an otherwise timorous Latin teacher, who once told me how, in a modest Toyota Corolla, he had defiantly "stuck it" to the driver of an eighteen-wheeler who he felt was hogging the road. Some mysterious force had turned this gentle suburban scholar into the Travis Bickle of the turnpike. (Are you tailgatin' me?) Was it traffic, or had the beast always been lurking within?

The more you think about it—or, rather, the more time you spend in traffic with time to think about it—the more these sorts of puzzling questions swim to the surface. Why can one sit in traffic jams that seem to have no source? Why does a ten-minute "incident" create one hundred minutes of gridlock? Do people really take longer to vacate a parking spot when someone else is waiting, or does it just seem so? Do the car-pool lanes on highways help fight congestion or cause more of it? Just how dangerous are large trucks? How does what we drive, where we drive, and with whom we drive affect the *way* we drive? Why do so many New Yorkers jaywalk, while hardly anyone in Copenhagen does? Is New Delhi's traffic as chaotic as it seems, or does a beautiful order lurk beneath the frenzied surface?

Like me, you may have wondered: What could traffic tell us, if someone would just stop to listen?

The first thing you hear is the word itself. *Traffic*. What did you think of when you read that word? In all likelihood you pictured a crowded highway, filled with people obstructing your progress. It was not a pleasant thought. This is interesting, because for most of its long life the word *traffic* has had rather positive connotations. It originally referred (and still does) to trade and the movement of goods. That meaning slowly expanded to include the people engaging in that trade and the dealings among people themselves—Shakespeare's prologue to *Romeo and Juliet* describes the "traffic of our stage." It then came to signify the movement itself, as in the "traffic on this road." At some point, people and things became interchangeable. The movement of goods and people were intertwined in a single enterprise; after all, if one was going somewhere, it was most likely in pursuit of commerce. This is still true today, as most traffic problems occur during the times we are all going to work, but we seem less likely to think of traffic in terms of motion and mobility, as a great river of opportunity, than as something that makes our lives miserable.

Now, like then, we think of traffic as an abstraction, a grouping of things rather than a collection of individuals. We talk about "beating the traffic" or "getting stuck in traffic," but we never talk—in polite company, at least—about "beating people" or "getting stuck in people." The news lumps together "traffic and weather" as if they were both passive forces largely outside our control, even though whenever we complain about it, we do so because we're part of the traffic. (To be fair, I suppose we are now part of the weather as well, thanks to the atmospheric emissions of that same driving.) We say there is "too much traffic" without exactly knowing what we mean. Are we saying there are too many people? Or that there are not enough roads for the people who are there? Or that there is too much affluence, which has enabled too many people to own cars?

One routinely hears of "traffic problems." But what is a traffic problem? To a traffic engineer, a "traffic problem" might mean that a street is running below capacity. For a parent living on that street, the "traffic problem" could be too many cars, or cars going too fast. For the store owner on that same street, a "traffic problem" might mean there is not enough traffic. Blaise Pascal, the renowned seventeenth-century French scientist and philosopher, had perhaps the only foolproof remedy for traffic: Stay home. "I have discovered that all the unhappiness of men arises from one single fact," he wrote. "That they cannot stay quietly in their own chamber." Pascal, as it happens, is credited with inventing history's first urban bus service. He died a mere five months later. Was Parisian traffic his undoing?

Whatever "traffic problem" means to you, it may give you some comfort to know that traffic problems of all variety are as old as traffic itself. Ever since humans began to propel themselves artificially, society has struggled to catch up with the implications of mobility, to sort out technical and social responses to the new demands.

Visitors to the ruins of Pompeii, for example, will see rutted streets marked by the tracks of chariot wheels. But many are wide enough for only one set of wheels. The tourist wonders: Was it a one-way street? Did a lowly commoner have to reverse himself out of the way when a member of the imperial legions came trotting along in the other direction? If two chariots arrived at an intersection simultaneously, who went first? These questions were neglected for years, but recent work by the American traffic archaeologist Eric Poehler has provided some answers.

By studying the wear patterns on curbstones at corners, as well as the stepping stones set up for pedestrians to cross the "rutways," Poehler was able to discern not just the direction of traffic but the direction of turns onto two-way streets at intersections. It seems, based on the "directionally diagnostic wear patterns" on the curbstones, that Pompeii drivers drove on the right side of the street (part of a larger cultural preference for right-handed activities), used primarily a system of one-way streets, and were banned from driving on certain streets altogether. There seemed to be no traffic signs or street signs. It may please the reader to know, however, that Pompeii did suffer from its share of road construction and detours (as when the building of baths forced the reversal of the Vico di Mercurio).

In ancient Rome, the chariot traffic grew so intense that Caesar, the self-proclaimed *curator viarum*, or "director of the great roads," declared a daytime ban on carts and chariots, "except to transport construction materials for the temples of the gods or for other great public works or to take away demolition materials." Carts could enter the city only after three p.m. And yet, as one so often finds in the world of traffic, there is very rarely an action without an equal and opposite reaction. By making it easier for the average Roman to move around during the day, Caesar made it harder for them to sleep at night. The poet Juvenal, sounding like a second-century version of a contemporary Roman complaining about scooter traffic, lamented, "Only if one has a lot of money can one sleep in Rome. The source of the problem lies in the carts passing through the bottlenecks of the curved streets, and the flocks that stop and make so much noise they would prevent . . . even a devil-fish from sleeping."

By the time we get to medieval England, we can see that traffic was still a problem in search of a solution. Towns tried to limit, through laws or tolls, where and when traveling merchants could sell things. Magistrates restricted the entry of "shod carts" into towns because they damaged bridges and roads. In one town, horses were forbidden to drink at the river, as children were often found playing nearby. Speeding became a social problem. The *Liber Albus,* the rule book of fifteenth-century London, forbade a driver to "drive his cart more quickly when it is unloaded than when it is loaded" (if he did, he would be looking at a forty-pence speeding ticket or, more drastically, "having his body committed to prison at the will of the Mayor").

In 1720, traffic fatalities from "furiously driven" carts and coaches were named the leading cause of death in London (eclipsing fire and "immoderate quaffing"), while commentators decried the "Controversies, Quarreling, and Disturbances" caused by drivers "contesting for the way." Meanwhile, in the New York of 1867, horses were killing an average of four pedestrians a week (a bit higher than today's rate of traffic fatalities, although there were far fewer people and far fewer vehicles). Spooked runaways trampled pedestrians underfoot, "reckless drivers" paid little heed to the 5-mile-an-hour speed limit, and there was little concept of right-of-way. "As matters now stand," the *New York Times* wrote in 1888, "drivers seem to be legally justified in ignoring crossings and causing [pedestrians] to run or dodge over vehicles when they wish to pass over."

The larger the cities grew, and the more ways people devised to get around those cities, the more complicated traffic became, and the more difficult to manage. Take, for instance, the scene that occurred on lower Broadway in New York City on the afternoon of December 23, 1879, an "extraordinary and unprecedented blockade of traffic" that lasted five hours. Who was in this "nondescript jam," as the *New York Times* called it? The list is staggering: "single and double teams, double teams with a tandem leader, and four-horse teams; hacks, coupes, trucks, drays, butcher carts, passenger stages, express wagons, grocers' and hucksters' wagons, two-wheeled 'dog carts,' furniture carts and piano trucks, and jewelers' and fancy goods dealers' light delivery wagons, and two or three advertising vans, with flimsy transparent canvas sides to show illumination at night."

Just when it seemed as if things could not get more complicated on the road, along came a novel and controversial machine, the first new form of personal transportation since the days of Caesar's Rome, a newfangled contrivance that upset the fragile balance of traffic. I am talking, of course, about the bicycle.

After a couple of false starts, the "bicycle boom" of the late nineteenth century created a social furor. Bicycles were too fast. They threatened their riders with strange ailments, like *kyphosis bicyclistarum*, or "bicycle stoop." They spooked horses and caused accidents. Fisticuffs were exchanged between cyclists and noncyclists. Cities tried to ban them outright. They were restricted from streets because they were not coaches, and restricted from sidewalks because they were not pedestrians. The bicycle activists of today who argue that cars should not be allowed in places like Brooklyn's Prospect Park were preceded, over a hundred years ago, by "wheelmen" fighting for the right for bicycles to be allowed in that same park. New bicycle etiquette questions were broached: Should men yield the right-of-way to women?

There is a pattern here, from the chariot in Pompeii to the Segway in Seattle. Once humans decided to do anything but walk, once they became "traffic," they had to learn a whole new way of getting around and getting along. What is the road for? Who is the road for? How will these streams of traffic flow together? Before the dust kicked up by the bicycles had even settled, the whole order was toppled again by the automobile, which was beginning to careen down those same "good roads" the cyclists themselves, in a bit of tragic irony, had helped create.

When driving began, it was like a juggernaut, and we have rarely had time to pause and reflect upon the new kind of life that was being made. When the first electric car debuted in mid-nineteenth-century England, the speed limit was hastily set at 4 miles per hour—the speed at which a man carrying a red flag could run ahead of a car entering a town, an event that was still a quite rare occurrence. That man with the red flag racing the car was like a metaphor of traffic itself. It was probably also the last time the automobile existed at anything like human speed or scale. The car was soon to create a world of its own, a world in which humans, separated from everything outside the car but still somehow connected, would move at speeds beyond anything for which their evolutionary history had prepared them.

At first, cars simply joined the chaotic traffic already in the street, where the only real rule of the road in most North American cities was "keep to the right." In 1902, William Phelps Eno, a "well-known yachtsman, clubman, and Yale graduate" who would become known as "the first traffic technician of the whole world," set about untangling the strangling miasma that was New York City's streets. (Deaths by automobile were already, according to the *New York Times*, "every-day occurrences" with little "news value" unless they involved persons of "exceptional social or business prominence.") Eno was every bit the WASP patrician as social reformer, a familiar character then in New York. He thundered at "the stupidity of drivers, pedestrians and police" and bluntly wielded his favorite maxim: "It is easy to control a trained army but next to impossible to regulate a mob." Eno proposed a series of "radical ordinances" to rein in New York's traffic, a plan that seems hopelessly quaint now, with its instructions on the "right way to turn a corner" and its audacious demands that cars go in only *one* direction around Columbus Circle. But Eno, who became a global celebrity of sorts, boating off to Paris and São Paulo to solve local traffic problems, was as much a social engineer as a traffic engineer, teaching vast numbers of people to act and communicate in new ways, often against their will.

In the beginning this language was more Tower of Babel than Esperanto. In one town, the blast of a policeman's whistle might mean stop, in another go. A red light indicated one thing here, another thing there. The first stop signs were yellow, even though many people thought they should be red. As one traffic engineer summed up early-twentieth-century traffic control, "there was a great wave of arrow lenses, purple lenses, lenses with crosses, etc., all giving special instructions to the motorist, who, as a rule, hadn't the faintest idea of what these special indications meant." The systems we take for granted today required years of evolution, and were often steeped in controversy. The first traffic lights had two indications, one for stop and one for go. Then someone proposed a third light, today's "amber phase," so cars would have time to clear the intersection. Some engineers resisted this, on the grounds that vehicles were "amber rushing," or trying to beat the light, which actually made things more dangerous. Others wanted the yellow light shown before the signal was changing to red *and* before it was changing from red back to green (which one sees today in Denmark, among other places, but nowhere in North America). There were strange regional one-offs that never caught on; for example, a signal at the corner of Wilshire and Western in Los Angeles had a small clock whose hand revealed to the approaching driver how much "green" or "red" time remained.

Were red and green even the right colors? In 1923 it was pointed out that approximately one in ten people saw only gray when looking at a traffic signal, because of color blindness. Might not blue and yellow, which almost everyone could see, be better? Or would that create catastrophic confusion among all those who had already learned red and green? Despite all the uncertainty, traffic engineering soon hoisted itself onto a wobbly pedestal of authority, even if, as the transportation historian Jeffrey Brown argues, engineers' neutral-sounding Progressive scientific ideology, which compared "curing" congestion to fighting typhoid, reflected the desires of a narrow band of urban elites (i.e., car owners). Thus it was quickly established that the prime objective of a street was simply to move as many cars as quickly as possible—an idea that obscured, as it does to this day, the many other roles of city streets.

After more than a century of tinkering with traffic, plus years of tradition and scientific research, one would think all these issues would have been smoothed out. And they have been, largely. We drive in a landscape that looks virtually the same wherever we go: A red light in Morocco means the same thing as it does in Montana. A walk "man" that moves us across a street in Berlin does the same in Boston, even if the "man" looks a bit different. (The beloved jaunty, hat-clad *Ampelmännchen* of the former German Democratic Republic has survived the collapse of the Berlin Wall.) We drive on highways that have been so perfectly engineered we forget we are moving at high speeds—indeed, we are sometimes barely aware of moving at all.

For all this standardized sameness, though, there is much that is still simply not known about how to manage the flows of all those people in traffic—drivers, walkers, cyclists, and others—in the safest and most efficient manner. For example, you may have seen, in some cities, a "countdown signal" that indicates, in seconds, exactly how much time you have before the "Walk" signal will change to "Don't Walk." Some people in the traffic world think this innovation has made things better for pedestrians, but it is just as easy to find others who think it offers no improvement at all. Some people think that marked bicycle lanes on streets are the ideal for cyclists, while others prefer separated lanes; still others suggest that maybe having no bicycle lanes at all would be best for bike riders. For a time it was thought that highway traffic would flow better and more safely if trucks were forced to obey a slower speed limit than cars. But "differential speed limits" just seemed to swap out one kind of crash risk for another, with no overall safety benefit, so the "DSLs" were gradually rolled back.

Henry Barnes, the legendary traffic commissioner of New York City in the 1960s, reflecting on his long career in his charmingly titled memoir *The Man with the Red and Green Eyes*, observed that "traffic was as much an emotional problem as it was a physical and mechanical one." People, he concluded, were tougher to crack than cars. "As time goes on the technical problems become more automatic, while the people problems become more surrealistic."

That "surrealistic" side of traffic will be the focus of this book. I began my research with the intention of stopping to take a look around at an environment that has become so familiar we no longer see it; I wanted to slow down for a moment and think about what's going on out there as we drive, walk, cycle, or find some other way to get around. (Look out for the SKATEBOARD ROUTE signs the next time you're in Portland, Oregon.) My aim was to learn to read between the dotted lines on the highway, sift through the strange patterns that traffic contains, interpret the small feints, dodges, parries, and thrusts between vehicles. I would study not only the traffic signals we obey but also the traffic signals we send.

Many of us, myself included, seem to take driving a car fairly lightly, perhaps holding on to some simple myths of independence and power, but it is actually an incredibly complex and demanding task: We are navigating through a legal system, we are becoming social actors in a spontaneous setting, we are processing a bewildering amount of information, we are constantly making predictions and calculations and on-the-fly judgments of risk and reward, and we're engaging in a huge amount of sensory and cognitive activity—the full scope of which scientists are just beginning to understand.

Much of our mobile life is still shrouded in mystery and murk. We welcome into our vehicles new technologies like cell phones, in-car navigation systems, and "radio display system" radios (which show song titles) before we have had time to understand the complicated effects those devices might have on our driving. Opinion is often divided on the most fundamental aspects of how we should do things. Should hands be at ten a.m. and two p.m. on the steering wheel, as we were once taught—or have air bags made that a dangerous proposition? When changing lanes, is it sufficient to simply signal and check the mirrors? Or should you turn your head and glance over your shoulder? Relying on mirrors alone leaves one open to blind spots, which engineers say can exist on any car (indeed, they almost seemed designed to occur at the most inconvenient and dangerous place, the area just behind and to the left of the driver). But turning your head means not looking forward, perhaps for that vital second. "Head checks are one of the most dangerous things you can do," says the research director of a highway safety agency.

So what do we do? If these issues aren't complicated enough, consider the right side-view mirror itself. In the United States, the driver will notice that their passenger side-view mirror is convex; it usually carries a warning such as "Objects in mirror are closer than they appear." The driver's side mirror is not. In Europe, both mirrors are convex. "What you have today is this clearly pretty wrong situation," says Michael Flannagan, a researcher at the University of Michigan who specializes in driver vision. "It's wrong in the sense that Europe does one thing, the U.S. does another. They can't *both* be optimal. These are both entrenched traditions, neither of which is fully based on rational, explicit argument." The mirror, as with so many things in traffic, is more complicated than it might appear.

And so we drive around with vague ideas of how things work. Every last one of us is a "traffic expert," but our vision is skewed. We see things only through our own windshields. It is a repeated truism, borne out by insurance company surveys, for example, that most accidents happen very close to home. On first glance, it makes statistical sense: You're likely to take more trips, and spend more time in the car, in your immediate surroundings. But could there be something deeper at work? Habits, psychologists suggest, provide a way to reduce the amount of mental energy that must be expended on routine tasks. Habits also form a mind-set, which gives us cues on how to behave in certain settings. So when we enter a familiar setting, like the streets around our house, habitual behavior takes over. On the one hand, this is efficient: It frees us from having to gather all sorts of new information, from getting sidetracked. Yet on the other hand, because we are expending less energy on analyzing what is around us, we may be letting our mental guard down. If in three years there has never been a car coming out of the Joneses' driveway in the morning, what happens on the first day of the fourth year, when suddenly there is? Will we see it in time? Will we see it at all? Our feeling of safety and control is also a weakness. A study by a group of Israeli researchers found that drivers committed more traffic violations on familiar routes than on unfamiliar routes.

Surely you have had a moment when you were driving down the road and suddenly found yourself "awake at the wheel," unable to remember the last few minutes. In a way, much of the time we spend in traffic is like that, a kind of gauzy dream state of automatic muscle movements and half-remembered images. Traffic is an in-between time in which we are more likely to think about where we are going than where we are at the moment. Time and space are skewed in traffic; our vision is fragmented and often unclear, and we take in and then almost immediately forget hundreds, perhaps thousands of images and impressions. Every minute we are surrounded by a different group of people, people we will share space with but never talk to, never meet.

Considering that many of us may spend more time in traffic than we do eating meals with our family, going on vacation, or having sex, it seems worth probing a bit deeper into the experience. As an American in the early twenty-first century, I live in the most auto-dependent, car-adapted, mileage-happy society in the history of the planet. We spend more on driving than on food or health care. As of the last census, there were more cars than citizens. In 1960, hardly any household had three vehicles, and most had only one. Now more own three than own one.

Even as the size of the average North American family has fallen over the past several decades, the number of homes with multicar garages has almost doubled—one in five new homes has a three-car garage.

To pay for all that extra space, commute times have also been expanding. One of the fastest-growing categories in the last "commuting census" in the United States was that of "extreme commuters," people who spend upward of two hours a day in traffic (moving or otherwise). Many of these are people pushed farther out by higher home prices, past the billboards that beckon "If you lived here, you'd be home by now," in a phenomenon real estate agents call "drive till you qualify"—in other words, trading miles for mortgage. The average American, as of 2005, spent thirty-eight hours annually stuck in traffic. In 1969, nearly half of American children walked or biked to school; now just 16 percent do. From 1977 to 1995, the number of trips people made on foot dropped by nearly half. This has given rise to a joke: In America, a pedestrian is someone who has just parked their car.

Traffic has become a way of life. The expanding car cup holder, which became fully realized standard equipment only in the 1980s, is now the vital enabler of dashboard dining, a "food and beverage venue" hosting such products as Campbell's Soup at Hand and Yoplait's Go-Gurt. In 2001, there were 134 food products that featured the word *go* on the label or in ads; by 2004, there were 504. Accordingly, the number of what the industry calls "on-the-go eating occasions" in the United States and Europe combined is predicted to rise from 73.2 billion in 2003 to 84.4 billion in 2008. Fast-food restaurants now clock as much as 70 percent of their sales at drive-through windows. (Early in our romance with the car, we used to go to "drive-in" restaurants, but those now seem relics of a gentler, slower age.) An estimated 22 percent of all restaurant meals are ordered through a car window in America, but other places, like Northern Ireland—where one in eight people are said to eat in the car at least once per week—are getting into the act too. McDonald's has added a second lane to hundreds of its restaurants in the United States in order to speed traffic, and at its new drive-throughs in China, dubbed De Lai Su (for "Come and Get It Fast"), the company is pitching retooled regional offerings like "rice burgers" to its burgeoning drive-through customers. Starbucks, which initially resisted the drive-through for its fast-food connotations, now has drive-throughs at more than half of its new company-owned stores. The "third place" that Starbucks espouses, the place for community and leisure between home and work, is, arguably, the car.

Traffic has even shaped the food we eat. "One-handed convenience" is the mantra, with forkless foods like Taco Bell's hexagonal Crunchwrap Supreme, designed "to handle well in the car." I spent an afternoon in Los Angeles with an advertising executive who had, at the behest of that same restaurant chain, conducted a test, in actual traffic, of which foods were easiest to eat while driving. The main barometer of success or failure was the number of napkins used. But if food does spill, one can simply reach for Tide to Go, a penlike device for "portable stain removal," which can be purchased at one of the more than twelve hundred (and growing) CVS drugstores that feature a drive-through window. The "audiobook," virtually unheard of before the 1980s, represents a business worth $871 million a year, and wouldn't you know it, "traffic congestion" gets prominent mention in sales reports from the Audio Publishers Association. Car commuting is so entrenched in daily life that National Public Radio refers to its most popular segments as "driveway moments," meaning the listener is so riveted to a story they cannot leave their car. In Los Angeles, some synagogues have been forced to change the time of their evening services from eight p.m. to six p.m. in order to capture commuters on their way home, as going home and then returning to services is too much to bear in L.A. traffic. So much time is spent in cars in the United States, studies show, that drivers (particularly men) have higher rates of skin cancer on their left sides—look for the opposite effect in countries where people drive on the left.

Americans have long been fabled for their love of mobility. The nineteenth-century French visitor Alexis de Tocqueville wrote of millions "marching at once toward the same horizon," a phrase that springs to mind today when I'm flying over any large city and look at the parallel strings of red and white lights, draped like glittering necklaces over the landscape.

But this is not just a book about North America. While the United States may still have the world's most thoroughgoing car culture, traffic has become a universal condition, inflected with regional accents. In Moscow, the old images of Russians waiting in line have been replaced by images of idling cars stuck in heavy congestion. Ireland has seen its car-ownership rates double since 1990. The once tranquil Tibetan capital of Lhasa now has jams and underground parking garages. In Caracas, Venezuela, traffic is currently ranked "among the world's worst," thanks in part to an oil-fueled economic boom—and in part to cheap gas (as low as seven cents a gallon). In São Paulo, the wealthy shuttle between the city's more than three hundred helipads rather than brave the legendary traffic. In Jakarta, desperate Indonesians work as "car jockeys," hitchhikers of a sort who are paid to help drivers meet the passenger quota for the faster car-pool lanes.

Another traffic-related job has emerged outside Shanghai and other Chinese cities, according to Jian Shou Wang, the head of Kijiji (the eBay of China). There, one can find a new type of worker: *Zhiye dailu*, or professional road guides, who for a small fee will jump into one's car and provide directions in the unfamiliar city—a human "nav system." But with opportunity comes cost. In China, the number of people being killed on the road every year is now greater than the total number of vehicles the country was manufacturing annually as recently as 1970. By 2020, the World Health Organization predicts, road fatalities will be the world's third-leading cause of death.

We are all traveling the same road, if each in our own peculiar way. I invite you to join me on that road as I try, over the din of passing cars, to hear what traffic has to say.

## Chapter One

# Why Does the Other Lane Always Seem Faster? How Traffic Messes with Our Heads

### Shut Up, I Can't Hear You: Anonymity, Aggression, and the Problems of Communicating While Driving

HORN BROKEN. WATCH FOR FINGER.

—bumper sticker

In *Motor Mania*, a 1950 Walt Disney short, the lovably dim dog Goofy stars as "Mr. Walker," a model pedestrian (on two legs). He is a "good citizen," courteous and honest, the sort who whistles back at birds and wouldn't "step on an ant." Once Mr. Walker gets behind the steering wheel of his car, however, a "strange phenomenon takes place." His "whole personality changes." He becomes "Mr. Wheeler," a power-obsessed "uncontrollable monster" who races other cars at stop lights and views the road as his own personal property (but still "considers himself a good driver"). Then he steps out of his car, and, deprived of his "personal armor," reverts to being Mr. Walker. Every time he gets back into his car, despite the fact that he knows "how the other fellow feels," he is consumed by the personality of Mr. Wheeler.

What Disney was identifying, in his brilliantly simple way, was a commonplace but peculiar fact of life: We are how we move. Like Goofy, I, too, suffer from this multiple personality disorder. When I walk, which as a New Yorker I often do, I view cars as loud, polluting annoyances driven by out-of-town drunks distracted by their cell phones. When I drive, I find that pedestrians are suddenly the menace, whacked-out iPod drones blithely meandering across the street without looking. When I ride a bike, I get the worst of both worlds, buffeted by speeding cars whose drivers resent my superior health and fuel economy, and hounded by oblivious pedestrians who seem to think it's safe to cross against the light if "only a bike" is coming but are then startled and indignant as I whisk past at twenty-five miles per hour.

I am guessing this sort of thing happens to you as well. Let us call it a "modal bias." Some of this has to do with our skewed perceptual senses, as I will discuss in Chapter 3. Some of it has to do with territoriality, like when bicyclists and pedestrians sharing a path yell at each other or someone pushing a triplet-sized stroller turns into the pedestrian version of the SUV, commandeering the sidewalk through sheer size. But something deeper and more transformative happens when we move from people who walk to people who drive. The "personal armor" described by Disney is perhaps not so far-fetched. One study of pedestrian fatalities by French researchers showed that a significant number were associated with a "change of mode"—for example, moving from car to foot—as if, the authors speculated, drivers leaving their vehicles still felt a certain invulnerability.

Psychologists have struggled to understand the "deviant driver," creating detailed personality profiles to understand who's likely to fall prey to "road rage." An early mantra, originally applied to what was called the "accident-prone driver," has long held sway: "A man drives as he lives." This is why car insurance premiums are tied not only to driving history but, more controversially, to credit scores; risky credit, the thinking goes, correlates with taking risks on the road. The statistical association between lower credit scores and higher insurance losses is just that, however; the reasons why how one lives might be linked to how one drives are less clear. And as inquiries into this question typically involve questionnaires, they're open to various self-reported response biases. How would *you* answer this sample question: Are you a raving psychopath on wheels? (Please check "never," "sometimes," or "always.") Generally, these inquiries come to what hardly seem earth-shattering conclusions: that "sensation-seeking," "risk-seeking," "novelty-seeking," and "aggressive" individuals tend to drive in a riskier, more aggressive manner. You weren't going to bet your paycheck on daredevil drivers being the risk-averse people who crave quiet normalcy and routine, were you?

Even using a phrase like "road rage" lends a clinical legitimacy to what might simply be termed bad or boorish behavior elsewhere. "Traffic tantrums" is a useful alternative, nicely underscoring the raw childishness of aggressive driving. The more interesting question is not whether some of us are more prone to act like homicidal maniacs once we get behind the wheel but why we *all* act differently. What is going on seems to have less to do with a change in personality than with a change in our entire being. In traffic, we struggle to stay human.

Think of language, perhaps the defining human characteristic. Being in a car renders us mostly mute. Instead of complex vocabularies and subtle shifts in facial expression, the language of traffic is reduced—necessarily, for reasons of safety and economy—to a range of basic signals, formal and informal, that convey only the simplest of meanings. Studies have shown that many of these signals, particularly informal ones, are often misunderstood, especially by novice drivers. To take one example, the Reverend David Rowe, who heads a congregation in the wealthy Connecticut suburb of Fairfield and, improbably, is a great fan of the neopunk band Green Day, told me he was once driving down the road when he spotted a car with a Green Day bumper sticker. He honked to show his solidarity. For his efforts he was rewarded with a finger.

Even formal signals are sometimes hazy: Is that person who keeps driving with their right turn signal on actually going to turn or have they forgotten it's still blinking? Unfortunately, there's no way to ask the driver what they mean. This may lead to a rhetorical outburst: "Are you going to turn or not?" But you can't ask; nor would there be a way to get an answer back. Frustrated by our inability to talk, we gesture violently or honk—a noise the offending driver might misinterpret. At some point you may have been the recipient of an unsolicited honk, to which you immediately responded with defensive anger—*What?!*—only to learn that the honker was trying to convey to you that you left your gas cap open. *Thanks! Have a good one!*

Traffic is riddled with such "asymmetries" in communication, as Jack Katz, a sociologist at the University of California in Los Angeles and the author of *How Emotions Work*, describes them. "You can see but you can't be heard," he told me. "In a very precise way, you're made dumb. You can shout as much as you want but nobody's going to hear you."

Another way to think about this "asymmetry" is that while you can see a lot of other drivers making mistakes, you are less likely to see yourself doing so. (A former mayor of Bogotá, Colombia, had a wonderful solution to this, hiring mimes to people the city's crosswalks and silently mock drivers and pedestrians who violated traffic laws.) Drivers also spend much of their time in traffic looking at the rear ends of other cars, an activity culturally associated with subordination. It also tends to make the communication one-way: You're looking at a bunch of drivers who cannot see you. "It's like trying to talk to someone who's walking in front of you, as opposed to someone who's face-to-face with you," Katz says. "We're looking at everybody's rear, and that's not how human beings were set up to maximize their communicative possibility."

This muteness, Katz argues, makes us mad. We are desperate to say something. In one study, in-car researchers pretended to be measuring the speed and distance perception of drivers. What they were really interested in was how their subjects would react to a honk from another driver. They made this happen by giving subjects instructions as they paused at a stop sign. They then had an accomplice pull up behind the stalled car and honk. More than three-quarters of the drivers reacted verbally, despite the fact they would not be heard by the honker.

When a driver is cut off by another driver, the gesture is read as rude, perhaps hostile. There is no way for the offending driver to indicate that it was anything but rude or hostile. Because of the fleeting nature of traffic, the act is not likely to be witnessed by anyone else. No one, save perhaps your passenger, will shake their heads in unison with you and say, "Can you believe he did that?" There are at least two possible responses. One is to speed ahead and cut the offending driver off in turn, to "teach them a lesson." But there is no guarantee that the person receiving the lesson is aware of what they have done—and so your lesson simply becomes a provocation—or that they will accept your position as the "teacher" in any case. And even if your lesson is successful, you're not likely to receive any future benefit. Another response is to use an "informal" traffic signal, like the middle finger (or, as is gaining currency in Australia, the pinkie, after an ad campaign by the Road and Traffic Authority to suggest that the person speeding or otherwise driving aggressively is overcompensating for deficient male anatomy). This gains power, Katz says, if the person you give the finger to visually registers that you're giving him the finger. But what if that person merely gives the finger back?

Finally, it is often impossible to even send a message to the offending driver in the first place. Yet still we get visibly mad, to an audience of no one. Katz argues that we are engaging in a kind of theatrical storytelling, inside of our cars, angrily "constructing moral dramas" in which we are the wronged victims—and the "avenging hero"—in some traffic epic of larger importance. It is not enough to think bad thoughts about the other driver; we get angry, in essence, to watch ourselves get angry. "The angry driver," Katz argues, "becomes a magician taken in by his or her own magic." Sometimes, says Katz, as part of this "moral drama," and in an effort to create a "new meaning" for the encounter, we will try to find out something after the fact about the driver who wronged us (perhaps speeding up to see them), meanwhile running down a mental list of potential villains (e.g., women, men, teenagers, senior citizens, truck drivers, Democrats, Republicans, "idiots on cell phones," or, if all else fails, simply "idiots") before finding a suitable resolution to the drama.

This seems an on-road version of what psychologists call the "fundamental attribution error," a commonly observed way in which we ascribe the actions of others to who they are; in what is known as the "actor-observer effect," meanwhile, we attribute our own actions to how we were forced to act in specific situations. Chances are you have never looked at *yourself* in the rearview mirror and thought, "Stupid #$%&! driver." Psychologists theorize that the actor-observer effect may stem from one's desire to feel more in control of a complex situation, like driving in traffic. It also just might be easier to chastise a "stupid driver" for cutting you off than to fully analyze the circumstances that caused this event to occur.

On a larger scale, it might also help explain, more than actual national or civic chauvinism, why drivers the world around have their own favorite traffic targets: "The Albanians are terrible drivers," say the Greeks. "The Dutch are the worst drivers," say the Germans. It's best not to get New Yorkers started about New Jersey drivers. We even seem to make the fundamental attribution error in the way we travel. When bicyclists violate a traffic law, research has showed it is because, in the eyes of drivers, they are reckless anarchists; drivers, meanwhile, are more likely to view the violation of a traffic law by another driver as somehow being required by the circumstances.

At least some of this anger seems intended to maintain our sense of identity, another human trait that is lost in traffic. The driver is reduced to a brand of vehicle (a rough stereotype at best) and an anonymous license-plate number. We look for glimpses of meaning in this sea of anonymity: Think of the curious joy you get when you see a car that matches your own, or a license plate from your home state or country when you are in another. (Studies with experimental games have shown that people will act more kindly toward someone they have been told shares their birth date.) Some drivers, especially in the United States, try in vain to establish their identities with personalized vanity plates, but this raises the question of whether you really want your life summed up in seven letters—let alone why you want to tell a bunch of people you don't know who you are! Americans seem similarly (and particularly) predisposed to putting cheap bumper stickers on their expensive cars—announcing the academic wizardry of their progeny, jocularly advising that their "other car is a Porsche," or giving subtle hints ("MV") of their exclusive vacation haunts. One never sees a German blazing down the autobahn with a PROUD TO BE GERMAN sticker.

Trying to assert one's identity in traffic is always going to be problematic, in any case, because the driver yields his or her identity to the cars. We become, Katz says, cyborgs. Our vehicle becomes our self. "You project your body way out in front of a vehicle," says Katz. "When somebody's changed lanes a hundred yards ahead, you instantly feel you've been cut off. They haven't touched you physically, they haven't touched your car physically, but in order to adjust the wheel and acceleration and braking, you have projected yourself." We say, "Get out of my way," not "Get out of my and my car's way."

Identity issues seem to trouble the driver alone. Have you ever noticed how passengers rarely seem to get as worked up about these events as you do? Or that they may, in the dreaded case of the "backseat driver," even question your part in the dispute? This may be because the passenger has a more neutral view. They do not feel that their identity is bound up with the car. Studies that have examined the brain activity of drivers and passengers as they engaged in simulated driving have shown that different neural regions are activated in drivers and passengers. They are, in effect, different people. Studies have also shown that solo drivers drive more aggressively, as measured by such indices as speed and following distance. It is as if, lacking that human accompaniment—and thus any sense of shame—they give themselves over to the car.

Like many everyday travails, this whole situation is succinctly illustrated in a hit country song, Chely Wright's "The Bumper of My S.U.V." The song's protagonist complains that a "lady in a minivan" has given her the finger because of a United States Marines Corps bumper sticker on her SUV. "Does she think she knows what I stand for / Or the things that I believe," sings Wright, just because the narrator has a bumper sticker for the U.S. Marines on the aforementioned bumper of her S.U.V.? The first issue here is the struggle over identity; the narrator is upset that her identity has been defined by someone else. But the narrator may be protesting too much: How *else* would we know the things that you stand for or believe if you did not have a bumper sticker on your SUV? And if you are resentful at having your identity pigeonholed, why put a pigeonholing sticker on your bumper in the first place?

In the absence of any other visible human traits, we do draw a lot of information from bumper stickers. This point was demonstrated by an experiment conducted in 1969 at California State College, a place marked by violent clashes between the Black Panther Party and the police. In the trial, fifteen subjects of varying appearance and type of car affixed a bright BLACK PANTHER sticker to their auto's rear bumper. No one in the group had received traffic violations in the past year. After two weeks with the bumper sticker, the group had been given thirty-three citations. (The idea that people with distinguishing marks on their vehicle will be singled out for abuse or cause other disruptions of smooth traffic is just one of the problems with proposals to add scarlet letter–style designations to license plates; suggestions have ranged from identifying sex offenders in Ohio to marking the cars of the reckless drivers known as "hoons" in Australia.)

In being offended, the SUV driver has made several huge assumptions of her own. First, she has presumed that the finger had something to do with the bumper sticker, when in fact it could have been directed at a perceived act of aggressive driving on her part. Or could it have been the fact that this single driver was tooling around in a large SUV, inordinately harming the environment, putting pedestrians and drivers of cars at greater risk, and increasing the country's dependence on foreign oil? Secondly, by invoking a "lady in a minivan," later echoed by references to "private schools," she is perpetuating a preemptive negative stereotype against minivans: that their drivers are somehow more elitist than the drivers of SUVs—which makes no sense as SUVs, on average, cost more than minivans. The narrator is guilty of the same thing she accuses the minivan driver of.

In traffic, first impressions are usually the only impressions. Unlike the bar in *Cheers*, traffic is a place where no one knows your name. Anonymity in traffic acts as a powerful drug, with several curious side effects. On the one hand, because we feel that no one is watching, or that no one we know will see us, the inside of the car itself becomes a useful place for self-expression. This may explain why surveys have shown that most people, given the choice, desire a *minimum* commute of at least twenty minutes. Drivers desire this solitary "me time"—to sing, to feel like a teenager again, to be temporarily free from the constricted roles of work and home. One study found that the car was a favored place for people to cry about something ("grieving while driving"). Then there's the "nose-pick factor," a term used by researchers who install cameras inside of cars to study drivers. They report that after only a short time, drivers will "forget the camera" and begin to do all sorts of things, including nasal probing.

The flip side of anonymity, as the classic situationist psychological studies of Philip Zimbardo and Stanley Milgram have shown, is that it encourages aggression. In a well-known 1969 study, Zimbardo found that hooded subjects were willing to administer twice the level of electric shock to others than those not wearing hoods. Similarly, this is why hooded hostages are more likely to be killed than those without hoods, and why firing-squad victims are blindfolded or faced backward—not for their sake, but to make them look less human to the executioners. Take away human identity and human contact and we act inhuman. When the situation changes, we change.

This is not so different in traffic. Instead of a hood, we have the climate-controlled enclosure of the car. Why not cut that driver off? You do not know them and will likely never see them again. Why not speed through this neighborhood? You don't live here. In one study, researchers planted a car at an intersection ahead of a series of various convertibles, and had the blocking car intentionally not move after the light changed to green. They then measured how quickly the driver behind the plant vehicle honked, how many times they honked, and how long each honk was. Drivers with the top down took longer to honk, honked fewer times, and honked for shorter durations than did the more anonymous drivers with the tops up. It could have been that the people who put their tops down were in a better mood to begin with, but the results suggest that anonymity increases aggressiveness.

Being in traffic is like being in an online chat room under a pseudonym. Freed from our own identity and surrounded by others known only by their "screen names" (in traffic, license plates), the chat room becomes a place where the normal constraints of life are left behind. Psychologists have called this the "online disinhibition effect." As with being inside the car, we may feel that, cloaked in electronic anonymity, we can at last be ourselves. The playing field has been leveled, all are equal, and the individual swells with exaggerated self-importance. As long as we're not doing anything illegal, all is fair game. This also means, unfortunately, that there is little incentive to engage in normal social pleasantries. And so the language is harsh, rude, and abbreviated. One faces no consequences for one's speech: Chat room visitors aren't speaking face-to-face, and do not even have to linger after making a negative comment. They can "flame" someone and sign off. Or give someone the finger and leave them behind a cloud of exhaust.

## Are You Lookin' at Me? Eye Contact, Stereotypes, and Social Interaction on the Road

> GEORGE: This guy's giving me the stare-ahead.
>
> JERRY: The stare-ahead. I hate that. I use it all the time.
>
> GEORGE: Look at me! I am man! I am you!
>
> —*Seinfeld*

The movie *Crash* opens with the voice of the narrator, a driver in Los Angeles, speaking over a scene of a collision. "In L.A., nobody touches you. We're always behind this metal and glass. I think we miss that touch so much that we crash into each other, just so we can feel something."

The statement is absurd, but not without truth. Sometimes, we do come across little moments of humanity in traffic, and the effect is powerful. A classic case you have no doubt experienced is when you are trying to change lanes. You catch someone's eye, they let you in, and you wave back, flushed with human warmth. Now, why did that feel so special? Is it just because traffic life is usually so anonymous, or is something else going on?

Jay Phelan, an evolutionary biologist who works a few buildings over from Jack Katz at UCLA, often thinks about traffic as he pilots his motorcycle through Los Angeles. "We evolved in a world in which there were about a hundred people in the group you were in," he says. "Every person you saw you had an ongoing relationship with." Was that person good to you? Did they return the spear they borrowed last week? This way of getting along is called "reciprocal altruism." You scratch my back, I'll scratch yours; we each do it because we think it will benefit us "down the road." What happens in traffic, Phelan explains, is that even though we may be driving around Los Angeles with hundreds of thousands of anonymous others, in our ancient brains we are Fred Flintstones (albeit not driving with our feet), still inhabiting our little prehistoric village. "So when someone does something nice for you on the road, you're processing it like, 'Wow, I've got an ally now.' The brain encodes it as the beginning of a long-term reciprocal relationship."

When someone does something good or something bad, Phelan suggests, we keep score in our heads—even though the chances are infinitely small that we will ever see that person again. But our big brains, which are said to have evolved to help manage relatively large social networks, might be getting a powerful signal from that encounter. So we get angrier than we really should over minor traffic slights, or feel much better than we should after moments of politeness. "I feel like that happens a lot on the road," says Phelan. "Somebody waves you over to get in the turn lane. I get these unjustified warm feelings about the world, that there's kindness in it and everybody's looking out for each other." Or someone cuts you off, and the world is a dark, nasty place. In theory, neither should matter all that much, but we seem to react strongly either way.

These moments seem like traffic versions of the "ultimatum game," an experiment used by social scientists that seems to reveal an inherent desire for reciprocal fairness in humans. In the game, one person is given a sum of money and an instruction to share it with another person as they see fit. If the second person accepts the offer, both keep their share; if he or she rejects it, neither gets anything. Researchers have found that people will routinely reject offers that are less than 50 percent, even though this means they walk away with nothing. The cost is less important than the sense of fairness, or perhaps the bad feeling of being on the "losing end." (One study showed that people who did more rejecting had higher testosterone levels, which probably also explains why I tend to get more worked up about people who cut me off than my wife does.)

This sense of fairness might cause us to do things in traffic like aggressively tailgate someone who has done the same to us. We do this despite the costs to our own safety (we might crash, they might be homicidal) and the fact that we will never see the person we are punishing again. In small towns, it makes sense to be polite in traffic: You might actually see the person again. They might be related to you. They might learn not to do that to you again. But on the highway or in large cities, it is a puzzle why drivers try to help or hurt each other; those other drivers are not related to you (or even an immediate threat to your "kinship group"), and you are not likely to ever see those other drivers again. Have we been fooled into thinking our altruistic gesture might be returned, or are we just inherently nice? This traffic behavior is simply one part of the larger puzzle of why humans—who, unlike ants, are not all brothers and sisters working for the queen—get along (give or take your occasional war), something that scientists are still working to explain.

The Swiss economist Ernst Fehr and his colleagues have proposed a theory of "strong reciprocity," which they define as "a willingness to sacrifice resources for rewarding fair and punishing unfair behavior *even if this is costly and provides neither present nor future material rewards for the reciprocator.*" This is, after all, what we are doing when we go out of our way to scold someone on the road. In experimental games that involve people donating money into a communal investment pot, the best outcome for all players is achieved when everyone pools their resources. But a single player can do best if they contribute nothing, skimming off everyone else's profits instead. (This is like the person who drives to the front of a lengthy queue waiting to exit the highway and jumps in at the last minute.) Gradually, players stop contributing to the pool. Cooperation breaks down. When players in Fehr's game are given an option to punish people for *not* investing, however, after a couple of rounds most people give everything they have. The willingness to punish seems to ensure cooperation.

So perhaps, as the economist Herbert Gintis suggests, certain forms of supposed "road rage" are good things. Honking at or even aggressively tailgating that person who cut you off, while not strictly in your best self-interest, is a positive for the species. "Strong reciprocators" send signals that may make would-be cheaters more likely to cooperate; in traffic, as with any evolutionary system, conforming to the rules boosts the "collective advantage" of the group, and thus helps the individual. Not doing anything raises the risk that the transgressor will harm the good-driving group. You were not thinking of the good of the species when you honked at a rude driver, you were merely angry, but your anger may have been altruistic all the same. (And, like a bird squawking to warn of an approaching predator, honking at a threatening driver does not consume much energy.) In other words: Honk if you love Darwin!

Whatever the evolutionary or cultural reasons for cooperation, the eyes are one of its most important mechanisms, and eye contact may be the most powerful human force we lose in traffic. It is, arguably, the reason why humans, normally a quite cooperative species in comparison with our closest primate relatives, can become so noncooperative on the road. Most of the time we are moving too fast—we begin to lose the ability to maintain eye contact around 20 miles per hour—or it is not safe to look. Maybe our view is obstructed. Often other drivers are wearing sunglasses, or their car may have tinted windows. (And do you really want to make eye contact with those drivers?) Sometimes we make eye contact through the rearview mirror, but it feels weak, not quite believable at first, as it is not "face-to-face."

Because eye contact is so absent in traffic, it can feel uncomfortable when it does happen. Have you ever been stopped at a light and "felt" someone in a neighboring car looking at you? It probably made you uneasy. The first reason for this is that it may violate the sense of privacy we feel in traffic. The second is that there is no purpose for it and no appropriate neutral reaction, a condition that can provoke a fight-or-flight response. So what did you do at the intersection when you saw someone looking at you? If you sped up, you were not alone. In one study, researchers had an accomplice drive up on a scooter next to cars waiting at a traffic signal and stare at the driver of a neighboring car. These drivers roared through the intersection faster than those who were not stared at. Another study had a pedestrian stare at a driver waiting at the light. The result was the same. This is why trying to make eyes at your neighboring driver is bound to fail, and it is the larger problem with in-car dating networks like Flirting in Traffic, which allow drivers to send messages (via an anonymous e-mail to a MySpace-style Web site) to people bearing a special sticker. Most people—except middle-aged guys in Ferraris—do not want to be stared at while driving.

When you need to do something like change lanes, however, eye contact is a key traffic signal. On television's *Seinfeld*, Jerry Seinfeld was on to something when he advised George Costanza, who was waving his hand while trying to negotiate a difficult New York City merge, "I think we're gonna need more than a hand. They have to see a human face."

Many studies have confirmed this: Eye contact greatly increases the chances of gaining cooperation in various experimental games (it worked for *Seinfeld*'s George, by the way). Curiously, the eyes do not even need to be *real*. One study showed that the presence of cartoon eyes on a computer screen made people give more money to another unseen player than when the eyes were not present. In another study, researchers put photographs of eyes above an "honor system" coffee machine in a university break room. The next week, they replaced it with a photograph of flowers. This cycle was repeated for a number of weeks. Consistently, more people made donations on "eye" weeks. The very design of our eyes, which contain more visible sclera, or "white," than those of any of our closest primate relatives, may have even evolved, it has been argued, to facilitate cooperation in humans. This greater proportion of white helps us "catch someone's eye," and we're particularly sensitive to the direction of one's gaze. Infants will eagerly follow your glance upward but are less likely to follow if you close your eyes and simply tilt your head up. The eyes, one might argue, help reveal what we would like; eye contact is also a tacit admission that we do not think we will be harmed or exploited if we disclose our intentions.

There are times when we do not want to signal our intentions. This is why some poker players wear sunglasses. It also helps explain another game: driving in Mexico City. The ferocity of Mexico City traffic is revealed by the *topes*, or speed bumps, that are scattered throughout the capital like the mysterious earthen mounds of an ancient civilization. Mexico City's speed bumps may be the largest in the world, and in their sheer size they are bluntly effective at curbing the worst impulses of *chilango* (as the capital's residents are known) motorists. Woe to the driver who hits one at anything but the most glacial creep. Older cars have been known to stall out at a bump's crest and be turned into a roadside food stand.

*Topes* are hardly the only traffic hazard in Mexico City. There are the *secuestros express*, or "express kidnappings," in which, typically, a driver stopped at a light will be taken, at gunpoint, to an ATM and forced to withdraw cash. Often the would-be criminal is more nervous than the victim, says Mario González Román, a former security official with the U.S. embassy and himself a kidnapping victim. Calmness is essential. "Most of the people dead in carjackings are people that send the wrong signal to the criminal," he explained while driving the streets of the capital in his 1976 Volkswagen Beetle (known as a *vocho*). "You have to facilitate the work of the criminal. If the car is all he wants, you are lucky."

Express kidnappings, thankfully, are fairly rare in Mexico City. The more common bane of driving in the Distrito Federal is the endless number of intersections without traffic lights. Who will go, who will yield—it is an intricate social ballet with rough, vague guidelines. "There is no order, it's whoever arrives first," according to Agustín Barrios Gómez, an entrepreneur and sometime politico, as he drove in the Polanco neighborhood in his battered Nissan Tsuru, a car that seemed a bit beneath his station. "Mexican criminals are very car-conscious and watch-conscious," he explained. "In Monterrey I wear a Rolex; here I wear a Swatch." At each crossing, he slowed briefly to assess what the driver coming from the left or right might be doing. The problem was that cars often seemed to be arriving at the same time. In one of these instances, he barreled through, forcing a BMW to stop. "I did not make eye contact," he said firmly, after clearing the intersection.

Eye contact is a critical factor at unmarked intersections in Mexico City. Look at another driver and he will know that you have seen him, and thus dart ahead of you. *Not* looking at a driver shifts the burden of responsibility to him (assuming he has actually seen you), which allows you to proceed first—if, that is, he truly believes you are not aware of him. There's always the chance that both drivers are not actually looking. In the case of Barrios Gómez, the perceived social cost of stopping might have been greater for the BMW, higher as it is in the social hierarchy than an old Nissan Tsuru; then again, the BMW had more to lose in terms of sheer car value by not stopping. Drivers not wanting to cooperate, unwilling to begin that relationship of "reciprocal altruism," simply do not look, or they pretend not to look—the dreaded "stare-ahead." It is the same with the many beggars found at intersections in Mexico City. It is easier not to give if one does not make eye contact, which is why one sees, as in other cities, so many drivers looking rigidly ahead as they wait for the light.

Your daily drive may not seem to have much to do with the strategies of the Cold War, but every time two cars approach an unmarked intersection simultaneously, or four cars sidle up to a four-way stop at about the same time, a form of game theory is being applied. Game theory, as defined by the Nobel Prize–winning economist Thomas Schelling, is the process of strategic decision making that occurs when, as in a nuclear standoff or a stop-sign showdown, "two or more individuals have choices to make, preferences regarding the outcomes, and some knowledge of the choices available to each other and of each other's preferences. The outcome depends on the choices that both of them make, or all of them if there are more than two."

Traffic is filled with these daily moments of impromptu decision making and brinksmanship. As Schelling has argued, one of the most effective, albeit risky, strategies in game theory involves the use of an "asymmetry in communication." One driver, like Barrios Gómez in Mexico City, makes himself "unavailable" to receive messages, and thus cannot be swayed from going first through the intersection. These sorts of tactics can be quite effective, if you feel like risking your neck to prove a bit of Cold War strategy. Pedestrians, for example, are told that making eye contact is essential to crossing the street at a marked crosswalk (the kind without traffic lights), but at least one study has shown that drivers were more likely to let pedestrians cross when they did *not* look at the oncoming car.

Drivers at intersections are acting from a complicated set of motives and assumptions that may or may not have anything to do with traffic law. In one study, researchers showed subjects a series of photographs of an intersection toward which two vehicles, equally distant from the intersection, were traveling. One had the legal right-of-way, and the other did not; the second driver also did not know if the first driver would take the right-of-way. Subjects were asked to imagine that they were one of the drivers and to predict who would "win" the right-of-way under a variety of conditions; whether they were making eye contact, whether they were a man or a woman, and whether they were driving a truck, a medium-sized car, or a small car. Eye contact mattered hugely. When it was made, most subjects thought the driver who had the legal right-of-way would claim it. Drivers were also more likely to yield when the approaching car was the same size. They were even more likely to yield when the driver was female—an artifact, the researchers suggested, of a belief that women drivers were less "experienced," "competent," or "rational." Or was it just chivalry?

Traffic is thus a living laboratory of human interaction, a place thriving with subtle displays of implied power. When a light turns green at an intersection, for example, and the car ahead of another driver has not moved, there is some chance that a horn will be sounded. But *when* that horn will be sounded, for how *long* and how many *times* it will be sounded, *who* will be sounding the horn, and who the horn will be sounded *at* are not entirely random variables.

These honks follow observed patterns that may or may not fit your preexisting notions. We've already seen that drivers in convertibles with their tops down, less cloaked in anonymity, were less likely to honk than other drivers. For a similar reason, drivers in New York City, surrounded by millions of strangers, are likely to honk more, and sooner, than a driver in a small town in Idaho, where a car that has not moved might not be a random nuisance but the stalled vehicle of a friend. What the driver ahead is doing also matters. One study showed that when a car was purposely held as the light changed to green, drivers were more likely to honk—more often and for a longer time—if the nonmoving driver was quite obviously having a cell phone conversation than if they were not. (Men, it turned out, were more likely to honk than women, though women were just as likely to visibly express anger.)

All kinds of other factors—everything from gender to class to driving experience—also come into play. In another classic American study, replicated in Australia, the status of the car that did not move was the key determinant. When the "blocking car" was "high-status," the following drivers were less likely to honk than when a cheaper, older car was doing the blocking. A study in Munich reversed the equation, keeping the car doing the blocking the same (a Volkswagen Jetta) and looking instead at who did the honking; if you guessed Mercedes drivers were faster to the horn than Trabant drivers, you guessed right. A similar study tried in Switzerland did not find this effect, which suggests that cultural differences, like the Swiss reserve and love of quiet, may have been at work. Another study found that when the driver of the blocking car was a woman, more drivers—*including* women—would honk than when it was a man. An experiment in Japan found that when the blocking drivers drove cars with mandatory "novice driver" stickers, the cars behind were more likely to honk than when they did not (perhaps the horn was just a driving "lesson"). A study across several European countries found that drivers were more likely to honk, and honk sooner, when the stalled driver ahead had an identity sticker indicating that they were from another country than when they were fellow nationals.

Men honk more than women (and men and women honk more *at* women), people in cities honk more than people in small towns, people are more reluctant to honk at drivers in "nice" cars—perhaps you already suspected these things. The point is that as we are moving around in traffic, we are all guided by a set of strategies and beliefs, many of which we may not even recognize as we act upon them. This is one of the themes guiding a fascinating series of experiments by Ian Walker, a psychologist at the University of Bath in England. In a complex system such as traffic, Walker says, where myriad people with a loose sense of the proper traffic code are constantly interacting, people construct "mental models" to help guide them. "They just develop their own idea of how it works," Walker told me over lunch in the village of Salisbury. "And everyone's got different ideas."

Take the case of a car and a bicycle at an intersection. As it happens, studies consistently show intersections to be one of the most dangerous places for cyclists (not to mention cars) in traffic. Some of the reasons have to do with visibility and other perceptual problems; these will be addressed in Chapter 3. But even when drivers do see cyclists, things are not so simple. In one study, Walker showed "drivers" (i.e., qualified drivers in a lab) a photograph of a cyclist stopped at an intersection who was gazing toward the cross street but not making a turn signal with their arm. When drivers were asked to predict the cyclist's next move, 55 percent said the cyclist was not going to turn, but 45 percent said the opposite. "This is what I mean about the informality of people's mental models," he said. "There are a lot of informal signals on the road that are being used. In that study you've actually got half the population taking it to mean one thing and half the population taking it to mean another thing—which is crying out for accidents."

But there's something even more interesting than mere misinterpretation going on here, Walker suggests. In another study, Walker presented subjects (again, qualified drivers in a lab) with photographs of a brightly clad bicyclist in a number of different traffic situations in a typical English village. Using a computer, the subjects were asked to "stop" or "go" depending on what they thought the cyclist was going to do at various intersections. Cyclists were shown making a proper turn signal with the arm, giving a glance or a look over the shoulder, or not signaling at all. Results were tallied on the number of "good outcomes" (when the driver made the right choice), "false alarms" (the driver stopped when they did not have to), and what Walker predicted would be collisions. As might be expected (or hoped), drivers tended to sound false alarms most often when a cyclist looked over their shoulder or gave no signal at all. As they did not know what the cyclist was going to do, they behaved overcautiously. But when Walker studied the "collisions," he found that these happened most often when the cyclist had given the most clear indication of all, an arm turning signal. What's more, when drivers made the correct decision to stop, their reaction times were slowest when they were confronted with the arm signal.

Why should proper signaling, even when it's seen and understood by the driver, be more linked to danger in this study than lack of signaling? The answer may be that the cyclists are guilty of simply looking like humans, rather than anonymous cars. In a previous study, Walker had subjects look at various photographs of traffic and describe what was going on. When subjects saw a photograph with a car, they were more likely to refer to the photo's subject as a thing. When subjects looked at a picture that showed a pedestrian or a cyclist, they were more likely to use language that described a person. It somehow seems natural to say "the bicyclist yielded to the car," while it sounds strange to say "the driver hit the bicycle." In one photograph Walker showed, a woman was visible in a car, while a man on a bike waited behind. Although the woman could be clearly seen in the car, she was never referred to as a person, while the cyclist almost always was. Even when she was visible she was rendered invisible by the car.

In theory, this is good news for bicycle riders: What cyclist does not want to be considered human? The problem may come from the inhuman environment of traffic I have already described. Vehicles are moving at velocities for which we have no evolutionary training—for most of the life of the species we did not try to make interpersonal decisions at speed. So, when we're driving and along comes a *person* on wheels, we cannot help but look at their face and, again, their eyes. In another study Walker performed, using photographs of cyclists and subjects hooked up to eye-tracking software, he found that the subjects' gazes went instinctively toward the cyclists' faces and lingered there longest, no matter what other information was in the picture.

Eyes are the original traffic signals. Walker has a good demonstration of this. On his laptop are two photographs of himself. In one, he is looking straight at the camera (i.e., the viewer). In another, he's looking almost imperceptibly askance, but I could still feel, quite powerfully, that something had changed. How much had his eyes moved so that I knew he was no longer looking at me? A mere two pixels (out of 640 pixels across the width of the screen). What Walker is suggesting is that when we view a cyclist's eyes, or even their arm motion, we begin—perhaps automatically—a chain of cognitive processing. We cannot help but look for those things we seek out when we see another person. This seems to take longer than looking at mere things, and it seems to involve more mental effort (studies have shown that electroencephalographic, or EEG, readings spike when two people's eyes meet). We may be trying to gauge more from them than simply which direction they are going to turn. We may be looking for signs of hostility or kindness. We may be looking for reciprocal altruism. We may look where they are looking rather than see what their arm is signaling.

Whether or not we realize it, we are always making subtle adjustments in traffic. A kind of nonverbal communication is going on. Walker revealed this in a powerful way when he moved from the lab setting to the actual road. As a cyclist himself, he was curious about the anecdotal accounts from cyclists who said, in effect, that the more road space they took up, the more space passing cars gave them. He was also curious about survey reports that hinted that drivers tended to view cyclists wearing helmets as more "serious, sensible and predictable road users."

Did any of this matter on the road, or did cars simply pass cyclists *as* cyclists, more or less randomly? To find out, Walker mounted a Trek hybrid bicycle with an ultrasonic distance sensor and set out on the roads of Salisbury and Bristol. He made trips wearing a helmet and not wearing a helmet. He made trips at different distances from the edge of the road. And he made trips dressed as a man and dressed as a woman, wearing, as a rough signifier of gender, a "long feminine wig." After he had crunched the data, the numbers revealed an interesting set of patterns. The farther he rode from the edge of the road, the *less* space cars gave him. When he wore a helmet, vehicles tended to pass closer than when he did not wear a helmet. Passing drivers may have read the helmet as a sign that there was less risk for the cyclist if they hit him. Or perhaps the helmet dehumanized the rider. Or—and more likely, according to Walker—drivers read the helmet as a symbol of a more capable and predictable cyclist, one less likely to veer into their path. In either case, the helmet changed the behavior of passing drivers.

Finally, drivers gave Walker more space when he was dressed as a woman than as a man. Was this a "novelty effect" based on the fact there are statistically fewer female cyclists on England's roads? Or were drivers simply thinking, "Who is this crazy man-cyclist wearing that terrible wig?" Or were drivers (whose gender Walker was not able to record) giving women cyclists more room out of some sense of politeness or, perhaps, as he suggests, because they were operating with a stereotypical idea of women cyclists as less predictable or competent?

Interestingly, the possible gender bias, however misguided, echoes the intersection study mentioned earlier, in which drivers were more likely to yield the right-of-way if a female driver was approaching. Drivers, whether aware of it or not, seem to rely on stereotypes (a version of Walker's "mental models"). Indeed, stereotypes seem to flourish in traffic. One reason, most simply, is that we have little actual information about people in traffic, as with the "Bumper of My S.U.V." dilemma. The second reason is that we rely on stereotypes as "mental shortcuts" to help us make sense of complex environments in which there is little time to develop subtle evaluations. This is not necessarily bad: A driver who sees a small child standing on the roadside may make a stereotypical judgment that "children have no impulse control" and assume that the child may dash out. The driver slows.

It does not take a great leap to imagine, however, the problems of seeing something that does not conform to our expectations. Consider the results of one well-known psychological study. People were read a word describing a personal attribute that confirmed, countered, or avoided gender stereotypes. They were then given a name and asked to judge whether it was male or female. People responded more quickly when the stereotypical attribute matched the name than when it did not; so people were faster to the trigger when it was "strong John" and "gentle Jane" than when it was "strong Jane" and "gentle John." Only when subjects were actively asked to try to counter the stereotype and had a sufficiently low "cognitive constraint" (i.e., enough time) were they able to overcome these automatic responses.

Similarly, the drivers passing Walker on his bicycle seemed to be making automatic judgments. But did the stereotype of the helmet-wearing Walker as a competent, predictable cyclist help or hurt in the end? After all, motorists drove more closely to him. Would he have been better off wearing a wig, a Darth Vader mask, or anything else that sent a different "traffic signal" to the driver? The answer is unclear, but Walker came away from the experiment with a positive feeling about what looking human can mean in traffic. "You can stick a helmet on and it will lead to measurable changes in behavior. It shows that as a driver approaches a given cyclist, they can make an individual judgment on that person's perceived needs. They are judging each person as individuals. They're not just invoking some default behavior for passing cyclists. That's *got* to be encouraging."

Our traffic lives are ruled by anonymity, but this doesn't mean we give up trying to infer things about the people we encounter, or acting on those things in ways we may not even register.

## Waiting in Line, Waiting in Traffic: Why the Other Lane Always Moves Faster

> When people are waiting, they are bad judges of time, and every half minute seems like five.
>
> —Jane Austen, *Mansfield Park*

When was the last time you were angry at something that seemed out of your control? There is a very good possibility it was in one of three situations: being stuck in a traffic jam; waiting in line at a bank, an airport, a post office, or some such place; or being placed on hold for a "customer service representative."

In all three cases, you were in a queue. Of course, you were probably *more* angry in the first and third cases, because you were most likely in the privacy of your car or home. But there is ample opportunity for you to get angry in a public queue, which is why corporations have spent a lot of money, and thought long and hard, not only about how to reduce queues but how to make them feel shorter.

In traffic, we wait in several kinds of queues. Traffic lights cause the most traditional kind. The traffic light takes the place of the "server." A particularly slow server, like a particularly slow traffic light, bears the brunt of our frustration. As with traditional queues, traffic engineers try to estimate the flow of "arrivals." Do cars arrive in a random way, or in a "Poisson" process (after the French mathematician Siméon-Denis Poisson), as in a bank queue? Or is it non-Poisson, nonrandom (think of immigration queues at airports, which are periodically flooded by "platoons" of deplaning passengers)? Traffic engineers extend the "cycle time" during peak hours in the same way a Starbucks might add employees during the morning rush.

There are also "moving queues," as when you're in the faster left-lane on a highway, stuck behind what engineers call a "platoon" of vehicles. As some vehicles shift to slower lanes, you can "move up" the queue. If someone is in your way you might flash your lights or crowd their tail, which is roughly the equivalent of lightly coughing or tapping the shoulder of someone who is daydreaming in line ahead of you and has forgotten to move. You may have noticed how we tend to do this even when it clearly will not change the overall wait time, as if the sight of empty space makes us anxious.

Traffic congestion baffles traditional queue logic. We are waiting in a queue, but we often do not know where it begins or ends. How are we to measure our progress? Whether or not traffic always acts like a traditional queue, what's interesting is that it seems to affect us in exactly the same way. David Maister, an expert in "the psychology of queuing," has come up with a series of propositions about waiting in line. Strikingly, they all seem to hold true for traffic.

Take proposition no. 1: "Unoccupied time feels longer than occupied time." This is why grocery stores put magazines near the cashiers, and why we listen to radios or talk on cell phones in our cars. Or proposition no. 3: "Anxiety makes waits seem longer." Ever been stuck in traffic on your way to an important meeting or when you were low on gas? Or proposition no. 4: "Uncertain waits are longer than known, finite waits." This is why highway engineers use CMS, or "changeable message signs," to tell us how long a stretch of commute will take. Studies suggest that when we know the exact time of a wait, we devote less attention to thinking about it. Traffic engineers in Delhi, India, have put up "countdown signals" on a number of traffic lights, marking the number of seconds until the light turns green, for this very reason.

Also worth considering is proposition no. 6: "Unfair waits are longer than equitable waits." Think of ramp meters, those signals that delay drivers' entrance onto the freeway. Drivers fume: Why should I have to wait on the ramp while the freeway is moving? One study found that people thought of waiting on the ramp as 1.6 to 1.7 times "more onerous" than waiting on the highway itself. The more people understand the purpose of ramp meters (which I will discuss in Chapter 4), the less bothersome the wait becomes. This relates to proposition no. 5: "Unexplained waits are longer than explained waits." Hence our frustration when we find no "cause" for a traffic jam. If we know there is an accident or construction, the delay is easier to process. Proposition no. 8 is appropriate, too: "Solo waiting feels longer than group waiting." One study found that solo drivers placed the highest value on saving time in traffic. The implication is that they are more affected by delays than people not traveling alone, which is ironic, considering that under HOV lane schemes people traveling in groups often move faster.

Queues, wherever they occur, play strange games with our perception of time, our feeling of satisfaction, even our sense of "social justice." Studies have shown that people routinely overestimate the amount of time they have actually spent in a queue, and thus are less satisfied when they get served. (This is why Disney World inflates the posted waiting times for their attractions.) And while you might think that the most important factor of a queue is how many people are in front of the person waiting, research suggests that the number *behind* is significant as well. One study, at a Hong Kong post office, found that the more people there were behind a person waiting in a queue, the less likely they were to "renege," or quit. The queue might have suddenly seemed more valuable. Another theory is that when people are anxious—as is common in queues—they're more likely to make "downward" comparisons than "upward": instead of "Look how far along they are," at the front of the line, they think, "At least I'm better off than you at the back."

What really seems to rankle us is seeing people get ahead. This is why, says Richard Larson, director of the Center for Engineering Systems Fundamentals at the Massachusetts Institute of Technology and one of the world's leading authorities on queues, any number of companies—from banks to fast-food chains—have switched from systems in which multiple lines feed multiple servers to a single, serpentine line. "There's a theorem in queuing theory that says the average wait in either configuration is the same," Larson explains. Yet people prefer the single line, so much so that they have said they would be willing to wait in a longer line at Wendy's, the hamburger chain where a single line is used, than at a shorter line at McDonald's, which uses multiple lines. Why? Social justice, says Larson. "If you have the single serpentine line, you're guaranteed first come, first served. If you have the multiple lines, you have what happens at McDonald's at lunchtime. You have the stress of joining a line with high likelihood that somebody who's joined a queue next to you will get served before you. People get really irritated with that."

This happens all the time in traffic, in which reneging on the queue is often impossible. It's why I changed lanes and became a "late merger," and why people get frustrated with late mergers. (I will explain shortly why they shouldn't.) Sometimes, changing lanes (i.e., moving to a different queue) is actually a useful strategy. Often, however, it gets us no real gain. A Canadian television news program had two drivers commute along the same route on a highway at the same time. One was told to make as many lane changes as possible, the other to avoid changing lanes. The chronic lane changer saved a mere four minutes *out of an eighty-minute drive,* which hardly seems worth it. The stress involved in making all those changes probably took more than four minutes off the driver's life.

One reason why many people constantly change lanes was demonstrated in a fascinating experiment conducted by Donald Redelmeier and Robert Tibshirani, a clinical epidemiologist in Toronto and a statistician at Stanford University, respectively. Using a simple computer simulation of two lanes' worth of congested traffic obeying typical traffic behavior, as well as a video of an actual congested highway, the researchers found an illusion when looking at a sample driver: Even though the subject car had as many "passing events" as it had "overtaking events"—meaning it was maintaining the same overall relative pace as the next lane—the car spent more *time* being passed by cars than it did passing them.

Traffic, for reasons I will later explain, tends to act like an accordion: As traffic slows in a jam, it compresses; as congestion eases, the accordion "opens" and cars begin to speed up. Because of the uneven nature of stop-and-go traffic, these shifts happen in different lanes at different times. A driver in a temporarily opening lane may very quickly pass a cluster of compressing cars in the next lane. But then he will find himself in the compressing lane. And what happens? He spends more time watching those vehicles zip by in the next lane. To make matters worse, the researchers found that the closer a driver drove to the car in front of him, and the more glances he made to the next lane, the worse the illusion seemed.

Something else might also be helping to create the illusion. Drivers spend most of their time—anywhere from 80 percent to over 90 percent, studies have found—looking at the forward roadway. This includes, of course, the adjacent lane; estimates are that for every two glances we make at our own lane, we make one glance at the next lane—simply so we can actually stay in our lane. This means we are highly aware of vehicles passing us. We spend only about 6 percent of our driving time looking in the rearview mirror. In other words, we're much more aware of what is passing us than what we have passed.

The fact that we spend more time seeing losses than gains while driving in congestion plays perfectly into a well-known psychological theory called "loss aversion." Any number of experiments have shown that humans register losses more powerfully than gains. Our brains even seem rigged to be more sensitive to loss. In what psychologist Daniel Kahneman has called the "endowment affect," once people have been given something, they are instantly more hesitant to give it up.

Do you remember the childlike glee you felt the last time you found a parking spot at the mall on a crowded day? You may have left the spot with a certain reluctance, particularly if someone else was waiting for it. Studies have shown that people take longer to leave a parking spot when another driver is waiting, even though they predict they will not. It's as if the space suddenly becomes more valuable once another person wants it. In strict terms it does, even though it is no longer of intrinsic value to the person leaving it. This sensitivity to loss might also help explain the late-merger dilemma described in the Prologue. What really triggers the decision to change lanes is not so much the coolly rational assessment of underused transportation capacity but the fact that people kept passing while the early mergers stood still. The late merger's gain is perceived as the early merger's loss.

But what's the harm in merely changing lanes, anyway? One study, by the National Highway Traffic Safety Administration, found that almost 10 percent of all crashes involved lane changes. How many of those lane changes were necessary, and how many were discretionary? Do we really understand what is involved in the choices we are making? It is this last question that was at the heart of Redelmeier and Tibshirani's lane-changing study, for Redelmeier, a soft-spoken, sober doctor who spends a third of his time seeing patients at the Sunnybrook Health Sciences Centre in Toronto, has a privileged window on to the consequences of the decisions we make while driving.

"I mostly look at individuals that get seriously damaged in the aftermath of a crash," he told me in his office. "For many of them, their lives are ruined forever. For many of them, there's also this tremendous sense of remorse or chagrin—you know, if only they had behaved slightly differently, they would have never ended up in the hospital. There's a real element of almost counterfactual thinking that goes on in the aftermath of a crash. When someone comes down with pancreatic cancer there's a lot of suffering that's going on, but they usually don't start second-guessing themselves about how things could have been done differently in order to avoid this terrible predicament, whereas with motor vehicle crashes it's a very strong theme. That got me thinking how complicated driving is."

We may be doing things in traffic for reasons we do not even understand as we are acting. But how we can resist things like the next-lane-is-faster illusion? Redelmeier suggests, if not completely seriously, that we might feel better if we spent more time looking in the rearview mirror. Then we could make "downward comparisons," as with the poor saps in the Hong Kong post office, and not feel so bad. But we would also quite likely collide with the vehicle in front of us, and then cars in the next lane really would be going faster. The very nature of driving, posited as a constant progress along an endless queue, defeats us. Traffic messes with our heads in a strangely paradoxical way: We act too human, we do not act human enough.

## Postscript: And Now, the Secrets of Late Merging Revealed

People are afraid to merge on freeways in Los Angeles.
—Bret Easton Ellis, *Less Than Zero*

We humans have achieved great things. We have unlocked the once-unfathomable human genetic sequence, sent space probes to the far reaches of the solar system, and even managed to freeze a beam of light. But there's one scientific conquest that has largely eluded us. It's all the more puzzling because, on the face of it, it seems so mundane: We have not found a way to make drivers merge with the most efficiency and safety on the highway.

The situation described in the Prologue that I encountered on the Jersey highway is known in the traffic-engineering world as a "work-zone merge." Work zones, it turns out, are among the most complex and dangerous areas on the highway. Despite the signs often warning of large penalties for striking a worker (or pleas like SLOW DOWN, MY DADDY WORKS HERE), they are much more dangerous for the drivers passing through them than for the workers—some 85 percent of people killed in work zones are drivers or passengers. The reasons are not difficult to imagine. Drivers moving from an incredibly fast, free-flowing environment are suddenly being asked, sometimes unexpectedly, to come to a crawl or even a full stop, perhaps change lanes, and pass through a narrow, constricted space filled with workers, heavy machinery, and other objects of visual fascination.

And then there's the inevitable point at which two lanes of traffic will be forced to become one (or three to become two, etc.), when the early mergers, the late mergers, and everyone in between are suddenly introduced to one another. This can get sticky. It seems that even though (or maybe because) we're all tossed together on the road, drivers are not all that comfortable with interacting; a survey undertaken by the Texas Transportation Institute found that the single most common cause of stress on the highway was "merging difficulties."

Traffic engineers have spent a lot of time and money studying this problem, but it is not as simple as you might think. The "conventional merge" site, the sort I experienced on the highway in New Jersey, works reasonably well when traffic is light. Drivers are warned in advance to move into the correct lane, and they do so at a comfortable distance and speed, without a "conflict" with a driver in the other lane. But the very nature of a work zone means that traffic is often *not* light. A highway going from two lanes to one, or experiencing a "lane drop," loses at least half of its capacity to process cars—even more if drivers are slowing to see what is going on in the work zone itself. Because the capacity is quickly exceeded by the arriving cars, a "queue" soon forms. The queue, inevitably, is longer in the lane that will remain open, probably because signs have told drivers to move there.

This causes more problems. As the queue grows, it may move far back up the highway—engineers call this "upstream"—perhaps even past the signs warning of the lane closure. This means that newly arriving drivers will be encountering an unexpected queue of cars. Seeing no reason for it, they will be unaware that they're in a lane that is due to close. Once they learn this, they will have to "force" their way into the queued line, whose drivers may view the new arrivals, fairly or not, as "cheaters." As the entering drivers slow or even stop to merge, they create a temporary second queue. Drivers who grow frustrated in the queued line might similarly force their way into the faster open lane. This is all a recipe for rear-end collisions, which, as it happens, are among the leading types of crashes in work zones.

To improve things, North American engineers have responded in two basic ways. First, there is the school of Early Merge. To tackle the "forced merge" problem, Early Merge spreads out the whole merging zone. Drivers are warned by a sign several miles in advance of the "taper" that a lane drop is coming, rather than the twelve hundred feet or so in the conventional merge. "No Passing Zones" signs are often placed in the lane that will close. The earlier notice, in theory, means drivers will merge sooner and with less "friction," as engineers politely say, and will be less surprised by a sudden queue of stopped cars. Indeed, a 1997 study of an Indiana construction site using this system showed very few forced merges, few "traffic conflicts," and few rear-end collisions.

Early Merge suffers from a critical flaw, however. It has not been shown to move vehicles through the work zone more quickly than the conventional merge. One simulation showed that it actually took vehicles *longer* to travel through the work zone, perhaps because faster-moving cars were being put behind slower-moving cars in a single lane sooner than they might naturally have gotten there, thus creating an artificial rolling traffic jam. An Early Merge system would also seem to require some kind of active law enforcement presence to make sure drivers do not violate the concept. As we all know, the presence of a police car on the highway has its own unique effects on traffic.

The second school, Late Merge, was rolled out by traffic engineers in Pennsylvania in the 1990s in response to reports of aggressive driving at merge locations. In this system, engineers posted a succession of signs, beginning a mile and a half from the closure. First came USE BOTH LANES TO MERGE POINT, then a ROAD WORK AHEAD or two, and finally, at the lane drop: MERGE HERE TAKE YOUR TURN.

The beauty of the Late Merge system is that it removes the insecurity or anxiety drivers may feel in choosing lanes, as well as their annoyance with a passing "cheating" driver. The Late Merge compresses what may normally be thousands of feet of potential merging maneuvers to a single point. There is, presumably, no lane jumping or jockeying, as the flow or speed should be no better or worse in one lane than another—hence there are fewer chances for rear-end collisions. Because cars are using both lanes to the end point, the queue is cut in half.

The most surprising thing about the Late Merge concept is that it showed *a 15 percent improvement in traffic flow* over the conventional merge. It turns out that the Live Free crowd was right. Merging late, that purported symbol of individual greed, actually makes things better for everyone. As one of my Live Free responders had succinctly put it: "Isn't it obvious that the best thing to do is for both lanes to be full right up to the last moment, and then merge in turn? That way, the full capacity of the road is being used, and it's fair on everyone, rather than a bunch of people merging early and trying to create an artificial one-lane road earlier than necessary." (Note: This does *not* apply to people "late-merging" their way to the head of queues at off-ramps and the like, as those late mergers may temporarily block an otherwise free-flowing lane of traffic, not to mention greatly irritating those already queued.)

It's not just North Americans who have problems with merging. The United Kingdom's Transport Research Laboratory, in an internal report looking at new work-zone merging treatments, noted the "poor utilization of the closed lane well in advance of the taper," which it partially attributed to "vehicles blocking this lane deliberately to prevent others from 'queue jumping.' " In the 1990s, U.K. road authorities began experimenting with new signs and the so-called zipper merge, used in Germany since the 1970s. Rather than simply warn of an impending lane closure, the signs, beginning well in advance of the lane drop, advised drivers, WHEN QUEUING USE BOTH LANES and MERGE IN TURN. But the TRL, in trials on Scottish motorways, found that while the system reduced queue lengths, it didn't make traffic flow any more smoothly through the work zone. (Part of the problem may be that drivers are often still unclear about exactly where to merge: where the sign tells them to, or where the two lanes become one, or somewhere in between?) Most European traffic engineers try to avoid merging problems wherever possible by simply eliminating the need to merge. Instead, they carve out extra lanes by making the remaining lanes much smaller; this not only preserves multiple lanes, it forces drivers to slow, which is also safer.

One important caveat of the American Late Merge is that it achieved its superior performance in congested conditions—the time, of course, when work-zone merging becomes most problematic. When traffic is flowing freely, there are obvious logistical problems with driving at 75 miles per hour to the end of a lane and then "taking your turn" at the last moment. That is why traffic engineers began working on a refinement, the "Dynamic Late Merge." This employs "changeable message signs" and flashing warnings that are activated when the traffic volume reaches the point at which late merging would be more desirable. When traffic is light, the signs call for a conventional merge.

But as a Dynamic Late Merge trial undertaken by the Minnesota Department of Transportation on Interstate 10 in the summer of 2003 showed, the best-laid plans of traffic engineers often run aground on the rocky shoals of human behavior. While the experiment was able to reduce the length of queues by 35 percent, it found that vehicle volume through the merge actually decreased.

What happened? It seemed that many drivers, despite the instructions urging them to USE BOTH LANES, either did not understand the command or refused it. Only a few drivers in the lane to be closed actually made it to the sign that said, quite plainly, MERGE HERE. Some vehicles simply merged early into the "continuous lane," while others found themselves blocked by trucks and other self-appointed, lane-straddling "traffic cops" who, despite the messages, seemed intent on preserving a single queue—often to the point of aggressively weaving to block a vehicle from passing. Perhaps because they have the most difficulty accelerating and merging at work zones, truck drivers often seem intent on preserving a single queue. Some drivers in the ending lane were observed "pacing" themselves next to a car in the open lane, as if they thought it rude to go faster than anyone else (this was Minnesota, after all, the home of that Paul Bunyan–sized politeness they call "Minnesota Nice"). When this happened, the drivers following them seemed to simply give up and perform an early merge. None of this was what the DOT had in mind, as it bemoaned in a report: "These multiple merging locations created unnecessary disruptions in the traffic flow, slowing vehicles and creating more stop-and-go conditions than necessary."

The result was that drivers, whether acting out of perceived courtesy or a sense of vigilante justice, thought they were doing the right thing. In fact, they were slowing things down for everyone. One might be willing to forgive the loss in time if they were somehow making the work zone safer or less stressful, but this is not the case; rather, they created confusion by not following instructions or by acting hostile toward those who tried to do so. The Minnesota DOT seemed quite puzzled: "For some unknown reason, a small number of drivers were unwilling to change their old driving behaviors." Things got better over time—but by then the construction project was finished.

Beyond simple engineering, there seems to be a whole worldview contained in each of the merge strategies that have been tried. The Early Merge strategy implies that people are good. They want to do the right thing. They want to merge as soon as possible, and with as little negotiation as possible. They can eschew temptation in favor of cooperation. The line might be a little longer, but it seems a small price for working toward the common good. The Late Merge strategy suggests that people are not as good, or only as good as circumstances allow. Rather than having people choose among themselves where and when and in front of whom to merge, it picks the spot, and the rules, for them. Late Merge also posits that the presence of that seductively traffic-free space will be too tempting for the average mortal, and so simply removes it. And the conventional merge, the one that most of us seem to find ourselves in each day? This is strictly laissez-faire. It gives people a set of circumstances and only a vague directive of what to do and leaves the rest up to them. This tosses the late mergers and the early mergers together in an unholy tempest of conflicting beliefs, expectations, and actions. Perhaps not surprisingly, it performs the worst of all.

I suggest the following: The next time you find yourself on a congested four-lane road and you see that a forced merge is coming, don't panic. Do not stop, do not swerve into the other lane. Simply stay in your lane—if there is a lot of traffic, the distribution between both lanes should be more or less equal—all the way to the merge point. Those in the lane that is remaining open should allow one person from the lane to be closed in ahead of them, and then proceed (those doing the merging must take a similar turn). By working together, by abandoning our individual preferences and our distrust of others' preferences, in favor of a simple set of objective rules, we can make things better for everyone.

## Chapter Two

# Why You're Not as Good a Driver as You Think You Are

### If Driving Is So Easy, Why Is It So Hard for a Robot? What Teaching Machines to Drive Teaches Us About Driving

> As you wish, Mr. Knight. But, since I sense we are in a slightly irritable mood caused by fatigue . . . may I suggest you put the car in the auto cruise mode for safety's sake?
>
> —K.I.T.T., *Knight Rider*

For those of us who aren't brain surgeons, driving is probably the most complex everyday thing we do. It is a skill that consists of at least fifteen hundred "subskills." At any moment, we are navigating through terrain, scanning our environment for hazards and information, maintaining our position on the road, judging speed, making decisions (about twenty per mile, one study found), evaluating risk, adjusting instruments, anticipating the future actions of others—even as we may be sipping a latte, thinking about last night's episode of *American Idol*, quieting a toddler, or checking voice mail. A survey of one stretch of road in Maryland found that a piece of information was presented every two feet, which at 30 miles per hour, the study reasoned, meant the driver was exposed to 1,320 "items of information," or roughly 440 words, per minute. This is akin to reading three paragraphs like this one while also looking at lots of pretty pictures, not to mention doing all the other things mentioned above—and then repeating the cycle, *every minute you drive.*

Because we seem to do this all so easily, we tend not to dwell on it. Driving becomes like breathing or an involuntary reflex. We just do it. It just happens. But to think anew about this rather astonishing human ability, it's worth pausing to consider what it actually takes to get a nonhuman to drive. This is a problem that Sebastian Thrun, director of the Artificial Intelligence Laboratory at Stanford University, and his team have dedicated themselves to for the last few years. In 2005, Thrun and his colleagues won the Defense Advanced Research Projects Agency's Grand Challenge, a 132-mile race through a tortuous course in the Mojave Desert. Their "autonomous vehicle," a Volkswagen Touareg named Stanley, using only GPS coordinates, cameras, and a variety of sensors, completed the course in just under seven hours, averaging a rather robust 19.1 miles per hour.

Stanley won because Thrun and his team, after a series of failures, changed their method of driving instruction. "We started teaching Stanley much more like an apprentice than a computer," Thrun told me. "Instead of telling Stanley, 'If the following condition occurs, invoke the following action,' we would give an example and train him." It would not work, for example, to simply tell Stanley to drive at a certain speed limit. "A person would slow down when they hit a rut," Thrun said. "But a robot is not that smart. It would keep driving at thirty miles per hour until its death." Instead, Thrun took the wheel and had Stanley record the way he drove, carefully noting his speed and the amount of shock the vehicle was absorbing. Stanley watched how Sebastian responded when the road narrowed, or when the shock level of his chassis went beyond a certain threshold.

Stanley was learning the way most of us learn to drive, not through rote classroom memorization of traffic rules and the viewing of blood-soaked safety films but through real-world observation, sitting in the backseats of our parents' cars. For Thrun, the process made him begin "questioning what a rule really is." The basic rules were simple: Drive on this road under this speed limit from this point to this point. But giving Stanley rules that were too rigid would cause him to overreact, like the autistic character played by Dustin Hoffman in the film *Rain Man*, who stops while crossing an intersection because the sign changes to DO NOT

WALK. What about when the conventions are violated, as they so often are in driving? "Nothing says that a tumbleweed has to stay outside the drivable corridor," Thrun explained. In other words, stuff happens. There are myriad moments of uncertainty, or "noise." In the same way we do things like judge whether the police car with the flashing lights has already pulled someone else over, Stanley needs to decipher the puzzling world of the road: Is that a rock in the middle of the street or a paper bag? Is that a speed bump in the road or someone who fell off their bike? The restrictions on a New York City "No Parking" sign alone would bring Stanley to his knees.

If all this seems complicated enough, now consider doing all of it in the kind of environment in which most of us typically drive: not lonely desert passes but busy city and suburban streets. When I caught up with Thrun, this is exactly what was on his mind, for he was in the testing phase for DARPA's next race, the Urban Challenge. This time the course would be in a city environment, with off-roading Stanley retired in favor of sensible Junior, a 2006 VW Passat Wagon. The goal, according to DARPA, would be "safe and correct autonomous driving capability in traffic at 20 mph," including "merging into moving traffic, navigating traffic circles, negotiating busy intersections, and avoiding obstacles."

We do not always get these things right ourselves, but most drivers make any number of complex maneuvers each day without any trouble. Teaching a machine to do this presents elemental problems. Simply analyzing any random traffic scene, as we constantly do, is an enormous undertaking. It requires not only recognizing objects, but understanding how they relate to one another, not just at that moment but in the future. Thrun uses the example of a driver coming upon a traffic island versus a stationary car. "If there's a stationary car you behave fundamentally differently, you queue up behind it," he says. "If it's a traffic island you just drive around it. Humans take for granted that we can just look at this and recognize it instantly. To take camera data and be able to understand this is a traffic island, that technology just doesn't exist." Outside of forty meters or so, Junior, according to Thrun, does not have a clue about what the approaching obstacle is; he simply sees that it is an obstacle.

In certain ways, Junior has advantages over humans, which is precisely why some robotic devices, like adaptive cruise control—which tracks via lasers the distance to the car in front and reacts accordingly—have already begun to appear in cars. When calculating the distance between himself and the car ahead, as with ACC, Junior is much more accurate than we are—to within one meter, according to Michael Montemerlo, a researcher at Stanford. "People always ask if Junior will sense other people's brake lights," Montemerlo said. "Our answer is, you don't really have to. Junior has the ability to measure the velocity of another car very precisely. That will tell you a car's braking. You actually get their velocity instead of this one bit of information saying 'I'm slowing down.' That's much more information than a person gets."

Driving involves not just the fidelity of perception but knowing what to do with the information. For Stanley, the task was relatively simple. "It was just one robot out in the desert all by himself," Montemerlo said. "Stanley's understanding of the world is very basic, actually just completely geometric. The goal was just to always take the good terrain and avoid the bad terrain. It's not possible to drive in an urban setting with that limited understanding of the world. You actually have to take and interpret what you're seeing and exhibit a higher-level understanding." When we approach a traffic signal that has just gone yellow, for example, we engage in a complex chain of instantaneous processing and decision making: How much longer will the light be yellow? Will I have time (or space) to brake? If I accelerate will I make it, and how fast do I have to go to do so? Will I be struck by the tailgater behind if I slam on the brakes? Is there a red-light camera? Are the roads wet? Will I be caught in the intersection, "blocking the box"?

Engineers call the moment when we're too close to the amber light to stop and yet too far to make it through without catching some of the red phase the "dilemma zone." And a dilemma it is. Judging by crash rates, more drivers are struck from the rear when they try to stop for the light, but more serious crashes occur when drivers proceed and are hit broadside by a car entering the intersection. Do you take the higher chance of a less serious crash or the lower chance of a more serious crash? Engineers can make the yellow light last longer, but this reduces the capacity of the intersection—and once word gets out on the generous signal timing, it may just encourage more drivers to speed up and go for it.

Some people have even proposed signs that warn the driver in advance that the light is about to turn amber, a sort of "caution for the caution" that extends what is called the "indecision zone." But a study in

Austria that looked at intersections where the green signal flashes briefly before turning yellow found mixed results: Fewer drivers went through the red light than at intersections without the flashing green, but more drivers stopped *sooner* than necessary. The danger of the latter result was shown in a study of intersections in Israel where the "flashing green" system had been introduced. There were more rear-end collisions at those intersections than at those without the flashing green. The longer the indecision zone, the more cars that are in it, the more decisions about whether or not to go or suddenly stop, and thus the more chances to crash.

In traffic, these sorts of dilemma zones occur all the time. There are no pedestrians present in the Grand Challenge ("Thank God," said Montemerlo); they would represent a massive problem for Junior. "I've thought a lot about what would happen if you let Junior loose in the real world," Montemerlo said. Driving at Stanford is relatively sedate, but what if there is a pedestrian standing on the curb, just off the crosswalk? As the pedestrian isn't in the road, he's not classified as an obstacle. But is he waiting to cross or just standing there? To know this, the robot would somehow have to interpret the pedestrian's body language, or be trained to analyze eye contact and facial gestures. Even if the robot driver stopped, the pedestrian might need further signals. "The pedestrian is sometimes wary to walk in front of someone even if they have stopped," Montemerlo said. "Often they wait for the driver to wave, 'You go first.' " Would you feel comfortable crossing in front of a driverless Terminator?

In some ways, however, a city environment is actually easier than a dusty desert track. "Urban driving is really constrained; there aren't many things you can do," said Montemerlo (who has clearly never driven on New York's FDR Drive). "This is actually how we're able to drive. We use the rules of the road and road markings to make assumptions about what might happen."

Traffic is filled with these assumptions: We drive at full speed through the green light because we're predicting that the other drivers will have stopped; we do not brace for a head-on collision every time a car comes our way in the opposite lane; we zoom over the crest of a hill because we do not think there is an oil truck stopped just on the other side. "We're driving faster than we would if we couldn't make these assumptions," Montemerlo said. What the Stanford team does is encode these assumptions into the 100,000 or so lines of code that make up Junior's brain, but not with such rigidity that Junior freezes up when something weird happens.

And weird things happen a lot in traffic. Let's say a traffic signal is broken. David Letterman once joked that traffic signals in New York City are "just rough guidelines," but everyone has driven up to a signal that was stuck on red. After some hesitation, you probably, and very carefully, went through the red. Or perhaps you came up behind a stalled car. To get around it would involve crossing a double yellow line, normally an illegal act. But you did it, and traffic laws usually account for exceptional circumstances. What about the question of who proceeds first at a four-way stop? Sometimes there is confusion about who arrived first, which produces a brief four-way standoff. Now picture four robot drivers who arrived at the *exact* same moment. If they were programmed to let the person who arrived first go first, two things might happen: They might all go first and collide or they might all sit frozen, the intersection version of a computer crash. So the Stanford team uses complex algorithms to make Junior's binary logic a bit more human. "Junior tries to estimate what the right time to go is, and tries to wait for its turn," Montemerlo said. "But if somebody else doesn't take their turn and enough time passes by, the robot will actually bump itself up the queue."

The Stanford team found that the best way for Stanley and Junior to learn how to drive was to study how humans drive. But might the robots have anything to teach us? In the very first Grand Challenge, Montemerlo said, Thrun was "always complaining that the robot slowed down too much in turns." Yet when a graduate student analyzed the race results, he came to the conclusion that the robot could have "cornered like a Ferrari" and still only shaved a few minutes off a seven-hour race—while upping the crash risk. The reason was that most of the course consisted of straight roads. Maintaining the highest average speed over these sections was more important than taking the relatively few turns (the most dangerous parts of the road) at the highest speed possible.

"Driving smarter," Montemerlo calls it. This is something he has thought a lot about for the Urban Challenge. "You might initially think, 'I'll take everything Junior does, and make it as fast as possible. I'll make it accelerate from the stop sign as fast as possible. I'll make it wait the minimum amount of time when it stops.' But it turns out it doesn't help that much. We all know it from traffic. You see the guy who speeds past you on a road, and then you see him again—you're stopped one car behind him at the next red light. The randomness of traffic overwhelms these tiny instances. At the same time, some of these little optimizations, like being a jerk at a stop sign, cause problems for everyone. They slow everyone down."

It took a group of some of the world's leading robotics researchers years of work to come up with an autonomous vehicle that, while clever and adept at certain driving tasks, would quickly go haywire in real traffic. That should be both a testament to the remarkable human ability that driving is as well as a cautionary reminder not to take this activity for granted. The advantage robots have in the long run is that the hardware and software keep getting better. We humans must use what we're born with. The human cognitive mechanism is powerful equipment, as the trials of teaching Stanley and Junior to drive show. But as we are about to see, it is not without bugs. And these are not the sort that are going to be fixed in Version 2.0.

## How's My Driving? How the Hell Should I Know? Why Lack of Feedback Fails Us on the Road

> There are two things no man will admit he cannot do well: drive and make love.
>
> —Stirling Moss, champion racer

A splashy television advertising campaign for the online auction site eBay came with the simple tagline "People Are Good." Interestingly, a number of the images it showed involved traffic: In one spot, people joined to help push a car stuck in the snow; in another, a driver slowed to let another driver in, with a wave of the hand. By tapping into these moments of reciprocal altruism, eBay was hoping to underscore the idea that you can buy something from somebody you have never met, halfway around the globe, and feel confident that the product will actually show up. This "everyday trust," as an eBay spokesperson described it, which "blossoms into millions of strangers transacting with each other and overwhelmingly comes off without a hitch," roughly describes what happens in traffic.

And yet people are not always good. Each month seems to bring some new form of scam to eBay, which the company duly investigates. Sophisticated software, for one thing, sniffs out suspicious bidding patterns. What keeps the site running, however, is not the prowess of its fraud squad—which would hardly have time to monitor more than a fraction of the many millions of daily auctions—but a more simple mechanism: feedback. The desire to get positive feedback and avoid negative feedback is, as anyone who has bought or sold on the site knows, a crucial part of the experience. This probably has less to do with people wanting to feel good than the fact that sellers with good reputations can, as one study found, make 8 percent more in revenue. Either way, feedback (provided it's authentic) is the social glue that holds eBay together.

What if there was an eBay-like system of "reputation management" for traffic? This idea was raised in a provocative paper by Lior J. Strahilevitz, a law professor at the University of Chicago. "A modern, urban freeway is a lot like eBay, without reputation scores," he wrote. "Most drivers on the freeway are reasonably skilled and willing to cooperate conditionally with fellow drivers, but there is a sizeable minority that imposes substantial costs on other drivers, in the form of accidents, delays, stress, incivility, and rising insurance premiums."

Inspired by the HOW'S MY DRIVING stickers used by commercial fleets, the idea is that drivers, when witnessing an act of dangerous or illegal driving, could phone a call center and lodge a complaint, using mandatory identification numbers posted on every driver's bumper or license plate. Calls could also be made to reward good drivers. An account would be kept and, at the end of each month, drivers would receive a "bill" tallying the positive or negative comments called in. Drivers exceeding a certain threshold could be punished in some way, such as by higher insurance premiums or a suspension of their license. Strahilevitz argues that this system would be more effective than sporadic law enforcement, which can monitor only a fraction of the traffic stream. The police are usually limited to issuing tickets based on obvious violations (like speeding) and are essentially powerless to do anything about the more subtle rude and dangerous moments we encounter—how often have you wished in vain for a police car to be there to catch someone doing something dangerous, like tailgating or texting on their BlackBerry? It would help insurance companies more effectively set rates, not to mention giving frustrated drivers a safer and more useful outlet to express their disapproval, and gain a sense of justice—than by responding in kind with acts of aggressive driving.

But what about false or biased feedback? What if your next-door neighbor who's mad at you for your barking dog phones in a report saying you were acting crazy on the turnpike? As Strahilevitz points out, eBay-style software can sniff out suspicious activity—"outliers" like one negative comment among many positives, or repeated negative comments from the same person. What about privacy concerns? Well, that's exactly the point: People are free to terrorize others on the road because their identity is largely protected. The road is not a private place, and speeding is not a private act. As Strahilevitz argues, "We should protect privacy if, and only if, doing so promotes social welfare."

Less ambitious and official versions of this have been tried. The Web site Platewire.com, which was begun, in the words of its founder, "to make people more accountable for their actions on the roadways in one forum or another," gives drivers a place to lodge complaints about bad drivers, along with the offenders' license plate numbers; posts chastise "Too Busy Brushing Her Hair" in California and "Audi A-hole" in New Jersey. Much less frequently, users give kudos to good drivers.

However noble the effort, the shortcomings of such sites are obvious. For one, Platewire, at the time of this writing, has a bit over sixty thousand members, representing only a minuscule fraction of the driving public. Platewire complaints are falling on few ears. For another, given the sheer randomness of driving, the chances are remote that I would ever come across the owner of New Jersey license plate VR347N—more remote even than the chance that they're reading this book—and, moreover, I'm unlikely to remember that they were the one a Platewire member had tagged for "reading the newspaper" while driving! Lastly, Platewire lacks real consequences beyond the anonymous shame of a small, disparate number of readers.

The call-center idea is aimed at countering the feeling of pervasive anonymity in traffic, and all the bad behavior it encourages. But it could also help correct another problem in traffic: the lack of feedback. As discussed earlier, the very mechanics of driving enable us to play spectator to countless acts of subpar driving, while being less aware of our own. Not surprisingly, if we were to ask ourselves "How's my driving?," research has shown that the answer would probably be a big thumbs-up—regardless of one's actual driving record.

In study after study, from the United States to France to New Zealand, when groups of drivers were asked to compare themselves to the "average driver," a majority inevitably respond that they were "better." This is, of course, statistically quite improbable and seems like a sketch from Monty Python: "We Are All Above Average!" Psychologists have called this phenomenon "optimistic bias" (or the "above-average effect"), and it is still something of a mystery why we do it. It might be that we want to make ourselves out to be better than others in a kind of downward comparison, the way the people in line in the first chapter assessed their own well-being by turning around to look at those lesser beings at the back of the queue. Or it might be the psychic crutch we need to more confidently face driving, the most dangerous thing most of us will ever do.

Whatever the reason, the evidence is strong that we self-enhance in all areas of life, often at our peril. Investors routinely claim they are better than the average investor at picking stocks, but at least one study of brokerage accounts showed that the most active traders (presumably among the most confident) generated the *smallest* returns. Driving may be particularly susceptible to the above-average effect. For one, psychologists have found that the optimistic bias seems stronger in situations we can control; one study found drivers were more optimistic than passengers when asked to rate their chances of being involved in a car accident.

The above-average effect helps explain resistance (in the early stages, at least) to new traffic safety measures, from seat belts to cell phone restrictions. Polls have shown, for example, that most drivers would like to see text messaging while driving banned; those same polls also show that most people have done it. We overestimate the risks to society and underestimate our own risk. It is the *other* person's behavior that needs to be controlled, not mine; this reasoning helps contribute to the long-standing gap, concerning evolving technology, between social mores and traffic laws. We think stricter laws are a good idea for the people who need them.

Another problem with our view of ourselves is that we tend to rank ourselves higher, studies have shown, when the activity in question is thought to be relatively easy, like driving, and not relatively complex, like juggling many objects at once. Psychologists have suggested that the "Lake Wobegon effect"—"where all the children are above average"—is stronger when the skills in question are ambiguous. An Olympic pole-vaulter has a pretty clear indication of how good she is compared to everyone else by the height of the bar she must clear. As for a driver who simply makes it home unscathed from work, how *was* their performance? A 9.1 out of 10?

Most important, we may inflate our own driving abilities simply because we are not actually capable of rendering an accurate judgment. We may lack what is called "metacognition," which means, as Cornell University psychologists Justin Kruger and David Dunning put it, that we are "unskilled and unaware of it." In the same way a person less versed in the proper rules of English grammar will be less able to judge the correctness of grammar (to use Kruger and Dunning's example), a driver who is not fully aware of the risks of tailgating or the rules of traffic is hardly in a good position to evaluate their own relative risk or driving performance compared to everyone else's. One study showed that drivers who did poorly on their driving exam or had been involved in crashes were not as good at estimating their results on a simple reaction test as the statistically "better" (i.e., safer) drivers. And yet, as mentioned earlier, people seem easily able to disregard their own driving record in judging the quality of their own driving.

So whether we're cocky, compensating for feeling fearful, or just plain clueless, the roads are filled with a majority of above-average drivers (particularly men), each of whom seems intent on maintaining their sense of above-averageness. My own unscientific theory is that this may help explain—in America, at least—why drivers polled in surveys seem to find the roads less civil with each passing year. In an 1982 survey, a majority of drivers found that the majority of other people were "courteous" on the road. When the same survey was repeated in 1998, the rude drivers outnumbered the courteous.

How does this tie into pumped-up egos? Psychologists suggest that narcissism, more than insecurity propelled by low self-esteem, promotes aggressive driving. Rather like the survey data that show a mathematical disconnect between the number of sexual partners men and women claim to have had, polls of aggressive driving behavior find more people seeing it than doing it. Someone is self-enhancing. And so narcissism, like road nastiness, seems to be on the rise. Psychologists who examined a survey called the Narcissistic Personality Inventory, which has for the past few decades gauged narcissistic indicators in society (measuring reactions to statements like "If I ruled the world, it would be a better place"), found that in 2006, two-thirds of survey respondents scored higher than in 1982. More people than ever, it seems, have a "positive and inflated view of the self." And over the same period that narcissism was growing, the road, if surveys can be believed, was becoming a less pleasant environment. Traffic, a system that requires conformity and cooperation to function best, was filling with people sharing a common thought: "If I ruled the road, it would be a better place."

When negative feedback does come our way on the road, we tend to find ways to explain it away, or we quickly forget it. A ticket is a rare event that one grumblingly attributes to police officers having to "make a quota"; a honk from another driver is a cause for anger, not shame or remorse; a crash might be seen as pure bad luck. But usually, for most people, there is no negative feedback. There is little feedback at all. We drive largely without incident every day, and every day we become just a little bit more above average. As John Lee, head of the Cognitive Systems Laboratory at the University of Iowa, explained, "As an average driver you can get away with a lot before it catches up to you. That's one of the problems. The feedback loops are not there. You can be a bad driver for years and never really realize it, because you don't get that demonstrated to you. You could drive for years with a cell phone and say, 'How can cell phones be dangerous, because I do it every day for two hours and nothing's happened?' Well, that's because you've been lucky."

Even the moments when we almost crash become testaments to our skill, notches on our seat belts. But as psychologist James Reason wrote in *Human Error*, "In accident avoidance, experience is a mixed blessing." The problem is that we learn how to avoid accidents precisely by avoiding accidents, not by being in accidents. But a near miss, as Reason described it, involves an *initial error* as well as a process of *error recovery*. This raises several questions: Are our near misses teaching us how to avoid accidents or how to prevent the errors that got us into the tight spot to begin with? Does avoiding a minor accident just set us up for having to get out of much bigger accidents? How, and what, do we learn from our mistakes?

. . .

What do we learn from mistakes? This last question was also raised by the technology of a company called DriveCam, located in an office park in suburban San Diego, where I spent a day watching video footage of crashes, near crashes, and spectacularly careless acts of driving. The premise is simple: A small camera, located around the rearview mirror, is constantly buffering images (the way TiVo does for your television shows) of the exterior view and the driver. Sensors monitor the various forces the vehicle is experiencing. When a driver brakes hard or makes a sudden turn, the camera records ten seconds before and after the event, for context. The clip is then sent to DriveCam analysts, who file a report and, if necessary, apply "coaching."

DriveCam, whose motto is "Taking the risk out of driving," has its cameras installed in everything from Time Warner Cable vans to Las Vegas taxicabs to rental-car shuttle buses at airports. Companies that have installed DriveCam have seen their drivers' crash rates drop by 30 to 50 percent. The company contends that it has several advantages over the traditional methods of trying to improve the safety records of commercial fleets. One earlier approach, as DriveCam CEO Bruce Moeller told me, was giving drivers spot safety drills. "They'd come in for the training. You're all hopped up, 'I'm going to do right.' But then over time, you start pushing the envelope. You didn't hit anybody and nobody yelled at you. So that's fine, you get away with it, and pretty soon you start lapsing back to your old ways." The widespread onset of "How's My Driving?" phone numbers in the 1980s created the potential for more constant feedback, but it was often late or of debatable quality, says Del Lisk, the company's vice president. "It's highly prone to very subjective consumer call-ins," he said. "Like, 'I'm mad about my phone bill so I'm going to call in that AT&T guy.' "

Given that the company car is the most statistically hazardous environment for workers, it seems appropriate that the thinking behind DriveCam is inspired by the work of H. W. Heinrich, an insurance investigator for the Travelers Insurance Company and the author of a seminal 1931 book, *Industrial Accident Prevention: A Scientific Approach.* After investigating tens of thousands of industrial injuries, he estimated that for every one fatality or major injury in the workplace, there were 29 minor injuries and 300 "near-miss" incidents that led to no injury. He arranged these in the so-called Heinrich's triangle and argued that the key to avoiding the one event at the top of the triangle lay in tackling the many small events at the bottom.

When I'd met Moeller, the first thing he'd told me, after introductory pleasantries, was: "If we were to put a DriveCam in your car, not knowing you at all, I guarantee you that you've got driving habits you're not even aware of that are an accident waiting to happen." He pointed to the Heinrich triangle he had drawn on a whiteboard. "You know about the twenty-nine and the one"—the crashes and the fatality—"because there's hard evidence that somebody got killed or somebody crashed," he said. "What we show you with the DriveCam monitoring this thing twenty-four/seven is that all the very same unsafe behaviors that are going on down here"—he pointed to the bottom of the triangle—"can result, or will result, in accidents, except for pure luck."

The key to reducing what DriveCam calls "preventable accidents," as Lisk sees it, lies at the bottom of the triangle, in all those hidden and forgotten near misses. "Most people would look at that triangle and use the top two tiers as their way of estimating how good a driver they are. The truth is, it's really the bottom tier that is the real evaluator." In other words, a driver thinks of their own performance in terms of crashes and traffic tickets. People riding along with a driver look at it differently. "All of us, as passengers," Lisk said, "will ride along and evaluate drivers from the bottom of the pyramid, squeezing the armrest and pushing our feet into the floorboards."

As I played virtual passenger on a number of DriveCam moments, a disturbing realization came to my attention. There is much careless driving, to be sure. In one clip, a man takes his hands off the steering wheel to jab at a boxer's speed bag suspended from the rearview mirror. In any number of clips, drivers struggle to keep their eyes open and their bobbing heads erect. "We've got one where a guy's driving a tanker truck full of gas for *eight* full seconds as he's asleep," Moeller said. (A dip on a Los Angeles freeway had triggered the camera.)

But what is most unsettling in a number of clips is not the event itself as much as what else was visible in the camera, just outside the frame. In one bit of footage, a man looks down to dial a cell phone as he drives down a residential street. His eyes are off the road for much of the nine seconds of the recorded event, and his van begins to drift off the road.

Startled by the vibration of the roadside, he swerves back onto the road. He grimaces in a strange mixture of shock and relief. Examining the image closely, however, one sees a child on a bicycle and the child's friend, standing just off the road, less than a dozen feet away from the triggered event. "Do you think he ever even saw the bike rider and other person?" Lisk asked. "It's just luck. It's that pyramid."

Not only was the driver unaware of the real hazards he was subjecting himself and others to in the way he was driving, he was not even aware that he was unaware. "This guy's probably a great guy, good family man, good employee," Lisk said. "He doesn't even know this is happening. If we told him it happened, with a black box or something, he wouldn't even believe it." Without the video, the driver would not have realized the potential consequences of his error. "I get reinforced more positively every day that I don't hit a kid because I'm not seeing that stuff," Moeller said. "I'm thinking I'm good, I can do this. I can look down at my BlackBerry, I can dial a phone, I can drink. We all get reinforced the wrong way."

Until the moment when we do not, of course, and something goes wrong. We commonly refer to these moments as "accidents," meaning that they were unintended or unforeseen events. *Accident* is a good word for describing such events as an otherwise vigilant driver being unable to avoid a tree that suddenly fell across the road. But consider the case of St. Louis Cardinals pitcher Josh Hancock, who was tragically killed in 2007 when his rented SUV slammed into the back of a tow truck that was stopped on the highway, lights flashing, at the scene of a previous crash. Investigators learned that Hancock (who days before had crashed his own SUV) had a blood alcohol concentration nearly twice the legal limit, was speeding, was not wearing a seat belt, and was on a cell phone at the time of the fatal crash.

Despite the fact that all these well-established risky behaviors were present, simultaneously, the event was still routinely referred to in the press as an "accident." The same thing happened with South Dakota congressman Bill Janklow. A notorious speeder who racked up more than a dozen tickets in the span of four years and had a poster of himself boasting that he liked to live in the "fast lane," in 2003 Janklow blazed through a stop sign and killed a motorcyclist. The press repeatedly called it an "accident."

The problem with this word, as the *British Medical Journal* pointed out in 2001 when it announced that it would no longer use it, is that accidents are "often understood to be unpredictable," and thus unpreventable. Were the Hancock and Janklow crashes really unpredictable or unpreventable? They were certainly unintentional, but are "some crashes more unintentional than others"? Did they "just happen" or were there things that could have been done to prevent them, or at least greatly reduce the chances of their happening? Humans are humans, things will go wrong, there are instances of truly bad luck. And psychologists have argued that humans tend to exaggerate, in retrospect, just how predictable things were (the "hindsight bias"). The word *accident*, however, has been sent skittering down a slippery slope, to the point where it seems to provide protective cover for the worst and most negligent driving behaviors. This in turn suggests that so much of the everyday carnage on the road is mysteriously out of our hands and can be stopped or lessened only by adding more air bags (pedestrians, unfortunately, lack this safety feature).

Most crashes involve a violation of traffic laws, whether intentional or not. But even the notion of "unintentional" versus "intentional" has been blurred. In 2006, a Chicago driver reaching for a cell phone while driving lost control of his SUV, killing a passenger in another car. The victim's family declared, "If he didn't drink or use drugs, then it's an accident." As absurd as that statement may sound, given that the driver intentionally broke the law, the law essentially agreed: The driver was fined $200. Similarly strange distinctions are found with "sober speeders." There is a huge gulf in legal recrimination between a person who boosts his blood alcohol concentration way over the limit and kills someone and a driver who boosts his speedometer way over the limit and kills someone.

A similar bias creeps into news reports, which are often quick to note, when reporting fatal crashes, that "no drugs or alcohol were involved," subtly absolving the driver from full responsibility—even if the driver was flagrantly exceeding the speed limit. Car companies would rightly be castigated if they advertised the joys of drinking and driving. But as a survey of North American car commercials by a group of Canadian researchers showed, it is quite acceptable to show cars being driven, soberly, in ways that a panel of viewers labeled "hazardous." Nearly half of the more than two hundred ads screened (always carrying careful, if duplicitous, disclaimers) were considered by the majority of the panel to contain an "unsafe driving sequence," usually marked by high speeds. Ads for SUVs were the most frequent offenders, and across all commercials, when drivers were shown, the majority were men.

What the video footage at DriveCam showed, more often than not, is not that unforeseen things happen on the road for no good reason but that people routinely do things to make crashes "unpreventable." If the van driver had struck the child by the side of the road, it would have been reasonably "accidental" only in the sense that he did not intend to do it. Would this have just been "bad luck"? The psychologist Richard Wiseman has demonstrated in experiments that people are also capable of making their own "luck." For example, people who know lots of people are more likely to have seemingly lucky "small-world" encounters than those who do not (and those who did not have many such chance meetings more often viewed themselves as "unlucky").

We cannot entirely prevent "bad luck" from landing on our doorstep, but the van driver dialing his cell phone, the one who narrowly missed the kids in the DriveCam video, was virtually throwing open his door and inviting it inside. DriveCam's hindsight does make it glaringly easy to see all the things drivers were doing wrong. The question is, Why didn't they? Why do people act in ways that put themselves and others at unnecessary risk? Are they being negligent, ignorant, overconfident, just plain dumb—or are they just being human? Can we actually learn from our mistakes *before* they have real consequences?

Psychologists have demonstrated that our memory, as you might expect, is tilted in favor of more recent things. We also tend to emphasize the ends of things—as, for example, when told a series of facts and later asked to recall the entire series. Studies have confirmed that people are less likely to remember traffic accidents the further back in time they happened. In this same way, a near crash or a crash might loom more vividly than the things that led up to it. "Almost rear-ending someone will stick in your mind, but that freezing it and remembering it comes at the cost of losing the precipitating events," Rusty Weiss, director of DriveCam's consumer division, explained. Time also takes its toll. A study led by Peter Chapman and Geoff Underwood at the University of Nottingham in England found that drivers forgot about 80 percent more of their near crashes if they were first asked about them two weeks later than if they were asked at the end of their trip. This is exactly the point with DriveCam: It does not let you forget the precariousness of your existence on the road.

Weiss, who came to DriveCam after setting up a program to put the camera in the cars of teenage drivers in a trial in Minnesota, theorizes that this amnesia for what helped lead up to a crash, something we are all subject to, troubles beginning drivers in particular. They are the ones, ironically, who are constantly finding themselves moving in and out of risky situations. "These kids should be learning rapidly," he says. "There's lots of learning opportunities, yet they continue making mistakes. At the moment they say it wasn't their fault, but then they see the video and go, 'Oh my God.' It's like video feedback for your golf swing. It makes you aware of things you're not aware of when you're there in the moment."

The problem may be that they are simply forgetting the moments they should be learning from. Another study by Chapman and Underwood found that when drivers were shown videos of hazardous driving situations, novice drivers were less likely to remember details from the event than were more experienced drivers.

One reason may have been that they were not looking in the right places. Researchers have long known that inexperienced drivers have much different "visual search" patterns than more experienced drivers. They tend to look overwhelmingly near the front of the car and at the edge markings of the road. They tend not to look at the external mirrors very often, even while doing things like changing lanes. Knowing where to look—and remembering what you have seen—is a hallmark of experience and expertise. In the same way that eye-tracking studies have shown reliable differences in the way artists look at paintings versus the way nonartists do (the latter tend to zero in on things like faces, while artists scan the whole picture), researchers studying driver behavior can usually tell by a driver's glance activity how experienced they are.

Teenage drivers were, in many ways, the perfect next step for DriveCam. Like the drivers of commercial vehicles, teens are often driving someone else's car, and they are driving under the supervision of a higher authority—in this case, Mom and Dad. A trial in Iowa put DriveCams in the cars of twenty-five high school students for eighteen weeks. Triggered events were sent to parents, and the scores (using an anonymous ID) were posted so the drivers could judge exactly where they stood in relation to their peers. According to Daniel McGehee, the trial's head and director of the Human Factors and Vehicle Safety Research Program at the University of Iowa's Public Policy Center, teenagers in Iowa, because of its agricultural character, can begin driving to school at fourteen. "That crash rate is absolutely out of sight," he said. Teenagers in Iowa also drive a lot: In thirteen months of driving, the twenty-five drivers put over 360,000 miles on the odometer, many of them on the statistically most dangerous roads: rural two-lane highways.

The early clips he showed were indeed troubling: drivers sailing heedlessly through red lights, or singing and looking around absentmindedly before flying off a curve into a cornfield. Admittedly, I felt a bit uneasy peering into this little cocoon of privacy during these moments of raw, unfiltered emotion. Apparently the teens, in this age of reality television, were not so shy. The DriveCam contains a button that drivers can press to add a comment about a triggered event. Some teens used it to record diary entries, a sort of dashboard confessional about events in their lives outside the car. Driving also provided a rather unique window on to the social lives of teens, McGehee told me. "We could tell when someone got a new girlfriend or boyfriend. They would drive more aggressively to show off."

But it was the safety effects, not the video confessions or dating habits, that interested the researchers. When I spoke to McGehee later, he was in the sixteenth week of the trial. "The riskiest drivers dropped their safety-relevant behaviors by seventy-six percent," he said. "The farther we get into this, the risky behaviors are just drying up." Whereas before, the riskiest drivers had been triggering the device up to ten times a day, McGehee said, they were now triggering it only once or twice a week. "Even the magnitude of those triggers is pretty benign relative to their early days," he noted. "They still might be taking a corner a little too fast but it might be right above the threshold."

What was really happening to the teens? Were they afraid of getting in trouble with their parents? Were they just seeing their own mistakes for the first time? Or were they simply gaming the system, trying to crack the code like they do with their SATs? "I think what you see is that drivers in this pure behavioral psychology loop are becoming sensors themselves," McGehee said. "This little accelerometer in there—they start to sense over time what the limit is." As DriveCam's Weiss put it, "One kid said, 'I figured out how to beat the system. I just look way ahead and anticipate traffic and slow down for corners, and I haven't set it off in a month.' " He was, whether he realized it or not, acting like a good driver.

But what happens when the DriveCam is gone? "I don't pretend to represent DriveCam as anything but an extrinsic motivation system," Moeller had said. He admits that in the early days of a DriveCam trial, the mere presence of the camera is enough to get drivers to act more cautiously, in a version of the famous "Hawthorne effect," which says that people in an experiment change their behavior simply because they know they are in an experiment. But without any follow-up coaching, without "closing the feedback loop," results begin to erode. "The driver starts to think, 'The camera's not intrusive at all. Nothing's ever going to happen—this is just there so in case I get in a crash this will record who was at fault,' " Moeller said. "When you inject coaching in, then he realizes there is an immediate and certain consequence for his risky driving behavior. That twenty-second loss of privacy is enough for most people."

The things that DriveCam finds itself coaching drivers on most often do not involve actual driving skills per se—like cornering ability or obstacle avoidance—but mistakes that are born from overconfidence. The most striking example of this came in a trial that Weiss, then with the Mayo Clinic in Minnesota, did with an ambulance company that was trying to improve the "ride experience" for patients. One might think the DriveCam would have been triggered quite regularly in emergency situations, when the drivers, with lights and sirens, were speeding their patients to the hospital, careening around corners, and slaloming through red lights. That was not the case. "It's actually smoother when you have the red lights and siren on, is how it turned out," Weiss explained. "We triggered more events—we had harder cornering and more erratic driving—when they were just doing their own thing." Weiss, himself a former ambulance driver and paramedic, suspected he knew why. "The big difference between running lights and a siren and your normal driving is that you're focused. They're seeing the hazards that are out there and they're slowing sooner when someone can't see them. Smoother is quicker when you're running lights and a siren."

Since most of us don't have sirens and lights, our driving is of the everyday variety. As the sense of routine begins to take over, we begin to ratchet up our sense of the possible—how close we can follow, how fast we can take curves—and become conditioned to each new plateau. We forget those things that the Stanford researchers were learning as they tried to teach their robot to drive: It is not as easy as it appears. Lisk, who had that morning reviewed a sheaf of collision reports, said that "the large majority were just people who didn't have enough space, or were not attentive enough. A lack of good old-fashioned basic driving skills was a huge part of it."

He showed one clip, of a driver moving rather quickly down an open lane toward a tollbooth, flanked on either side by queues of cars. "The driver's thinking it's wide open. It's a football mentality—I've got all my blockers and I can go," Lisk said. It's as if the driver has already imagined himself to have passed through the lines of cars and past the open tollbooth. There is just one problem: All those other drivers are eagerly salivating over that same space. "Because they're boxed in they've got to come in a pretty abrupt angle and at low speed," Lisk said. "We see a lot of collisions where the driver hasn't slowed down enough when they're approaching that high-risk, open-lane situation."

This may help explain why EZ Pass–style automated payment lanes at tollbooths, which should theoretically help reduce crashes at these statistically risky areas—drivers no longer have to fumble for change—have been shown to increase crash rates. Drivers approach at a higher speed, with nothing to stop them from zooming through the toll plaza, while other cars, finding themselves in the "wrong" lanes, dart out and jockey among lanes more than they would have under the old system, in which there was less chance of finding a shorter queue.

Each month, DriveCam receives more than fifty thousand of these triggered clips, making it, Moeller said, the world's largest "repository of risky driving behavior." The technology of the camera is allowing glimpses into what has been, for most of the automobile's existence, a kind of closed world: the inner life of the driver.

"Driver behavior" has previously been teased out through things like driving simulators, test tracks, or actually having a researcher sit in the car, clipboard in hand—none of which is quite like real-world driving. Cars could be watched from the outside, via cameras or lab assistants on highway overpasses, but that did not give any glimpse into what the driver was doing. The study of crashes was based largely on police investigations and witness reports, which are both prone to distortion—the latter particularly so.

People are more likely to assign blame to one person or another when a crash is severe, research has shown, than when it is minor. In another study, a group of people were shown films of car crashes. When the subjects were asked, a week later, to gauge the speed of various cars in the films, they estimated higher speeds when the questions used the word "smash," versus words like "hit" or "contacted." More subjects remembered seeing broken glass when the word "smash" was used, even though no glass was broken. A driver's own memory of events is usually clouded by a desire to lessen their own responsibility for an event (perhaps so as to not conflict with their enhanced self-image or to avoid legal liability). "Baker's law," named after crash reconstructionist J. Stannard Baker, notes that drivers "tend to explain their traffic accidents by reporting circumstances of lowest culpability compatible with credibility"—that is, the most believable story they can get away with.

Most elusive of all, before Drivecam-style devices, were the crashes that *almost* happened. There was no way to determine why and how they nearly occurred (or did not), nor how often these near misses took place. If the top of the triangle was murky, the bottom of the triangle was as vast a mystery as the deepest ocean floor.

That has now changed, and large-scale studies, using technology like DriveCam's, are providing new clues into how drivers behave and, most important, new insight into just why we encounter trouble on the road. The answer is not so much all the things that the road signs warn us about—the high winds on bridges or the deer crossing the highway. Nor is it mostly tire blowouts, faulty brakes, or the mechanical flaws that prompt car makers to issue recalls ("human factors" are said to account for 90 percent of all crashes). Nor does it seem to be "driver proficiency" or our ability to understand traffic signals.

What seems to gives us the most trouble, apart from our overconfidence and lack of feedback in driving, are the two areas in which Stanley and Junior, Stanford's clumsy robot drivers, have a decided edge. The first is the way we sense and perceive things. As amazing as this process is, we do not always interpret things correctly. More important, we aren't always aware of this fact. The second thing that separates us from Stanley and Junior on the road is that we are not driving machines: We cannot keep up a constant level of vigilance. Once we feel we have things under control, we begin to act differently. We look out the window or talk on a cell phone. Much of our trouble, as I will show in the next chapter, comes because of our perceptual limitations, and because we cannot pay attention.

Chapter Three

# How Our Eyes and Minds Betray Us on the Road

## Keep Your Mind on the Road: Why It's So Hard to Pay Attention in Traffic

> Any man who can drive safely while kissing a pretty girl is simply not giving the kiss the attention it deserves.
>
> —Albert Einstein

Here is a common traffic experience: You are driving, perhaps down a mostly empty highway, perhaps on the quiet streets around your house, when you suddenly find yourself "awake at the wheel." You realize, with a mixture of wonder and horror, that you cannot remember what you have been doing for the past few moments—nor do you know how long you have been "out." You may find yourself sitting in your driveway and asking, as the Talking Heads once did, "How did I get here?"

This phenomenon has been called everything from "highway hypnosis" to the "time-gap experience," and while it has long puzzled people who study driving, it is still not fully understood. What is known is that it usually happens in fairly monotonous or familiar driving situations. Some scientists suggest that it's related to drowsiness, and that we may even be taking what are called "microsleeps" at the wheel.

What is also unclear is how much attention we were actually paying to the road while under the spell of highway hypnosis versus to what extent we have simply forgotten everything that happened during that period. You may have wondered why you did not drift off the side of the road. Perhaps you were lucky; one study that had subjects drive for several (boring) hours in a driving simulator found that the roughly one in five drivers who succumbed to "driving without awareness"—as measured by EEG readings and eye movements—drifted out of their lane one-third of the time. You may have wondered what would have happened if a car (or bike or small child) had veered into the lane while you were zoning out. Would you have responded in time? *Did* a near accident almost happen during that period, one that you have since forgotten about?

Think back to the blank stares of drivers monitored by DriveCam. Why is it so hard to pay attention while we are driving? How and why do our eyes and mind betray us on the road?

Driving, for most of us, is what psychologists call an "overlearned" activity. It is something we're so well practiced at that we're able to do it without much conscious thought. That makes our life easier, and it is how we become good at things. Think of an expert tennis player. A serve is a complex maneuver with many different components, but the better we become at it, the less we think of each individual step. This example comes from Barry Kantowitz, a psychologist and "human factors" expert at the University of Michigan; he has spent years studying the safest and most efficient ways for humans to interact with machines, working with everyone from NASA pilots to operators of nuclear power plants. "One of the interesting things about learning and attention is that once something becomes automated, it gets executed in a rapid string of events," he says. "If you try to pay attention, you screw it up." This is why, for example, the best hitters in baseball do not necessarily make the best hitting coaches. Coaches need to be able to explain what to do; Charley Lau, the legendary batting coach and author of the classic book *The Art of Hitting .300*, never actually hit .300 himself.

The more overlearned an activity becomes, the less cognitive workload it imposes—though studies suggest that even the most mundane activities, like switching gears, never become fully automatic. The task always costs something. Having less workload is, on the one hand, a good thing. If, while driving, we were to really process every potential hazard, carefully analyze every motion and decision, and break down each maneuver into its component parts, we would quickly become overwhelmed. People who bring test subjects into driving simulators find something like this happening. "We're not going to get a driver to be one hundred percent vigilant to the driving task, because we would all get out of the car sweating," according to Jeffrey Muttart, a crash investigator and researcher at the University of Massachusetts. "If you see people get out of a driving simulator test, almost the first thing they do is take a deep, cleansing breath. Because I'm frying their brains. This is a ten-minute drive, and they want to try hard to do well."

Too little workload has its own problems. We get bored. We get tired. We lapse into highway hypnosis. We may make errors. Anyone who has (like me) put on mismatched socks or run the coffeemaker without adding coffee or water will be aware of this phenomenon. The absolute ease of the activity allows the mind to wander. A classic psychological principle, the Yerkes-Dodson law, posits that the ability to learn is harmed by too little—or too much—"arousal." This idea applies as well to human performance. Driving in North Dakota is on the low side of the curve, driving in Delhi on the high side. The ideal conditions presumably lie somewhere in between.

But where? Most driving rarely requires our full workload. So we listen to the radio, look out the window, or, increasingly, talk on the cell phone or read text messages—in the case of one fatal crash in California, the driver may have been operating a laptop computer as he drove. Or we may change the way we drive—we speed up because driving does not seem overly taxing. To the extent that this keeps us in the middle of the Yerkes-Dodson curve, it's a good thing. But the problem with driving is that we never know for sure when things are going to change very quickly, when that nice empty road—seemingly safe for a cell phone conversation—is going to turn into an obstacle course. We may also be unaware of just how much workload our secondary activity is consuming.

"Let's say you're driving on a straight road. It's relatively easy. I could ask you to do arithmetic at the same time and it wouldn't mess up your driving," Kantowitz said. "If you're driving on a curved road, especially if it's sharp curve, that takes more attention if you're to keep the car operating safely within the lane. If I ask you to do mental arithmetic on a curve you'll do it more slowly and you'll screw it up. Or if you do it well you'll screw up the driving." A study by a Danish researcher found that those same types of arithmetic problems took longer to do when driving in a village than on a highway.

This raises another point: Researchers look at how driving is affected when people do other things, but research also shows that secondary tasks suffer as well. We become worse drivers *and* worse talkers. This is obvious to anyone who has listened to the wandering, interrupted musings of a driver talking on a phone (journalists know that people calling from their cars give terrible interviews). As Kantowitz put it, "There's no free lunch."

"My basic belief after almost forty years of studying this stuff is that people can't time-share at all," Kantowitz told me. "You only get the appearance. It's like speed-reading. You think you can read really fast but your comprehension disappears. You can give the illusion of time-sharing if it's simple information, but in general we're not built for time-sharing." Think of the annoying crawl type found on the bottom of the screen on CNN and other news networks. We are led to believe that this is how people now process information, as if we are suddenly genetically programmed to multitask. Studies have shown, however, that the more information there is on the screen, the less we actually remember.

The relative ease of most driving lures us into thinking we can get away with doing other things. Indeed, those other things, like listening to the radio, can help when driving itself is threatening to cause fatigue. But we buy into the myth of multitasking with little actual knowledge of how much we can really add in or, as with the television news, how much we are missing. As the inner life of the driver begins to come into focus, it is becoming clear not only that distraction is the single biggest problem on the road but that we have little concept of just how distracted we are.

In the largest study to date of the way we actually drive today, the Virginia Tech Transportation Institute, working with NHTSA, equipped one hundred cars in the Washington, D.C., and northern Virginia area with cameras, GPS units, and other monitoring devices, and then set about recording a year's worth of what it calls "pre-crash, naturalistic driving data." After poring over forty-three thousand hours of data and more than two million miles of driving, the study found that almost 80 percent of crashes and 65 percent of the near crashes involved drivers who were not paying attention to traffic for up to three seconds before the event.

That period of time is critical. "A total time of two seconds looking away from the forward roadway is when people start to get in trouble," explained Sheila "Charlie" Klauer, a researcher at VTTI and the study's project manager. "That's when they get to the point when they are starting to lose track of what's going on in front of them." The two-second window is not technically related to the "two-second rule" for following distance, but the comparison is instructive. The point is that a lot can happen in two seconds—like colliding with the car in front if it came to a stop or slowed—but drivers, lulled by the expectancy that it will not stop, drive as if the world will not have changed when they return their eyes to the road after that two seconds. They drive as if the world is a television show viewed on TiVo that can be paused in real time—one can duck out for a moment, grab a beer from the fridge, and come back to right where they left off without missing a beat. For many of the crashes, Klauer found that "the eye glance happened to be at exactly the wrong time. If they had not chosen to look away at that very second they would have probably been okay."

The sources of distraction inside a car have been painstakingly logged by researchers. We know that the average driver adjusts their radio 7.4 times per hour of driving, that their attention is diverted 8.1 times per hour by infants, and that they search for something—sunglasses, breath mints, change for the toll—10.8 times per hour. Research has further revealed just how many times we glance off the road to do these things and how long each glance takes: In general, the average driver looks away from the road for .06 seconds every 3.4 seconds. "On average, radio tuning takes seven glances plus or minus three," said Linda Angell, a safety researcher at General Motors, in a conference room at the Technical Center in Warren, Michigan. "That's for an oldish radio. We do better with the modern radio, which zeroes you in on the right region." Most of these glances, Angell noted, do not take our eyes off the road for longer than 1.5 seconds. But there are exceptions, such as "intense displays" (e.g., lots of features) or looking for a button you have not pressed in a while. The iPod is changing the equation yet again: Studies have shown that scrolling for a particular song takes our eyes off the road for 10 percent longer than simply pausing or skipping a song—plenty of time for something to go wrong.

Even a succession of very short glances, less than two seconds each, can cause problems. Researchers talk of the "fifteen-second rule," which indicates the maximum amount of time a driver should spend operating any kind of in-car device, whether navigation or radio, even as they are (at least occasionally) looking at the road. "What we believe is that task time is very important," Klauer said. "The longer the task time, the more dangerous the task is, and the greater the crash risk." And so a fifteen-second task might require only short glances at the device, but, Klauer said, "that risk increases every time the driver looks away."

The study found that while dialing a cell phone put drivers at a greater crash risk, talking on a cell phone presented only a slightly higher risk than normal driving. "When a driver is talking or listening on their cell phone, at any given moment within that conversation what our odds ratio is telling us is they're only at a slightly higher crash risk than an alert driver. Statistically speaking, it's not different," Klauer said. Does that mean talking on a cell phone is safe? Maybe it's all that *dialing* we need to worry about. But the study also found that talking (or listening) on a cell phone was a contributing factor in as many crashes as dialing was. "We think that's probably true because while dialing is a much more dangerous task while the driver's doing it, the task is fairly short," Klauer told me. "But drivers typically talk on their cell phone for a long period of time. Over that long period of time a lot more crashes and near crashes are more apt to occur. That slight increase in crash risk is starting to add up." As more drivers talk for longer periods, Klauer said, "it's going to become a lot more dangerous."

The reason we talk for a long time on our cell phones is related to the reason we all think we are better drivers than we are, and to the thing that also makes us think we are better drivers on our cell phones than we are: lack of feedback. Cell phone users are not aware of the risk because, by all surface measures, they seem to be driving fine. Traffic affords us these illusions—until it does not, as the hundred-car study showed. "Cell phone conversations are particularly insidious because you don't notice your bad performance, particularly the cognitive side," John Lee argues. "So if you're dialing the phone, you get immediate feedback because you don't quite stay in the lane, because you're punching the buttons." Once the dialing is done, the driver can again look at the road. The weaving stops. They seem to be in control.

Drivers may confidently assume they can adequately compensate for talking on a cell phone or texting on a BlackBerry by lowering their speed or putting more space between their own car and the car ahead of them, but the evidence gleaned from the hundred-car survey suggests otherwise. One might think, for example, that rear-end collisions most commonly occur because the driver behind was following too closely. Yet the study found that the majority of rear-end crashes happened when the following car was *more* than two seconds away from the car it struck. "I think people compensated a little bit for their inattention," Klauer said. " 'I need to answer this cell phone, I need to look at these papers on the seat next to me.' So they back off the lead vehicle and give themselves some space. Then they start to engage in something else. Then something unexpected happens and they're in trouble."

The drivers were redistributing workload. With more of their attention devoted to a cell phone conversation, they may have had to work just a bit harder to stay in their lane; similarly, the narrower the lane, the more mental energy it takes to stay in that lane (my own theory is that cell phones in cars have contributed to the seeming death of signaling for turns). Driving closer to someone also requires more mental energy, as does driving fast. We can usually feel this starting to take a toll, so we do things like drop back from a car in front of us or slow down. Clearly we do not always compensate enough, and there is evidence to suggest that we hardly compensate at all for our cell phone impairment when we're doing things like changing lanes.

Something similar happens with very new drivers on highways: So much of their mental concentration is devoted to simply staying in the lane, they have trouble paying attention to their speed. And it is not only drivers who suffer, as anyone who has walked behind someone talking on a mobile phone has noticed. When psychologists have asked people to walk around a track while memorizing words that were shown to them, walking speeds slowed as the mental task got harder. Similarly, researchers in Finland have found that pedestrians using mobile devices walked more slowly *and* were less able to interact with the device, pausing occasionally to "sample the environment." But pedestrians on cell phones do not sample the environment as often as they should, as a study of a Las Vegas crosswalk showed: Those talking on cell phones were less likely to look at traffic while crossing *and* took longer to do so.

Our attention, like a highway dropping from three lanes to two lanes, suffers from a bottleneck, one theory claims: Only so much can get through at once. Trying to squeeze more mental "cars" past the bottleneck means we have to slow them all down, space them out—or it means that some of those cars might drive off the road. In the hundred-car study, something else was also happening when drivers got on their cell phones. They began to look almost exclusively straight ahead, much more so than they did when they were not on their cell phones. They were, by external measures, "paying attention." But keeping one's eyes on the road is not necessarily the same thing as keeping one's mind on the road.

Consider for a moment the incredibly complex question of what it even means to pay attention while driving. There are an infinite number of things we could notice if we chose to, or had the spare mental capacity. But through practice and habit we learn to expertly analyze complicated scenes and extract only the information we need, ignoring the rest. New drivers, as we have seen, look rather rigidly ahead and near the front of the car, using "foveal" rather than peripheral vision to help them stay in their lane. As drivers get more experienced, they cast their eyes farther out along the road, barely registering the pavement markings. This happens without their even noticing. Experiments have been done in which researchers pulled over drivers on the highway and asked them if they recalled having seen certain traffic signs. The recall rates were as low as 20 percent. Were drivers simply not seeing things? One study found that the remembered signs were not necessarily the most visible ones but the signs that drivers judged most important (e.g., speed limit). This suggests that drivers saw enough of the signs to process what they were, at some subconscious level, and then effectively forgot most of them.

We do this sort of thing all the time—and for good reason. Remembering traffic signs we have seen is not useful to our lives. Steven Most, a psychologist at the University of Delaware, compares the flow of information and images we get in daily life to a stream passing through our heads. Unless we stop to "scoop up" some of that water—or "capture" it with our attention—it will flow in and out of our minds. "Sometimes, you attend to things enough to be aware of them in the moment, but that encoding process isn't necessarily taking place," he told me. "The awareness is there but not the memory of the awareness. When attention is distracted enough, it's even questionable whether we have that momentary awareness."

The reason we notice things like signs while driving is not as simple as it might seem. The average driver, asked why he saw a stop sign, might say, "Because it was there" or "Because it's the color red, and humans are hard-wired to see red more easily." But often we see a sign simply because we know where to look for one. This curious fact was explained by Carl Andersen, a vision specialist with the Federal Highway Administration, in a laboratory filled with eye-catching prototype warning signs in bold new colors like "incident pink." "If drivers are in an area that they already know, they almost don't even see the sign, because they already know it's there," Andersen said. This is known as "top-down processing." We see something because we are looking for it. To see things that we are *not* looking for, like unexpected stop signs, we need to rely on "bottom-up processing." Something has to be conspicuous enough to catch our attention. "If you're on one of those divided state highways, the older highways, you're not expecting to stop," Andersen said. "You'd better have advance signing and reduce the speed to prepare people for it."

Drivers actually look at most traffic signs at least twice: once for "acquisition" and again for "confirmation." Curiously, we do not really read things like stop signs. "Studies have been done where they intentionally misspell 'stop,' " Andersen said. "Everybody stops and then they drive off. They query the people later and the vast majority never saw that it was misspelled." (In fact, they may not have even seen it; it's estimated that one-fifth of our viewing time is interrupted by blinks and what are known as saccades, or our eyes' rapid movements, during which we are, as one expert puts it, "effectively blind.") Other studies, in driving simulators, have done things like change "No Parking" signs briefly to stop signs, and then back again. When the signs were at intersections, where stop signs usually are, drivers were more likely to notice the change. When they popped up elsewhere (e.g., at mid-block), drivers hardly ever noticed the change. When the drivers did see the sign change from "No Parking" to "Stop" at the intersection, they did not see it change back to "No Parking." Their decision to stop, the researchers noted, had already been made.

What does this have to do with real driving? After all, traffic signs do not change capriciously. A lot of things in traffic *do* change, however, and the question of whether we will notice those things depends not just on how visible they are but, indeed, on whether or not we are looking for them and how much spare capacity we have to process them. In a now-famous psychological experiment, a group of researchers had subjects view a video that showed a circle of people passing a basketball around. Half wore white shirts, half wore black. The subjects were asked to count the number of passes. What at least half the subjects did not notice was that a person wearing a gorilla suit passed right through the middle of the circle of basketball players. They were suffering from what has been called "inattentional blindness."

The idea that people could not see something as striking as a gorilla in a group of basketball players, although their eyes were locked on the video screen, suggests just how unstable and selective attention is—even when we are giving something our "undivided" attention. "There's an unlimited amount of information in the world, but our capacity for attending to information is pretty limited," explained Daniel Simons, a psychologist at the University of Illinois and the coauthor of the gorilla study. "If you're limited in how many things you can pay attention to, and attention is a gateway to consciousness, then you can only be aware of a limited subset of what's out there."

Inattentional blindness, it has been suggested, is behind an entire category of crashes in traffic, those known as "looked but did not see accidents." As with the gorilla-experiment subjects, drivers were looking directly at a scene but somehow missed a vital part—perhaps because they were looking for something else, or perhaps because something came along that they were *not* looking for. All too often, for instance, cars collide with motorcycles. One of the most frequently cited reasons is "failure to see," and these events are so common that motorcyclists in England have taken to calling them SMIDSYs, for "Sorry, Mate, I Didn't See You."

Many people assume that "failure to see" means that the motorcycle itself was difficult to see, because of its size or its single headlight. But it may also be that car drivers tend to be on the lookout for other cars when entering an intersection or turning across a lane of oncoming traffic. They may be in a sense "looking through" the motorcycle, because it does not fit their mental picture of the things they think they should be seeing. This is why safety campaigns (e.g., "Watch for motorcycles" or the United Kingdom's "Take longer to look for bikes") stress the idea of drivers simply being aware that motorcycles are out on the road. "The common intuition is that we first see things in the world and then interpret the scene in front of us," said Most. "What this work shows is that it's possible that the idea you have in mind actually precedes the perception and affects what you see. Our expectations and knowledge of what's in a scene influence what we see in a scene."

These expectations might also help explain the troublingly high numbers of emergency vehicles that are struck on the highway, even as they sit on the shoulder with their lights flashing brightly (and despite the fact that most places have laws requiring drivers to change lanes or slow down in the presence of an ambulance). These incidents are so common that the term "moth effect" has been coined for them. The idea is that drivers are lured to the lights, like moths to a flame.

What could cause a moth effect? There are many theories, ranging from arguments that we tend to steer where we look (which raises the question of why we do not drive off the road every time we see something interesting) to the idea that humans instinctively look toward light (ditto). Other researchers have argued that the fixation of attention on the roadside leaves drivers less able to judge their position in the lane. Many moth effect crashes involve alcohol-impaired drivers, perhaps no surprise in light of work that suggests that alcohol has a particularly deleterious effect on our eyes' ability to perceive depth or direction while we are moving.

The simplest explanation may be that most drivers, upon seeing a car on the highway, assume that it is moving at the same high speed as everyone else—and cars with flashing lights are usually moving even *faster* than that. One study, conducted in a driving simulator, showed that drivers reacted more quickly when stopped police cars were parked at an angle to oncoming traffic, rather than straight ahead in the direction of traffic. As the two vehicles were essentially equally conspicuous, the reason the angled car was seen sooner had less to do with visibility than in how the drivers interpreted what they saw: a car that was obviously not moving in the direction of traffic. (This ability to interpret seemed to be a by-product of driving experience, as novice drivers had the same reaction times for both cars.)

Even when we see an unexpected hazard, the fact that it's outside our "attentional set" means we are slower to react to it. This is demonstrated in a classic psychological test for what is known as the "Stroop effect." Subjects are shown a list of color names; these words are printed in the same color as the name as well as in other colors. Naming the color a word is printed in, it turns out, typically takes longer when the word does not match the color; that is, it takes longer to say "red" when the word printed in red is "yellow" than when it's "red." One argument for why this happens is that while reading is for us an "automatic" activity, naming colors is not. The automatic gets in the way of the less automatic (as with the stereotyping studies in Chapter 1). But other theories suggest that attention is involved. That we can name the correct color when the word itself is "wrong" suggests that we can train our attention on certain things; yet the fact that it takes us longer to do it shows that we cannot always screen out the things on which we are not focused (i.e., the word itself).

What this means for traffic was highlighted in a study by Most and his colleague Robert Astur. Drivers on a computer driving simulator, navigating through an ersatz urban environment, were asked to look for an arrow at every intersection telling them where to turn. For some drivers, the arrow was yellow and for others it was blue. At one intersection, an approaching motorcycle, itself either blue or yellow, suddenly veered in front of the driver and stopped. Drivers' reaction times to slam on the brakes were slower—and their collision rates were higher—when the motorcycle was a different color than the arrow. In a purely bottom-up form of processing, we might expect the motorcycle to stand out because it is different; but because we are looking at the scene from a top-down perspective, the odd-colored motorcycle is less visible because it is different from those things for which we are searching.

This attention disorder could also help explain the "safety in numbers" phenomenon of traffic, as described by Peter Lyndon Jacobsen, a public-health consultant in California. You might think that as there are more pedestrians or cyclists on a street, the more chances there are for them to be hit. You are right. More pedestrians are killed by cars in New York City than anywhere else in the United States. But as Jacobsen found, these relationships are not linear. In other words, as the number of pedestrians or cyclists increases, the fatality rates per capita begin to drop. The reason, as Jacobsen points out, is not that pedestrians begin to act more safely when surrounded by more fellow pedestrians—in fact, in New York City, as a stroll down Fifth Avenue will reveal, the opposite is true. It is the behavior of *drivers* that changes. They are suddenly seeing pedestrians everywhere. The more they see, typically, the slower they drive; and, in a neatly perpetuating cycle, the more slowly they drive, the more pedestrians they effectually see because those pedestrians stay within sight for a longer period.

And so New York City, when one considers how many pedestrians it has, is actually one of the safest cities in the country for walkers. (One study, looking at 1997–98 figures, found the Tampa–St. Petersburg–Clearwater area to be the most dangerous for pedestrians.) To cite another instance, the Netherlands has a much lower fatality rate per mile traveled for cyclists than does the United States. It is not likely that Dutch cyclists are any more visible in terms of pure conspicuity; they rarely wear reflective clothing, favoring stylish black coats instead, and instead of flashing lights their bikes carry things like tulips. Nor do the Dutch more regularly wear helmets than American cyclists; the reverse is actually true. Perhaps the Dutch just have better bike paths, or maybe the flat landscape makes it easier for drivers to spot cyclists. But the most compelling argument is that Dutch cyclists are safer simply because there are more of them, and thus Dutch drivers are more used to seeing them. Dutch culture may be quite different from American culture, but the "safety in numbers" theory also holds for comparisons within the United States—in Florida, for example, Gainesville, a college town with the highest cycling rate in the state, is in fact the safest place to be a cyclist. The lesson: When you see more of something, you're more likely to *see* that thing.

In the gorilla experiment, an added condition made subjects less likely to see the gorilla: when their job got harder. Some subjects were asked to count not just passes but the types of passes—whether they were "bounce passes" or passes made in the air. "You've made the attention task that much harder, and used up more of your available resources," Simons said. "You're less likely to notice something unexpected."

In driving, you might protest, we do not do such things as tally basketball passes. Still, there may have been times when you were concentrating so much on looking for a parking spot that you did not notice a stop sign; or you might have almost hit a cyclist because she was riding against traffic, violating your sense of what you expected to see. And there is another activity, one that we increasingly often indulge in while driving, that closely resembles that very specific act of counting basketball passes: talking on a cell phone.

Let me ask you two questions: What route did you take to get home today? And what was the color of your first car? What just happened? Chances are, your eyes drifted away from the page. Humans, perhaps to free up mental resources, tend to look away when asked to remember something. (Indeed, moving the eyes is thought to aid memory.) The more difficult the act of remembering, the longer the gaze away. Even if your eyes had remained on the page, you would have been momentarily sent away in a reverie of thought. Now picture driving down a street, talking to someone on a mobile phone, and they ask you to retrieve some relatively complicated bit of information: to give them directions or tell them where you left the spare keys. Your eyes may remain on the road, but would your mind?

Studies show that so-called visual-spatial tasks, such as rotating a letter or a shape in one's mind, cause our eyes to fixate longer in one place than when we are asked to perform verbal tasks. The longer the fixation, the thinking goes, the more attention we are devoting to the task—and the less we're giving to other things, like driving. The mere act of "switching" tasks—like moving from solely driving to talking on the phone while driving or, say, to changing whom we're speaking to within the same cell phone call via call waiting—takes its toll on our mental workload. The fact that the audio information we are getting (the conversation) comes from a different direction than the visual information we are seeing (the road ahead) makes it harder for us to process things. Bad reception on the phone? Our struggle to listen more carefully consumes even more effort.

Now replace the gorilla of the basketball experiment with a car making an unexpected turn or a child on a bike standing near the side of the road. How many of us would see it? "Driving's already attention-demanding enough—if you add in the cognitive demands of talking on a cell phone, you're taking away whatever limited resources you had, and you're that much less likely to notice something unexpected," Simons said. "You might be able to stay on the road just fine, and you might be able to stay the same distance behind a car on the highway, but if something unexpected happens—a deer runs into the highway—you might not react as easily."

The notion that we could miss unexpected things while talking on a cell phone is powerfully demonstrated by our seeming failure to notice the expected things. Two psychologists at the University of Utah found, after running a number of subjects through a simulator test, that drivers not talking on a cell phone were able to remember more objects during the course of the drive than those who were. The objects ranged in their "driving relevance"; that is, the researchers ranked speed-limit signs and those warning about curves as more critical than Adopt-a-Highway signs. You might suspect that the cell phone drivers were just filtering out irrelevant information, but the study found no correlation between what was important and what was remembered. Most strikingly, the drivers using cell phones *looked* at the same number of objects as the drivers without cell phones—yet they still remembered fewer.

Drivers using a cell phone, as noted in the hundred-car study, tend to rigidly lock their eyes ahead, assuming a super-vigilant pose. But that stare may be surprisingly hollow. In a study with an admittedly small sample size, I took the wheel of a 1995 Saturn one day at the Human Performance Laboratory at the University of Massachusetts in Amherst, and got set for a virtual drive in the lab's simulator. While I drove down a four-lane highway, a series of sentences was read to me via a hands-free cell phone. My task was to first judge whether the sentences made sense or not (e.g., "The cow jumped over the moon") and then repeat (or "shadow," as researchers call it) the last word in the sentence. As I did this, the direction of my gaze (among other things) was being monitored via an eye-tracking device mounted to a pair of Bono-style sunglasses.

When I later watched a tape of my drive that plotted where my eyes had been looking, the pattern was striking. Under normal driving, my eyes danced around the screen, taking in signs, the speedometer, construction crews in a work zone, the video-game landscape. When I was on the phone, trying to discern whether the sentence made sense, my eyes seemed to train on a point very close to the front of the car—and they barely moved. Technically, I was looking ahead—my eyes were "on the road"—but they were gazing at a place that would not be useful in spotting any hazards coming from the side or even, say, determining whether the truck several hundred feet ahead might be stopping. Which is exactly why I smashed into its rear end. "You were driving like a sixteen-year-old" is how Jeffrey Muttart described it to me.

Our eyes and our attention are a slippery pair. They need each other's help to function, but they do not always share the load equally. Sometimes we send our eyes somewhere and our attention follows; sometimes our attention is already there, waiting for the eyes to catch up. Sometimes our attention does not think that everything our eyes are seeing is worth its time and trouble, and sometimes our eyes rudely interrupt our attention just as it's in the middle of something really interesting. Suffice it to say that what we see, or what we think we see, is not always what we get. "This is the reason the whole 'keep your eyes on the road, your hands upon the wheel, use the hands-free handset' idea is a silly thing," Simons said. "Having your eyes on the road doesn't do any good unless your attention is on the road too."

As with the subjects in the counting test who did not see the gorilla, drivers (and particularly drivers talking on cell phones) would be shocked to learn, later, what they missed—precisely those things the in-car cameras are now revealing. "It is striking that people miss this stuff," Simons said. "At some level it's even more striking how wrong our intuitions are about it. Most people are firmly convinced they would notice if something unexpected happened, and that intuition is just completely wrong."

Human attention, in the best of circumstances, is a fluid but fragile entity, prone to glaring gaps, subtle distortions, and unwelcome interruptions. Beyond a certain threshold, the more that is asked of it, the less well it performs. When this happens in a psychological experiment, it is interesting. When it happens in traffic, it can be fatal.

## Objects in Traffic Are More Complicated Than They Appear: How Our Driving Eyes Deceive Us

Try to picture, for a moment, the white stripes that divide the lanes on a major highway. How long would you guess they are? How much space would you say lies between each stripe? When first asked this question, I guessed about five feet, with maybe fifteen feet between the stripes. You might estimate six or even seven feet. While the exact length varies, the U.S. standard calls for ten feet, though depending on the speed limit of the road, the stripes may be as long as twelve or fourteen feet. Take a look at an overhead photo of a highway: In most cases, the stripe is as long as, or longer than, the cars themselves (the average passenger car is 12.8 feet). The spacing between the stripes is based on a standard three-to-one ratio; thus, for a twelve-foot stripe, there will be thirty-six feet between stripes.

I use this as a simple example of how what we see is not always what we get as we move in the unnaturally high speeds of traffic. You may be wondering how it is that humans can even do things like drive cars or fly planes, moving at speeds well beyond that ever experienced in our evolutionary history. As the naturalist Robert Winkler points out, creatures like hawks, whose eyes possess a much faster "flicker fusion rate" than humans', can track small prey from high above as they dive at well over 100 miles per hour. The short answer is that we cheat. We make the driving environment as simple as possible, with smooth, wide roads marked by enormous signs and white lines that are purposely placed far apart to trick us into thinking we are not moving as fast as we are. It is a toddler's view of the world, a landscape of outsized, brightly colored objects and flashing lights, with harnesses and safety barriers that protect us as we exceed our own underdeveloped capabilities.

What we see while driving is a visually impoverished view of the world. As Stephen Lea, a researcher at the University of Exeter, explains it, what matters is less the speed at which we or other things move than the rate at which images expand on our retinas. So in the same way that we easily observe a person 3 yards away jogging toward us at 6 miles per hour, we have little trouble tracking a car that is 30 yards away moving at 60 miles per hour. The "retinal speed" is the same.

While driving, we get a gently undulating forward view. Things are far away or moving at similar speeds, so they grow slowly in our eyes, until that moment when the car in front suddenly and jarringly "looms" into view (and you notice their bumper sticker: IF YOU CAN READ THIS, YOU'RE TOO CLOSE). But now picture looking directly down at the road while you're driving at a good speed. It is, of course, a blur. This is no less part of the actual environment in which we are driving, but we are physically unable to see it with any accuracy. Luckily, we do not usually *need* to see it to move safely—though, as we shall learn, there are other ways in which traffic puts our visual systems to severe tests.

Traffic illusions actually hit us before we even get in the car. You may have noticed how in movies or on television, the spokes on a car's wheels sometimes seem to be moving "backward." This so-called wagon-wheel effect happens in movies because they are composed of a flickering set of images (generally twenty-four frames per second), even though we perceive them to be smooth and uninterrupted. Like the dancers in a disco captured briefly by a strobe light, each frame of that movie captures an image of the spokes. If the frequency of the wheel's rotation perfectly matched the flicker rate of the film, the wheel would appear *not* to be moving. ("I replaced the headlights in my car with strobe lights," the comedian Steven Wright once joked, "so it looks like I'm the only one moving.") As the wheel moves faster, though, each spoke is "captured" at a different place with each frame (e.g., we may see a spoke at the twelve o'clock position on one sweep, but at eleven forty-five on the next). So it seemingly begins to move backward.

As the cognitive psychologists Dale Purves and Tim Andrews note, however, the wagon-wheel effect can happen in real life as well, under full sunlight, when the "stroboscopic" effect of movies does not apply. The reason we still see the effect, they suggest, is that, as with movies, we perceive the world not as a continuous flow but in a series of discrete and sequential "frames." At a certain point the rotation of the wheel begins to exceed the brain's ability to process it, and as we struggle to catch up, we begin to confuse the current stimulus (i.e., the spoke) in real time with the stimulus in a previous frame. The car wheel is not spinning backward, any more than disco dancers are moving in slow motion. But this effect should provide an early, and cautionary, clue to some of the visual curiosities of the road.

"Motion parallax," one of the most famous highway illusions, puzzled psychologists long before the car arrived. This phenomenon can be most easily glimpsed when you look out the side window of a moving car (though it can happen anywhere). The foreground whizzes past, while trees and other objects farther out seem to move by more slowly, and things far in the distance, like mountains, seem to move in the same direction as us. Obviously, we cannot make the mountains move, no matter how fast we may drive. What's happening is that as we fixate on an object in that landscape, our eyes, to maintain their fixation, must move in a direction opposite to the way we're going. Wherever we fixate in that view, the things we see before the point of fixation are moving quickly across our retina opposite to the direction we are moving in, while things past the point are moving slowly across our retina in the *same* direction as we're traveling. (See the notes for a quick demonstration of motion parallax.)

All this eye movement and the relative motion of the objects we are seeing, as confusing as it seems, help us judge how far away things are from us. As Mark Nawrot, a psychologist at North Dakota State University and an expert in motion parallax, describes it, this is why film directors like Peter Jackson like to move the camera around a lot. Because we are sitting, stationary, in a theater, and thus cannot get the sort of depth cues our eyes give us when we move, Jackson moves the camera instead, to make the film appear more realistic. But the price we pay for the depth cues that motion parallax provides us is the occasional illusion that we may or may not consciously notice. In traffic, motion parallax may trick us into thinking that an object is far and stationary when, in reality, it is *near and moving*.

The mind can play tricks on what we see, but motion parallax reminds us that what we see while driving plays tricks on our minds. Sense and perception are connected by a quite busy two-way street. The white stripes on the highway and the distance between them are designed precisely as an illusion, to make these high speeds seem comfortable. If both the stripes and the distance between them were short, the experience might feel nauseating. In fact, in some places, engineers have tried to exploit this by employing "illusory pavement markings" to make drivers think they are going faster than they are. In one trial, a series of arrowlike chevrons were painted, ever closer together, on a highway exit ramp. The theory was that as the drivers began to pass more chevrons for each moment they drove, it would appear as if they were going faster than they really were, and would thus slow down. That study did find that drivers reduced their speed, but in other trials the results have been mixed. Drivers may slow once or twice simply because there are strange markings on the pavement, but they may also quickly acclimate to the markings.

These experiments have been focused on exit ramps because they are a statistically dangerous part of the highway. One crucial reason involves a particular illusion we face in traffic: "speed adaptation." Have you ever noticed, when driving from a rural highway onto a village road with a lower speed limit, how absolutely slow it feels? When you again leave that town to rejoin the rural highway and its higher speed, does the disparity seem as noticeable? The longer we drive at high speeds, the harder it is for us to slow down. Studies have shown that drivers who drove for at least a few minutes at 70 miles per hour drove up to 15 miles per hour faster when they hit a 30-miles-per-hour zone than drivers who had not previously been traveling at the higher speed.

The reason, as Robert Gray, a cognitive psychologist at the University of Arizona, explained to me, is something that might be called the "treadmill effect." After running on a treadmill for a while, you may have noticed that the moment you stop you may briefly experience the sensation of moving backward. As Gray describes it with driving, neurons in the brain that track forward movement begin to become fatigued as a person looking ahead drives at the same speed for a time. The fatigued neurons begin to produce, in essence, a negative "output." When a person stops (or slows), the neurons that track backward motion are still effectively dormant, but the negative output of the forward neurons fools you into thinking you're moving backward—or, if you're changing from high speeds to lower speeds, it can fool you into thinking you have slowed more than you actually have. The illusion cuts both ways, studies suggest: We *underestimate* our speed when asked to slow down and *overestimate* our speed when asked to speed up. This helps explain why we often go too fast coming off a highway (and hence the chevron patterns); it might also explain why drivers entering a highway frequently fail to reach the speed of traffic by the time they're merging (frustrating those in the right-hand lane who are forced to slow).

We misjudge speed in all kinds of ways. Our general perception of how fast and in what direction we are moving—indeed that we are moving at all—comes largely, it is thought, from what has been called "global optical flow." When we drive (or walk), we orient ourselves via a fixed point on the horizon, our "target." As we move, we try to align that target so that it is always the so-called focus of expansion, the nonmoving point from which the visual scenes seem to flow, approaching us in a kind of radial pattern—think of the moment in *Star Wars* when the *Millennium Falcon* goes into warp speed and the stars blur into a set of lines streaming away from the center of the ship's trajectory. The "locomotor flow line"—or what you and I would call the road—is the most crucial part of the optic field in driving, and the "textural density" of what passes by us influences our sense of speed. Things like roadside trees or walls affect the texture as well, which is why drivers overestimate their speed on tree-lined roads, and why traffic tends to slow between noise-barrier "tunnels" on the highway. The finer the texture, the faster your speed will seem.

The fineness of the road texture is itself affected by the height at which it is viewed. We sense more of the road's optical flow the closer we are to it. When the Boeing 747 was first introduced, as the psychologist Christopher Wickens has noted, pilots seemed to be taxiing too fast, on several occasions even damaging the landing gear. Why? The new cockpit was twice as high as the old one, meaning that the pilots were getting half the optical flow at the same speed. They were going faster than they thought they were. This phenomenon occurs on the road as well. Studies have shown that drivers seated at higher eye heights but not shown a speedometer will drive faster than those at lower heights. Drivers in SUVs and pickups, already at a higher risk for rollovers, may put themselves at further risk by going faster than they intend to. Studies have shown, perhaps not surprisingly, that SUV and pickup drivers speed more than others.

The reason we have speedometers, and why you should pay attention to yours, is that drivers often do not have a clue about how fast they're actually traveling—even when they think they do. A study in New Zealand measured the speed of drivers as they passed children playing with a ball and waiting to cross the street. When questioned, drivers thought they were going at least 20 kilometers per hour (or about 12 miles per hour) more slowly than they really were (i.e., they thought they were going 18 to 25 miles per hour when they were really doing 31 to 37). Sometimes it seems as if we need someone standing on the side of the road, actually reminding us how fast we are really going. This is why we see "speed trailers," those electronic signs posted by the road that flash your speed. These plaintive appeals to conscience are usually effective, at least in the immediate vicinity, at getting drivers to slow down slightly—but whether drivers want to keep slowing down, day after day, is another issue. The speed trailers work, when they do, because they give us crucial feedback—which, as mentioned in the previous chapter, we so often lack on the road. Some highway agencies, responding to rising numbers of often-fatal rear-end crashes, have tried to put feedback of sorts right on the road, in the form of painted dots that inform drivers of the proper following distances (in one case, someone responded by painting a dot-eating Pac-Man on the highway). Drivers' following distances have tended to increase after dots are put down. Noise also gives feedback: We know we are going faster when the amount of road and wind noise picks up. The faster we go, the louder it gets. But have you ever found yourself listening to the radio at a high volume and then suddenly noticed you were speeding? A variety of studies have shown that when drivers lose auditory cues, they lose track of how fast they're going.

The robot car Junior, as you will recall, did not need to be able to "see" brake lights because he knew exactly how far the car ahead of him was, to within a few meters. For humans, however, distance, like speed, is something we often judge rather imperfectly (hence the Pac-Man dots). Unfortunately for us, driving is really all about distance and speed. Consider a common and hazardous maneuver in driving: overtaking a car on a two-lane road as another approaches in the oncoming lane. When objects like cars are within twenty or thirty feet, we're good at estimating how far away they are, thanks to our binocular vision (and the brain's ability to construct a single 3-D image from the differing 2-D views each eye provides). Beyond that distance, both eyes are seeing the *same* view in parallel, and so things get a bit hazy. The farther out we go, the worse it gets: For a car that is twenty feet away, we might be accurate to within a few feet, but when it is three hundred yards away, we might be off by a hundred yards. Considering that it takes about 279 feet for a car traveling at 55 miles per hour to stop (assuming an ideal average reaction time of 1.5 seconds), you can appreciate the problem of overestimating how far away an approaching car is—especially when they're approaching *you* at 55 miles per hour.

Since we cannot tell exactly how far away the approaching car might be, we guess using spatial cues, like its position relative to a roadside building or the car in front of us. We can also use the size of the oncoming car itself as a guide. We know it is approaching because its size is expanding, or "looming," on our retina.

But there are a few problems with this. The first is that viewing objects straight-on, as with an approaching car, does not provide us with a lot of information. Think of an outfielder catching a fly ball—a seemingly simple act, but one whose exact mechanics still elude scientists (and the occasional outfielder). One thing that's generally agreed upon, as University of Missouri psychology professor Mike Stadler notes, is that balls are harder to catch when they are hit directly at a fielder. Fielders often have trouble gauging distance and trajectory, and they find they need to move back or forth a bit to get a better picture; studies have shown that fielders have a harder time judging which balls can or cannot be caught when they are asked to stand still. Viewing a car head-on or directly from behind, as we almost universally do, is like viewing a baseball hit right at you: It doesn't give us a lot to go on.

Another problem is that the image of that car, when it does begin to expand in our eyes, does not do so in a linear, or continuous, way. The book *Forensic Aspects of Driver Perception and Response* gives this example: A parked car that an approaching driver sees 1,000 feet away will double on the retina by the time the driver is 500 feet away. Sounds about right, no? But it will double again in the next 250 feet, and again in the last 250 feet. It is nonlinear. To put it another way, we can tell the car is getting closer—although this itself may take as much as several seconds—but we have no idea of the rate at which it is getting closer. This difficulty in judging closing distance also makes passing the lead car a problem; studies have shown that it is struck in about 10 percent of overtaking crashes. Another way to think about this is to imagine what happens to skydivers. For much of their fall, they have little sense, looking downward, of how fast they are falling—or even that they're falling at all. But suddenly, as the distance to the ground begins to come within the limits of human perception, they experience what is called "ground rush," with the terrain suddenly exploding into their range of view.

If all this was not enough to worry about, there's also the problem of the oncoming car's *speed.* A car in the distance approaching at 20 miles per hour makes passing easy, but what if it is doing 80 miles per hour? The problem is this: We cannot really tell the difference. Until, that is, the car gets much closer—by which time it might be too late to act on the information. One study that looked at how and when cars decided to pass other cars on two-lane highways found that they were as likely to attempt a pass when an oncoming car was approaching at 60 miles per hour as when it was coming at 30 miles per hour. Why? Because when the passing maneuver began, the cars were about 1,000 feet apart—too far to tell the speed of the opposing car. At those distances we are not even really sure if the car is coming toward us or not; the fact that it's in the opposite lane, or that we can see its headlights, might be the only giveaway.

So at the crucial distance where one must make a decision, the driver has no idea of a key variable: the "closing rate" of the other car. This is why you may have been forced to rather suddenly abandon your attempted passing and make either a voluntary or a forced return to your own lane. We "cheat" like this regularly, relying on a car's perceived distance without taking into account its speed. One study, looking at drivers' left turns across oncoming traffic, found that when the speed of approaching cars was doubled, drivers' estimates of the safe "gap" in which they could cross, which you would guess should have also doubled, went up by only 30 percent. These small discrepancies are the stuff of crashes.

Evidence suggests that we are sometimes fooled into thinking things are not as far away as they appear (and not only the approaching objects in our mirrors!). Studies have shown that people think small cars are farther away than they really are, either because we maintain a mental image of a larger car or because there is less of the car to actually see. Large objects, though, also create problems. Researchers have long been puzzled about the relatively high number of drivers killed while crossing railroad tracks—often when visibility was clear and warning signals were in place. It raises an obvious question: How could a driver not see something as large (and as loud) as a train? One answer is that a driver may have crossed the same set of tracks three hundred times in the last year without ever seeing a train, even when the signals were flashing. Did they simply not expect it on the 301st trip across the tracks? Did they "look but not see"? The influential psychologist and vision expert H. W. Leibowitz, in what has become known as the "Leibowitz hypothesis," offered another possible explanation: biases in the drivers' perceptual systems.

Large objects often seem to move more slowly than small objects. At airports, small private jets seem to go faster than Boeing 767s, even when they are moving at the same speed. Even experienced pilots who are aware of the actual velocities fall for this illusion. The reason, Leibowitz argued, is that there are two different subsystems that influence the ways our eyes move. One system is "reflexive"—we do it without conscious thought—and is triggered by seeing contours. This system helps us continually see things while we ourselves are moving.

We also use, more actively, "pursuit" eye movements. This is how we view moving objects when we are stationary. We can tell how fast something is moving, Leibowitz said, by how much effort it takes this "pursuit" system to see it, and by how much object there is to see. The larger the object, the less our voluntary systems have to work, and the slower the object seems.

How much slower? Judging by a test of the Leibowitz hypothesis done by researchers at the University of California at Berkeley, a lot slower. Subjects looking at a computer screen were asked to estimate the speed of a series of large and small spheres that moved toward them. Despite the presence of stationary posts and lines on the ground that subjects could use as helpful cues to judge speed, the study found that most people still thought a smaller sphere was moving faster—even when a larger sphere was moving 20 miles per hour faster. It was not until a large sphere was moving *twice* as fast as a smaller one that subjects were no longer convinced that the latter was moving faster.

The problem with visual illusions—and it has been argued that all human vision is an illusion—is that we fall for them even when we know they are illusions. Imagine that you are not even aware of your visual shortcomings. This is what happens when we drive at night. We think we can see better than we actually can—and we drive accordingly. We "overdrive" our headlights, moving at speeds that would not allow us to stop in time for something we saw in the range of our lights. Why do we do this? Leibowitz's theory was that when the ambient light goes down, we lose the use of certain eye functions more than we lose others, in a process he called "selective degradation." Our "ambient vision," which happens mostly on the peripheral retina, helps us with things like walking down the sidewalk or staying on the road; this degrades less at night. Because of this, and because the roadside and the center lines are brightly illuminated by our headlights (studies show that we look at these lines much more at night), we essentially think we are seeing all there is to see.

But another element of our vision performs much worse at night, Leibowitz argued: the focal vision of the central retina. This is what we use to identify things, and it is the more conscious part of our vision. Most of the time, there is nothing to see on the road at night except the red taillights of cars, road signs (which we see and remember more at night), the brightly reflective pavement markings, and the section of road just in front of the car that is bathed in the full glow of our headlights.

Yet when a nonilluminated object enters the road—an animal, a stalled car, a piece of debris, or a pedestrian—we cannot see it as well as we might have thought we would based on how well we seem to be seeing everything else. We are blind to our blindness. Remember this the next time you are out walking. Studies have shown that pedestrians think drivers can see them up to *twice* as far away as drivers actually do. According to one expert, if we were to drive at night in a way that ensured we could see every potential hazard in time to stop—what is legally called the "assured clear distance"—we would have to drive 20 miles per hour.

Another kind of illusion bedevils us in fog. When fog rolls in on a highway, the result is often a huge, multicar chain-reaction crash. An incident that occurred in 1998 near Padua, Italy, involving more than 250 cars (and the death of four people), is an extreme example of a rather common condition. These sorts of events must be due to poor visibility, no? Obviously, it is harder to see in a fog. But the real problem may be that it is *even more difficult to see than we think it is.* The reason is that our perception of speed is affected by contrast. The psychologist Stuart Anstis has a clever demonstration of this; he shows that when a pair of boxes—one colored light, the other dark—are moved across a background of black-and-white stripes, the dark box seems to move faster when it crosses the white sections, while the light-colored box appears to go faster as it crosses the black sections. The higher the contrast, the faster the apparent motion, so even though the two boxes are moving at the exact same speed, they look as if they are taking alternating "steps" as they shuffle across the stripes.

In fog, the contrast of cars, not to mention the surrounding landscape, is reduced. Everything around us appears to be moving more slowly than it is, and *we* seem to be moving more slowly through the landscape. The idea that we are not aware of this discrepancy is suggested in studies showing that while drivers tend to slightly reduce their speed in foggy conditions, they do not do so by enough to ensure a safe margin—even when special temporary warning signs have been set up. Ironically, drivers may feel more comfortable staying closer to the vehicle ahead of them—so that they do not "lose" them in the fog—but given the perceptual confusion, this is exactly the wrong move. Similar things happen in the whiteout conditions of snow, in which it is not uncommon for drivers to crash into the back of orange-colored snowplow trucks with flashing lights. The culprit is not a slippery roadway but low contrast. Drivers may see the back of the truck "in time," but as they think it is going faster than it actually is they may not brake accordingly.

A simple object, present on every car, is a symbol of the complex interplay of what we see and what we think we see on the road: the side rearview mirror. This itself is a curious, and rather overlooked, device. We might think of it as an essential safety feature, but it is unclear to what extent, if any, it has actually reduced the number of crashes. Moreover, studies show that many drivers do not use it during lane changes, the time when it would be most helpful, relying instead on glances over the shoulder. Then there is the issue of exactly what we are seeing when we look in that mirror. Depending on where you are in the world, either both side mirrors or just the passenger-side one will be convex, or curved outward. Because of the natural blind spots that exist beyond the edges of any car mirror, the decision was made, beginning in the 1980s, to reveal more of the scene at the expense of the driver's ability to correctly judge distance. Better to see a car improperly than to not see it at all. This is why convex mirrors come with a familiar warning: "Objects in mirror are closer than they appear."

But Michael Flannagan, a researcher at the University of Michigan's Transportation Research Institute, has argued that something very strange is going on when we look in that mirror. Mirrors of any stripe tend to puzzle us. As a simple experiment, trace the outline of your head in a foggy bathroom mirror. People tend to think they are tracing the actual size, whereas actually it is *half*. The convex side-view mirror presents a particularly distorted and what he calls "impoverished" visual scene, with many of the typical visual cues we use to judge the world rendered more or less invisible. The only thing that reliably indicates distance, Flannagan says, is the retinal size of the image of the car we see. But the size of the car, like the entire "world" depicted, has been shrunk by the convex mirror. The curvature of the mirror means that everything is in essence being drawn closer to the viewer, which is why it is puzzling that things actually look *farther* away.

But it gets trickier still. Researchers can predict, by measuring the viewing angles and the geometry of the mirror, how much the mirror is distorting the image. (This distortion is greater when a driver looks over to the passenger-side mirror than when he looks at his own, closer mirror; thus, Flannagan notes, it's a bit of a mystery why in the United States we do not allow driver's-side convex mirrors.) In a number of studies, however, Flannagan and his colleagues have found that people's estimates of the distance of objects is not as far off as the models predict they should be. "The vehicle behind you looks less far away than it ought to based on the smallness of the image size, as if people were somehow correcting a bit," he says. "They're not going on just this retinal size; they know something is making them less susceptible to the distortion on paper than they ought to be."

These puzzles led Flannagan and his fellow researchers to a conclusion that might serve as a better warning label for side-view mirrors: "Objects in mirror are more complicated than they appear." The same could be said of driving, as well as our ability to drive, and probably us too. It is all more complicated than it appears. We would do well to drive accordingly.

Chapter Four

# Why Ants Don't Get into Traffic Jams (and Humans Do): On Cooperation as a Cure for Congestion

## Meet the World's Best Commuter: What We Can Learn from Ants, Locusts, and Crickets

> When insects can follow rules for laneing, why couldn't we the humans?
>
> —road sign in Bangalore, India

You may feel you have the worst commute in the world: the grinding monotony of sitting in congestion, alternately pressing your brake and accelerator like a bored lab monkey angling for a biscuit; the drivers who stymie you with their incompetence; the slow deadening of your psyche caused by the ritual of leaving home forty-five minutes sooner than you would like so you can arrive at work ten minutes later than your boss would like.

And yet, in spite of all this mental and physical anguish, there's at least small consolation awaiting you at the end of your daily slog: Your fellow commuters did not try to eat you.

Consider for a moment the short, brutish life of *Anabrus simplex*, or the Mormon cricket, so named for the species' devastating attack on Mormon settlers in Utah in the legendary 1848 "cricket war." Huge, miles-long migratory bands of flightless crickets, described as a "black carpet unrolling across the desert," are still a dreaded sight in the American West. They travel many dozens of miles, munching crops and carrion. They heedlessly spill across roads, causing death for themselves and headaches for another traveling species, *Homo sapiens*, whose cars may slip on the dense mat of pulsating crickets. "Crickets on Highway" signs have been posted in Idaho. It turns out the insects are actually katydids, but the point is well taken.

Viewed as a scurrying mass, the Mormon cricket band seems a well-organized, cooperatively driven collective search for food—a perfect swarm designed to ensure its own survival. But when a group of researchers took a closer look at a mass of Mormon crickets on the move in Idaho in the spring of 2005, they learned that something more complicated was going on. "It looks like this big cooperative behavior," says Iain Couzin, a research fellow at the Collective Animal Behaviour Laboratory in Oxford University's zoology department and a member of the Idaho team. "You can almost imagine it like a group of army ants, sweeping out to find food. But in actual fact we found out it's driven by cannibalism." What looks like cooperation turns out to be extreme competition.

Crickets choose food carefully based on their nutritional needs at the moment, and they often find themselves wanting in the protein and salt departments. One of a cricket's best sources for protein and salt, it turns out, is its neighbor. "They're getting hungry and they're trying to eat each other," says Couzin, an affable Scotsman wearing a faded "Death to the Pixies" T-shirt, in his small office. "If you're getting eaten, the best thing for you to do is to try and move away. But if you're also hungry and trying to eat, the best thing to do is move away from others that are trying to eat you, but also to move toward others to try and eat them." For crickets in the back of the pack, crossing over ground that has already been stripped of food by those in the front, another cricket may be the only meal in sight.

This seems a recipe for anarchy, not well-coordinated movement. What is actually happening is an example of the phenomenon known as "emergent behavior," or the formation of complex systems, like cricket bands, that "emerge," often unexpectedly and unpredictably, from the simple interactions of the individuals. Looking at the swarm as a whole, one might not easily see what is driving the movement. Nor could one necessarily predict by studying the local set of rules guiding each cricket's behavior—eat thy neighbor *and* avoid being eaten by thy neighbor—that this would all end up as a tight swarm.

For complex systems to work the way they do, they need all, or at least a good number, of their component parts to play by the rules. Think of the "wave" at football stadiums, which begins, studies have shown, on the strength of a few dozen people; nobody knows, however, how many waves simply died for lack of participation, or because they tried to go in the "wrong" direction. What if some crickets got tired of avoiding their neighbors' ravenous jaws and decided to leave the swarm? Some of Couzin's colleagues hooked up small radio transmitters to a number of individual crickets, which were then separated from the larger band. Roughly half of those separated were killed by predators within days. Among the radio-tagged crickets kept within the band, none died. So whatever the risk of being eaten by one's neighbors, no matter how stressful and unpleasant the experience, it's still a better option than going solo.

What's remarkable about the formation of these systems is how quickly the rules—and the form of the group—can change. Another insect Couzin has studied, both in the Oxford lab and in the wild in Mauritania, is the desert locust *(Schistocerca gregaria)*. These locusts have two personalities. In their "solitarius" phase, they're harmless. They live rather quietly, in small, scattered groups. "They're shy, cryptic green grasshoppers," Couzin says. But under certain conditions, such as after a drought, these Dr. Jekylls of the insect world, driven into closer contact by the search for food, will turn into a vast brown horde of marauding, "gregarious" Mr. Hydes. The impact is massive: Swarming locusts may invade up to 20 percent of Earth's land surface at a time, Couzin says, affecting the livelihood of countless people. Knowing why and how these swarms form might help scientists predict where and when they will form. And so the team assembled a large group of Oxford-raised locusts, put them in an enclosed space, and used custom tracking software to follow what was going on.

When there are few locusts, they keep to themselves, marching in different directions, "like particles in a gas," says Couzin. But when forced to come together, whether in a lab or because food has become scarce in the wild, interesting things start to happen. "The smell and sight of other individuals, or the touch on the back leg, causes them to change behavior," Couzin says. "Instead of avoiding one another, they'll start being attracted to each other, and this can cause a sort of cascade." Suddenly, once the locusts reach a "critical density," they will spontaneously start to march in the same direction.

Now what does all this have to do with traffic? you may be asking. The most obvious answer is that what the insects are doing looks a lot like traffic and that what we are doing on the road looks a lot like collective animal behavior. In both cases, simple rules govern the flow of the society, and the cost for violating those rules can be high. (Picture the highway police car or crashes in the role of predator.) Insects, like humans, are compelled to go on the move because they need to survive. Similarly, if we did not need to provide for ourselves, many of us would probably not choose to drive at the very same time everyone else is. Like insects, we have decided that moving in groups—even if most of us are alone in our own cars—makes the most sense. Virtually since traffic congestion began, plans have been put forward to stagger work schedules so that everyone is not on the roads at the same time, but even today, with telecommuting and flextime, traffic congestion persists because having a shared window of time during which we can easily interact with one another still seems the best way to conduct business.

In both insect and human vehicular traffic, large patterns contain all kinds of hidden interactions. A subtle change in these interactions can dramatically affect the whole system. To go back to the comparison between the Late and the Early Merge, if each driver simply adheres to one rule instead of another—merge only at the last moment instead of merge at your earliest opportunity—the merging system changes significantly. Like the pattern of locusts' movement, human traffic movement often tends to change at a point of critical density. In a reversal of the way that locusts go from disorder to order with the addition of a few locusts, with the addition of just a few cars, smoothly flowing traffic can change into a congested mess.

The locust or cricket commuter, by staying within a potentially cannibalistic traffic flow, is, as Couzin suggests, clearly making the best of a bad situation. And in many ways, we act like locusts. Our seeming cooperativeness can shift to extreme competition in the blink of a taillight. Sometimes, we may be those harmless Dr. Jekylls, minding our own business, keeping a safe distance from the car in front. But at a certain point the circumstances change, and our character changes. We become Mr. Hyde, furiously riding up to the bumper of the person in front of us (i.e., trying to eat them), angry at being tailgated (i.e., trying to avoid being eaten), wishing we could leave the main flow but knowing it is still probably the best way home. One study, taken from highways in California, showed a regular and predictable increase in the number of calls to a road-rage hotline during evening rush hours. Another study showed that on the same stretch of highway, drivers honked less on the weekend than during the week (even after the researchers adjusted for the difference in the number of cars).

Another creature does things differently, taking the high road in traffic. This is the New World army ant, or *Eciton burchellii,* and these insects may just be the world's best commuters. Army ant colonies are like mobile cities, boasting populations that can number over a million. Each dawn, the ants set out to earn their trade. The morning rush hour begins a bit groggily, but it quickly takes shape. "In the morning you have this living ball of ants, up to five feet high, perhaps living in the crevice of a tree," says Couzin, who has studied the ants in Panama. "And then the ants just start swarming out of the nest. Initially, it's like a big amoeboid, just seething bodies of ants. Then after a period of time they seem to start pushing out in one direction. It's unclear how they choose that direction."

As the morning commuters spread out, the earliest ones begin to acquire bits of food, which they immediately bring back to the nest. As other ants continue pushing into the forest, they create a complex series of trails, all leading back to the nest like branches to a tree trunk. Since the ants are virtually blind, they dot the trails with pheromones, chemicals that function like road signs and white stripes. These trails, which can be quite wide and long, become like superhighways, filled with dense streams of fast-moving commuters. There's just one problem: This is two-way traffic, and the ants returning to the nest are laden down with food. They often move more slowly, and often take up more space, than the outbound traffic. How do they figure out which stream will go where, who has right-of-way, on "roads" they have only just built?

Interested in the idea that ants may have evolved "rules to optimize the flow of traffic," Couzin, along with a colleague, made a detailed video recording of a section of army ant trail in Panama. The video shows that the ants have quite clearly created a three-lane highway, with a well-defined set of rules: Ants leaving the nest use the outer two lanes, while ants returning get sole possession of the center lane. It is not simply, says Couzin, that the ants are magically sticking to their own chemical-covered separate trails (after all, other types of ants do not form three lanes). Ants are attracted to the highest concentration of chemicals, which is where the highest density of ants tends to be, which happens to be the center lane.

A constant game of chicken ensues, with the outbound ants holding their ground against the returning ants until the last possible moment, then swiftly turning away from the oncoming traffic. There is the occasional collision, but Couzin says the three-lane structure helps minimize the subsequent delay. And ants are loath to waste time. Once finished with the evening commute, home by dusk, the entire colony moves, in the safety of darkness, to a new site, and the next morning the ants repeat the cycle. "These species have evolved for thousands of years under these highly dense traffic circumstances," says Couzin. "They really are the pinnacle of traffic organization in the actual world."

The secret to the ridiculous efficiency of army ant traffic is that, unlike traveling locusts—and humans—the ants are truly cooperative. "They really want to do what's best for the entire colony," says Couzin. As worker ants are not able to reproduce, they all labor for the queen. "The colony in a sense is the reproductive unit," Couzin explains. "To take a loose analogy, it's like the cells in your body, all working together for the benefit of you, to propagate your genes." The progress of each ant is integral to the health of the colony, which is why ant traffic works so well. No one is trying to eat anyone else on the trail, no one's time is more valuable than anyone else's, no one is preventing anyone else from passing, and no one is making anyone else wait. When bringing back a piece of food that needs multiple carriers, ants will join in until the group hits what seems to be the right speed. Ants will even use their own bodies to create bridges, making the structure bigger or smaller as traffic flow passing over it requires.

What about merging? I ask Couzin later, in the dining room at Balliol College. How are the ants at this difficult task? "There's definitely merging going on," he says with a laugh. "There seems to be something interesting going on at junctions. It's something we'd like to investigate."

## Playing God in Los Angeles

> Doesn't matter what time it is. It's either bad traffic, peak traffic, or slit-your-wrists traffic.
>
> —*The Italian Job* (2003)

"Sorry, the traffic was horrible." These five words rival "How are you?" as the most popular way to begin a conversation in Los Angeles. At times it seems like half the city is waiting for the other half to arrive.

But there is one night when being late simply will not do, when the world—or at least several hundred million inhabitants of it—wants everyone to get to the same place at the same time. This would be Oscar night, when eight hundred or so limousines, ferrying the stars, arrive in a procession at the corner of Hollywood and Highland, depositing their celebrity carriage at the Kodak Theater. On the red carpet, the media volley questions: "How are you feeling?" "Who are you wearing?" But on Oscar night no one ever asks a larger question: How did eight hundred cars get to the same party in a punctual fashion in Los Angeles?

The answer is found in the labyrinthine basement of City Hall in downtown L.A. There, in a dark, climate-controlled room with a wall-sized bank of glowing monitors, each showing strategic shots of intersections across the city, sits the brains of the Los Angeles Department of Transportation's Automated Traffic Surveillance and Control (ATSAC). Traffic centers like this one are essential in many modern cities, and one sees similar setups from Toronto to London (in Mexico City the engineers delightedly showed me footage of speeding drivers giving the finger to automatic speed-limit cameras).

The ATSAC room in Los Angeles would normally be empty on a Sunday, with only the quietly humming computers running the city's traffic lights—ATSAC will even call a human repairperson if a signal breaks down. But since it's Oscar night, an engineer named Kartik Patel has been in the "bunker" since nine a.m., working on the DOT's special Oscar package. Another man lurks at a desk and does not say much. Teams of engineers have also been deployed in the field at strategic intersections. On a desk sits a little statue of Dilbert at a computer, to which someone has attached a label: "ATSAC Operator."

Since the city cannot shut down the entire street network for the Oscars, the limos must be woven through the grid of Los Angeles in a complex orchestration of supply and demand. Normally, this is done by the system's powerful computers, which use a real-time feedback loop to calculate demand. The system knows how many cars are waiting at any major intersection, thanks to the metal-detecting "induction loops" buried in the street (these are revealed by the thin black circles of tar in the asphalt). If at three-thirty p.m. there are suddenly as many cars as there normally would be in the peak period, the computers fire the "peak-period plan." These area-wide plans can change in as little as five minutes. (For a quicker response, they could change with each light cycle, but this might produce overreactions that would mess up the system.) As ATSAC changes the lights at one intersection, it is also plotting future moves, like a traffic version of IBM's chess-playing computer Big Blue. "It's calculating a demand," says Patel. "But it needs to think ahead and say, 'How much time do I need for the next signal?' "

Over time, ATSAC amasses a profile of how a certain intersection behaves during a given time on a given day. Patel points to a computer screen, which seems to be running a crude version of the game SimCity, with computer renderings of traffic lights and streets but no people. An alert is flashing at one intersection. "This loop at three-thirty on a Sunday has a certain historical value, for a year's period of time," Patel explains. "Today it's abnormal, because it's not usually that heavy. So it'll flag that as out of the norm and post it up there as a possible incident." It will try to resolve the problem, says Patel, within the "confines of the cycling."

But on this occasion, the engineers want certain traffic flows—those conveying the stars' limos—to perform better than ATSAC would normally permit, without throwing the whole system into disarray. In the late afternoon, with the ceremony drawing near, it becomes apparent just how difficult this is. Harried requests are beginning to come in from field engineers, who are literally standing at intersections. "ATSAC, can you favor Wilcox at Hollywood?" asks a voice, crackling from Patel's walkie-talkie. Patel, on his cell phone, barks: "Man, did you happen to copy Highland and Sunset? There's quite a queue going northbound." At times Patel will have his cell phone in one hand, the walkie-talkie in another, and then the landline phone will ring. "The limos are starting to back up, almost at Santa Monica," someone cries through the static.

As Patel furiously taps on his keyboard, lengthening cycle times here, canceling a left-turn phase there, it becomes hard to resist the idea that being a traffic engineer is a little like playing God. One man pushing one button affects not just one group of people but literally the whole city, as the impact ripples through the system. It is chaos theory, L.A. style: A long red light in Santa Monica triggers a backup in Watts.

This is when it begins to look as if something odd is going on here this afternoon. Patel seems particularly concerned with the intersection of La Brea Avenue and Sunset Boulevard. "Yeah, Petey, what's up?" he shouts into his phone. "How many people are there? That's good." Patel then admits that his unit has a "labor problem." Some three hundred municipal engineers, on a sick-out, are picketing on the same streets on which the limos are trying to get to the Oscars. What better way to draw attention, and who better to know the streets on which to demonstrate? Some of the calls Patel receives are from engineers wondering why the limos have been held up, and some of the calls are from picketing engineers seeking updates about which intersections they should cross on foot. "Tell them to walk more slowly, they're going too damn fast," Patel says into his phone. Reports coming in say that police are hustling the picketers across the intersections, so as to not block traffic. "Oh my God, how can they kick you out? You have a legal right to cross. Any unmarked crosswalk, you can cross it . . . just keep on crossing there, moving slowly."

Patel is both trying to get the limos to their destination *and* coaching the picketers on how to best interrupt that progress. Does that mean he can give the sign-toting pedestrians more time, which would further their cause? A strange smile crosses Patel's face, but he says nothing. He later excuses himself and goes to an office in the back, where he takes phone calls. Is he a coconspirator? Or does his traffic-engineer side override his labor solidarity side? One cannot say for sure, but interestingly enough, Patel and another engineer were later charged with tampering with traffic lights at four key intersections as part of the ongoing labor dispute. In November 2008, the pair each pleaded guilty to a single felony count of illegal computer access, but were given a reduced sentence of community service.

Despite the picketers, the limos arrive on time. The winning picture, ironically, is *Crash*, a film about Los Angeles traffic on literal and metaphorical levels. Then the limos leave the Kodak Theater, rejoining the city's traffic, and head for the postevent parties.

. . .

That Oscar afternoon was a small but perfect illustration of how complicated human traffic is when compared to ant traffic. Ants have evolved over countless centuries to move with a seamless synchronicity that will benefit the entire colony. Humans, on the other hand, propel themselves around artificially, something they have done for only a few generations. They do not all move en masse with the same goal but instead travel with their own agendas (e.g., getting to the Oscars, staging a demonstration). Ants all move at roughly the same speed, while humans like to set their own speeds, ones that may or may not reflect the speed limit. And, crucially, ants move *as* ants. They can always feel their neighbors' presence. Humans separate themselves not only across space but into drivers and pedestrians, and tend to act as if they are no longer the same species.

Los Angeles, like all cities, is essentially a noncooperative network. Its traffic system is filled with streams of people who desire to move how they want, and where they want, when they want, regardless of what everyone else is doing. What traffic engineers do is to try to simulate, through technology and signs and laws, a cooperative system. They try to make us less like locusts and more like ants.

Take traffic signals. It's common to hear drivers in Los Angeles, as elsewhere, lament, "Why can't they time the signals so they're all green?" The obvious problem with so-called synchronized signals is that there is a driver moving in a different direction asking the same thing. Two people are competing for the same resource. The intersection, the fundamental problem of the traffic world, is an arena for clashing human desire. John Fisher, the head of the city's DOT, uses the analogy of an elevator in a tall building. "You get on the elevator, and it stops at every floor because someone presses the button. They want to get off or on. Now, it stops at every floor—is it synchronized or not synchronized? The reality is if there are many stops, it's going to take a while to get there. It's the same with signals."

Engineers can use sophisticated models to squeeze as much "signal progression" as possible out of a network, to give the driver the "green wave." Fisher says that when he came to the DOT in the 1970s, "we tried to hold the line and keep the signals at a quarter-mile spacing." By doing that, and setting the cycle time (or the time it takes to cycle through green, yellow, and red on the traffic light) at sixty seconds, vehicles traveling at 30 miles per hour could reasonably "expect to find a green."

But over time, as the city has grown more dense, so too has the pressure to add more traffic lights. In certain places there is now a light at every block, which means there is a potential demand to cross at every block. Engineers have been forced to expand the length of the cycle to ninety seconds—typically the maximum in cities. "Let's say you go to a ninety-second cycle," Fisher says. "Even if you have quarter-mile spacing it means your progressive speed is not thirty miles per hour anymore, but something like twenty miles per hour. If you complicate that further, and the signal spacing is every block or sixteenth of a mile, there's just no way you could progress from one end to the other. The best you could do is go a couple signals and stop, a couple signals and stop, in all directions." The green wave works well on major streets where the demand from side streets is small. But in Los Angeles, Fisher explains, "we have traffic going in all directions, and generally the same quantities." Some intersections receive so much competing demand that they are "oversaturated," as Fisher says, beyond the help of even ATSAC's computers.

To further complicate matters, there are, even in Los Angeles, pedestrians. Despite the hilarious scene in *L.A. Story* that showed Steve Martin driving to his next-door neighbor's house for dinner, people do walk, and not just to and from their parked car. As a profession, traffic engineering has historically tended to treat pedestrians like little bits of irritating sand gumming up the works of their smoothly humming traffic machines. With a touch of condescending pity, pedestrians are referred to as "vulnerable road users" (even though in the United States many more people die in cars each year, which leads one to wonder who exactly is more vulnerable). Engineers speak of things like "pedestrian impedance" and "pedestrian interference," which sound like nasty acts but really just refer to the fact that people sometimes have the gall to cross the street on foot, thus doing things like disrupting the "saturation flow rate" of cars turning at an intersection.

As a testament to the inherent bias of the profession, no engineer has ever written a paper about how "vehicular interference" disrupts the saturation flow rates of people trying to cross the street. In cities like New York, despite the fact that pedestrians vastly outnumber cars on a street like Fifth Avenue, traffic signals are timed to help move the fewer cars, not the many pedestrians—has anyone ever had an uninterrupted stroll up Fifth Avenue, a green wave for walking? Unlike in pedestrian-thronged New York City, where most push buttons to cross the street no longer work (even though they still tempt the impatient New Yorker), in Los Angeles the relative rarity of pedestrians means the buttons do work. The walker humbly asks the city's traffic gods for permission to cross the street, and, after a time, their prayers are answered. If you do not press the button, you will stand there until you're eventually ticketed for vagrancy.

Sometimes the traffic deities encounter even higher authorities. A curious fact of Los Angeles traffic life is that, at roughly seventy-five signals, in places ranging from Century City to Hancock Park, the button does *not* always have to be pressed to cross. These intersections run instead on what is known as Sabbath timing. As Sabbath-observant members of the Jewish faith are not supposed to operate machines or electrical devices from sundown on Friday to sundown on Saturday, or during a number of holidays, the act of pressing a button to cross the street is viewed as a violation of this tenet. With the only alternative rampant jaywalking, the city installed automated "Walk" signs at certain intersections (causing what Fisher jokingly calls "sacrificial interruptions" to traffic flow even when no pedestrians are present). "We have the Hebrew calendar programmed into our controller," Fisher told me.

When the DOT suggested installing "smart" devices that would sense the presence of a pedestrian at a crosswalk and activate a flashing signal, it was gently rebuffed by the Rabbinical Council of California, which opined that activating the light via a signal, even if it was done passively, violated the Sabbath restrictions. If pedestrians were *unaware* that their presence was triggering the device, the council noted, the smart device would be acceptable, but "people would quickly realize its presence and avoid using the crosswalk on the Sabbath."

These nuances pale before the overwhelming fact that Los Angeles is handling more traffic now than was ever thought possible. "A lot of major streets, like La Cienega and La Brea, carry sixty thousand vehicles a day," says Fisher. "Those streets were designed to carry thirty thousand vehicles a day." Years ago, engineers used capacity-expanding tricks like reversible lanes on Wilshire Boulevard and other major thoroughfares, changing the normal direction of one lane to help carry traffic in from the freeway in the morning and send it back out in the evening. That is no longer possible. "When you're getting a split like sixty-five percent of traffic one way, thirty-five percent the other way, reversible lanes work very well," Fisher says. "Today we rarely have that type of peaking anywhere in the city." The highways are no different. The San Diego Freeway, or I-405, was projected to carry 160,000 vehicles when it was completed at the end of the 1960s. It now carries almost 400,000 per day, and the junction where it connects to the Santa Monica Freeway is the most congested in the United States. The Santa Monica used to be a traditional sort of urban highway, with a heavier morning peak toward downtown and the reverse in the afternoon. "You try to go outbound in the morning and that often seems heavier than the inbound is," Fisher says.

"We used to have typical days where we would give volumes," notes Dawn Helou, an engineer with Caltrans, the sprawling and omnipresent agency in charge of California's highways. "A typical day is Tuesday, Wednesday, Thursday in a month preferably without holidays, in a week preferably without holidays. No rain, no holidays, no summer vacation, no incidents. We're running out of those typical days."

The thing that keeps the whole system from breaking down is precisely that advantage that humans have over ants: the ability to see, and direct, the whole traffic system at once. By making all these decisions for the drivers, by coordinating the complex ballet of wants and needs, supply and demand at intersections, engineers have been able to improve the city's traffic flow. A study a few years ago by the DOT showed that the area containing real-time traffic signals reduced travel times by nearly 13 percent, increased travel speeds by 12 percent, reduced delay by 21 percent, and cut the number of stops by 31 percent. Just by quickly alerting the DOT that signals have malfunctioned, the system squeezes out more efficiency. What the traffic engineers have done is added "virtual" capacity to a city that cannot add any more lanes to its streets.

The flow of information is crucial to maintaining the flow of traffic. With no spare capacity, irregularities in the system need to be diagnosed and addressed as soon as possible. Engineers at Caltrans say that as a rule of thumb, for every one minute a highway lane is blocked, an additional four to five minutes of delay are generated. The inductor loops buried in the highway can and do detect changes in the traffic patterns. But the highway loops are not in real time. There can be a gap of anywhere from a few minutes to a quarter of an hour before the information they're recording is processed. Often, visual confirmation by camera is needed to verify that there is a problem. In that time, a huge jam could develop. Or sometimes the loops in a particular section of highway are not working (Caltrans reports anywhere from 65 to 75 percent of its twenty-eight thousand statewide loops are working on a given day), or a section of highway will have no loops at all.

This is why, each day in Los Angeles, there is a frantic search for the truth. It's called the traffic report. Traffic news is the sound track of daily life in Los Angeles, a subliminal refrain of "Sig Alerts" and "overturned big rigs" always on the edge of one's consciousness. Occasionally the story is that there is no story, according to Vera Jimenez, who does the morning traffic on KCAL, the CBS affiliate in Los Angeles. "Sometimes it's funny," she said one morning at the Caltrans building. "The story is not that the traffic's really heavy, but, oh my gosh, it's surprisingly light. It's not a holiday, there isn't anything going on, it's just really light. Everyone's driving the way they should, everyone's merging, and believe it or not, look how nice it is."

No city in the world has more traffic reports or traffic reporters than Los Angeles, and to spend time with them is to see the city, and traffic, in a new way. Early one morning, I drive to Tustin, an Orange County suburb that is home to Airwatch, a Clear Channel subsidiary and one of America's largest traffic-reporting services. In a room filled with banks of televisions, computer monitors, and police scanners, Chris Hughes is several hours into the morning rush hour. Armed with a stopwatch and jittery from caffeine, Hughes rattles off a fast, well-calibrated, flow: "Heavy traffic in Long Beach this morning on the North 405 through Woodruff to the 710 then again from the 110 Freeway heading up to Inglewood . . . "

For each of the different radio stations for which Hughes reports, he must change the length of his report, as well as the way he says it. One station wants "upbeat and conversational," while another wants a precise robotlike diction they call "traffic formatics." Some stations have advertisements for Hooters Casino, but the Christian stations do not. Some stations actually want him to be someone else. "*Good* morning, I'm Jason Kennedy with AM 1150 traffic brought to you by Air New Zealand," I suddenly hear him say. "They're sort of competing stations," he explains sheepishly, "even though we own them both."

Hughes has an instinctual understanding of Los Angeles' highways. He can tell which way a rainstorm is moving by looking at the real-time traffic-flow highway maps. He knows Fridays heading east out of the city can be particularly bad. "Everyone's going to Las Vegas—all the way to ten p.m. that'll be backed up." He knows that people drive slower on highway stretches that have sound barriers to either side. He knows that mornings with heavy rains often lead to lighter afternoon traffic. "Maybe a lot of people got scared of the rain and disappeared," he says. He notes that while traffic information is easily available to the public, often the trick is in understanding it. "It's kind of like *The Matrix*," he says. "You're looking at the map and you can pick out what looks right and what doesn't. I can look at the map now and say, 'Hey, there's something wrong on the 101. A big-rig fire at Highland, probably.' "

There is no limit to the things that can disrupt the flow on Los Angeles highways. "Do you want to know the number one specific item dropped on the freeway?" asks Claire Sigman, another Airwatch reporter. "The most recorded item is ladders." Trucks, just like in the *Beverly Hills Cop* movies, also spill avocados and oranges. Portable toilets have been dumped in the middle of the freeway. In 2007, a house, replete with graffiti and a "For Rent" sign, sat for weeks on the Hollywood Freeway, abandoned during the course of its move after it struck an overpass (the owner had taken a detour onto an unauthorized route). People hold apocalyptic signs on overpasses, or threaten to jump. Wildfires break out. Out in the high desert, tumbleweeds cause problems. "People swerve out of the way, rather than just drive through it," Hughes says. A computer screen at the Airwatch office ticks off a steady flow of traffic incidents, ranging from the absurd to the horrifying, as recorded by the California Highway Patrol (CHP). Codes are used to disguise the presence of stalled female drivers, who might otherwise be preyed upon by unsavory men listening to police scanners. Not atypical of the stream is incident 0550, which describes a "WMA," or white male, wearing a plaid jacket and "peeing in middle of fwy." It adds a noteworthy detail: "No veh in sight." (Now, where was that wayward Porta Potti?)

CHP officers are the foot soldiers in the daily battle to keep Los Angeles' traffic from collapsing. The sophisticated computer modeling and fiber-optic cable that the traffic generals in the bunker have at their disposal are of little use when a car has stalled on Interstate 5, as I learned one afternoon when I went out for a patrol in a CHP cruiser with Sergeant Joe Zizi, an easygoing former trooper now doing public relations. CHP patrol officers begin each day by "cleaning their beat," or removing any abandoned vehicles or hazards from the road. "That way there's nothing that people have to look at when they're driving," Zizi says as he drives along the 101. Something as simple as a couch dumped in a roadside ditch can send minor shudders of curiosity through the traffic flow. A standard-issue black pump-action shotgun sits between the front seats. To enable drivers to carry out their traffic triage duties, patrol cars are outfitted with reinforced bumpers, designed to let them push cars off the road rather than wait for a tow truck. Their trunks are filled with a dizzying array of equipment for dealing with traffic contingencies, ranging from baby-delivering kits ("definitely a spectacle for rubberneckers") to dog snares.

"For some reason, dogs are attracted to the freeway," says Zizi. "They get on there, get completely freaked out, and start running down the center." According to CHP statistics, these Code 1125-As (traffic hazard—animal) peak on July 5, presumably from dogs scared by the previous night's fireworks. When traffic is moving, CHP officers pass the time by looking for stolen vehicles (screwdrivers in the ignition are a telltale sign) and, of course, writing traffic tickets. Does Zizi have any advice for beating tickets? "I have a lot of officers who say that women crying will get them out of tickets, while other officers say that if someone *does* cry they're getting the ticket," he says. "Of course we have a lot of men who cry trying to get out of tickets, but that really doesn't work on the heartstrings of officers."

For all the Caltrans cameras and loops wired into the road, for all the CHP officers flagging incidents, the highway system running through Los Angeles is so vast and incomprehensible that, sometimes, the only way to really understand what's happening is to pull way back and view the whole system from above. That is why there is still a place for people like Mike Nolan, KFI's "eye in the sky," a longtime L.A. traffic reporter who, twice daily, will take off in his Cessna 182 from Riverside County's Corona Airport and cover a swath of ground from Pasadena to Orange County.

"The learning curve is being able to read a freeway," he explains, banking his plane over a new subdivision carved into green hillside. "I know what's normal. I know where it should be slowing down and where it shouldn't. When I see something out of the ordinary, then I investigate it." Nolan, whose navigational mantra is "Keep the freeway to your left," knows traffic patterns like a grizzled fishing guide knows the best bass holes. A stalled Volkswagen in East Los Angeles is worse than an overturned oil truck in La Cañada ("More spectacular does not necessarily translate into worse," he says). Mondays, especially during *Monday Night Football*, tend to be a bit lighter. Thursday, congestion-wise, is now looking like the new Friday, traditionally the busy "getaway day." There are also strange blips in the pattern, like sunrise slowdowns. "The very first day of standard time, when we go from daylight saving time to darkness, everybody just locks up," he says. "The traffic goes from bad to horrendous." Rainy days can be bad, but the first rainy day in a while is even worse. "There's a buildup of oil and rubber if it hasn't rained in a while. It's like driving on ice, literally."

Nolan says people have long been predicting, because of ground sensors and in-vehicle probes that can detect the speed of traffic, that there will no longer be a need for aerial traffic reports. Indeed, on his instrument panel he has attached a TrafficGauge, a Palm Pilot–sized device fed by Caltrans data, that shows congestion levels on L.A. freeways. But he says that data rarely tell the whole story, or the correct story. "In my mind there's no substitute for looking out the window and telling people what you've got," he says. "The sensors in the road are delayed, they're inefficient. They're working half the time, not working half the time. There's no substitute for saying, 'It's in the right lane, I see it right there, right at the fill-in-the-blank overpass.' Or that the tow truck is in heavy traffic. The sensor can't tell you the tow truck's a block away, or ready to hook up and pull away. It can't give you the substantive info that comes from looking at it directly."

Indeed, that afternoon of flying around the city, accompanied by an Airwatch reporter receiving ground reports, seems to be an exercise in chasing ghosts. The jackknifed tractor-trailer on the 710 is not there, or never was there. The blockage on the 405 was a rumor. Nolan is the one who must try to make sense of the strange reports that come in, like the one that announced a dead dog was "blocking lanes one, two, three, and four." The most remarkable traffic event he ever saw was during the L.A. riots of 1992. "I remember seeing people stop at a stoplight in Hollywood. They would get out and loot a store. The light would turn green and they'd get back in and drive away. That was the most incredible thing I've ever seen."

Flying over a city like Los Angeles, it is easy to glance down and think, for a moment, that the people below, streaming along trails, look like ants. If only it were that simple.

## When Slower Is Faster, or How the Few Defeat the Many: Traffic Flow and Human Nature

> You hit the brakes for a second, just tap them on the freeway, you can literally track the ripple effect of that action across a two-hundred-mile stretch of road, because traffic has a memory. It's amazing. It's like a living organism.
>
> —*Mission: Impossible III*

At some point you may have come to a highway on-ramp, expecting to join the flow of traffic, only to be stopped by a red light. Such devices are called ramp meters, and they are found from Los Angeles to South Africa to Sydney, Australia.

Ramp meters often seem frustrating because the traffic on the highway appears to be moving just fine. "People ask me, 'How come you're stopping me at the ramp meter? The freeway is free-flowing,' " says Dawn Helou, the Caltrans engineer. "The freeway is free flowing because you're stopping."

This is one of the most basic, and often overlooked, facts about traffic: That which is best for an individual's interest may not be best for the common good. The game traffic engineers play to fight congestion involves fine-tuning this balance between what is "user optimal" and what is "system optimal." This happens on several different levels, both having to do with congestion: how traffic moves on roads and how larger traffic networks behave (an idea I'll return to in a later chapter).

The reason why highway ramp meters work is, on the face of it, simple once one knows a few basic facts about traffic flow. Engineers have been trying to understand, and model, traffic flow for many decades, but it is a huge and surprisingly wily beast. "Some puzzles remain unsolved," declares Carlos Daganzo, an engineer at the University of California, Berkeley. The first efforts merely tried to model the process known as "car following." This is based on the simple fact that the way you drive is affected by whether or not someone is in front of you, and how far away or close they are. Like ants responding to the presence of pheromones on the trail, you're influenced by the driver ahead, a constant, unsteady wavering between trying not to get too close and trying not to slip too far back. Now imagine those interactions, plus lane shifts and all the other driving maneuvers, a fluctuating mix of vehicle speeds and sizes, a wide range of driver styles and agendas, a dizzying spectrum of differing lighting and weather and road conditions; then multiply all this by the thousands, and you can begin to appreciate the higher-order complexities of traffic modeling.

Even the most sophisticated models do not fully account for human weirdness and all the "noise" and "scatter" in the system. Traffic engineers will offer caveats, like the disclaimer I saw at one traffic conference: "This model does not account for the heterogeneity of driver behavior." Do you feel uncomfortable driving next to someone else, and therefore speed up or slow down? Are you sometimes willing, for no apparent reason, to ride quite close to the car in front, before gradually drifting back? All kinds of strange phenomena lie outside easy capture by the traffic sensors. Car following, for instance, is filled with little quirks. A study that looked into how closely passenger-car drivers followed SUVs found that car drivers, contrary to what they said they did—and despite the fact that the SUV was blocking their view of the traffic ahead—actually drove closer to SUVs than when they followed passenger cars.

Or take what Daganzo has called the Los Gatos effect, after an uphill stretch of highway in California. You may have experienced this: Drivers seem reluctant to abandon the passing lane and join the lane of trucks chugging uphill, even when they are being pressured by other drivers, and even when the other lane is not crowded. What's going on? Drivers may not want to give up the fast lane for fear of having trouble returning to it. They may also be unsure whether the person behind truly wants to go faster or is just keeping a tight space to prevent someone else from passing. A tight "platoon" forms, but for how long? We all see these odd patterns. One of the idiosyncrasies I have noticed in traffic flow is something I call "passive-aggressive passing." You're in the passing lane when suddenly the driver behind you pressures you to move into the slower right-hand lane. After you have done so, they then move into your lane, in front of you, and slow down, thus forcing *you* to pass *them*.

The basic parameters of how highways perform have been gradually hammered out. One of the key performance measures is volume, also called flow, or the number of vehicles that pass a buried sensor or some other fixed point on the highway. At four a.m., before rush hour, cars may be zipping along a highway at 75 miles per hour. The volume is measured at 1,700 cars moving past a point in one hour. As rush hour begins, the volume quite naturally begins to rise in an upward curve, reaching a theoretical maximum of 2,400 cars traveling at 55 miles per hour. System-wise, this is traffic nirvana. Then, as additional vehicles enter the highway, the curve begins to drop. Suddenly, the volume is back at 1,700. This time the cars are going 35 miles per hour. "So you have the two 1,700s," Helou says. "Same volume, completely different situation."

Because traffic moves in time and space, measurements like volume can be deceiving, as can the highway itself. Solo drivers sitting in a highly congested lane may look to the HOV lane next to them and think that it's empty—a psychological condition so prevalent it even has a name, "empty lane syndrome." Many times it just seems empty because of the large headways between vehicles moving at much higher speeds. That lane may actually be achieving the same volume as the lane you are in, but the fact that the drivers might be going upward of 50 miles per hour faster creates an illusion that it's being underused. Of course, neither of these positive or negative individual outcomes—the driver whisking along at 80 miles per hour or the people stuck at 20 miles per hour in the congested lanes—are what's best for the entire system. The ideal highway will move the most cars, most efficiently, at a speed just about halfway.

Even as rush hour kicks in and the speed-flow curve begins to drop, traffic can perk along at what has been called "synchronized flow," heavy but steady. But as more vehicles pile onto the highway from on-ramps, the "density," or the number of cars actually found in a one-mile stretch (as opposed to passing a single spot), begins to thicken. At a certain point, the critical density (the moment, you will recall from before, when the locusts began their coordinated march), the flow begins to break down. Bottlenecks, fixed or moving, squeeze the flow like a narrowing pipe. There are simply too many cars for the road's capacity.

Ramp metering aims to keep the highway's "main-line flow" below the critical density by not letting the system be flooded with incoming on-ramp cars. "If you allow unimpeded access, then you have a platoon of vehicles that are entering the main line," says Helou. This means not only more cars but more cars jockeying to merge. Studies have shown that this is neither predictable nor always cooperative. "That [merging] eventually breaks down the right lane," she says. "This overflows to the next lane, because people try to merge left before they get to it. And then the people in the second lane try to merge to the next lane before they get to it, so you break down the whole freeway." A line of cars waiting to exit an off-ramp can trigger this same chain reaction, one study showed, even when all the other lanes were flowing nowhere near critical density.

If done properly, ramp metering, by keeping the system below the critical density, finds that sweet spot in which the most vehicles can move at the highest speed through a section of highway. Engineers call this "throughput maximization."

A simple way to see this in action involves rice. Take a liter of rice and pour it, all at once, through a funnel and into an empty beaker. Note how long it takes. Next, take the same rice and pour it not all at once but in a smooth, controlled flow, and time that process. Which liter of rice gets through more quickly? In a demonstration of this simple experiment by the Washington DOT, it took forty seconds for one liter of rice to pass through the funnel using the first method. The second method took twenty-seven seconds, nearly one-third less time. What seemed slower was actually faster.

Rice has more to do with traffic than you might think. Many people use water analogies when talking about traffic, because it's a great way to describe concepts like volume and capacity. One example, used by Benjamin Coifman, an engineering professor at Ohio State University who specializes in traffic, is to think of a bucket of water with an inch-wide hole in the bottom. If the inflow into the bucket is half an inch in diameter, no water will accumulate. Raise it to two inches, however, and the water rises, even though some water is still exiting. Whether we drive into a jam (or a jam drives into us) depends on whether the "water"—that is, the traffic trying to flow through a bottleneck—is draining or rising. "As a driver, the first thing you encounter is the end of the queue," Coifman told me. "The first thing you encounter is wherever the water level happens to be that day." The bucket metaphor also teaches us something else about traffic: No matter how much capacity there is in the rest of the bucket (or on the roads), the size of the hole (or the bottleneck) dictates what gets through.

At places like bottlenecks, however, traffic acts less like water (it does not speed up as highway "channels" narrow, for one) and more like rice: Cars, like grains, are discrete objects that act in peculiar ways. Rice is what's called a "granular media," a solid that can act like a liquid. Sidney Nagel, a physicist at the University of Chicago and an expert in granular materials, uses the analogy of adding a bit of sugar to a spoon. Pour too much, and the pile collapses. The sugar flows like a liquid as it collapses, but it's really a group of interacting objects that do not easily interact. "They do not attract one another," says Nagel. "All they can do is scatter off one another." Put a bunch of granular materials together, and it is not easy to predict how they will interact. This is why grain silos are the building type most prone to collapse, and it's also why my box of Cascadian Farm Purely O's cereal begins to bow outward at the bottom after several pours.

Why does the rice jam up as you pour it into the funnel? The inflow of rice exceeds the capacity of the funnel opening. The system gets denser and denser. Particles spend more time touching one another. More rice touches more rice. The rice gets "hung up" from the friction of the funnel walls. Sound familiar? "That's like cars on the highway," says Nagel. "And when you get narrowing of traffic, then that becomes very much stuff trying to flow through the hopper."

Pouring less rice at a time—or moving fewer cars—keeps more space, and fewer interactions, between the grains. Things flow faster. As intuitive as the "slower is faster" idea is, it's not always easy for a driver stuck in traffic to accept. In 1999, a state senator from Minnesota, claiming that ramp metering in the Twin Cities was doing more harm than good, launched a "Freedom to Drive" proposal that called for, among other things, shutting down the meters. The legislation died, but under another bill a ramp-meter "holiday" was declared. For two months the meters were turned off. Drivers could enter the highway at will, on so-called sane lanes, unfettered by troublesome red lights. And what happened? The system got worse. Speeds dropped, travel times went up. One study showed that certain highway sections had double the productivity with ramp meters than without. The meters went back on.

· · ·

The "slower is faster" idea shows up often in traffic. The classic example concerns roundabouts. Many people are under the mistaken impression that roundabouts cause congestion. But a properly designed roundabout can reduce delays by up to 65 percent over an intersection with traffic signals or stop signs. Sure, an individual driver who has a green light may fly through a signalized intersection much more quickly than through a roundabout. Roughly half the time, however, the light will not be green; and even if it is green there is often a rolling queue of vehicles just starting up from the previous red. Add to this such complications as left-turn arrows, which prevent the majority of drivers from moving, not to mention the "clearance phase," that capacity-deadening moment when *all* lights must be red, to make sure everyone has cleared the intersection. Drivers do have to slow down as they approach a roundabout, but under typical traffic conditions they rarely have to stop.

In the 1960s, experiments were made at the Holland Tunnel, one of the main arteries for traffic coming into and leaving New York City. When cars were allowed to enter the tunnel in the usual way, with no restrictions, the two-lane tunnel could handle 1,176 cars per hour, at an optimal speed of 19 miles per hour. But in a trial, the tunnel authorities capped the number of cars that could enter the tunnel every two minutes to 44. If that many cars got in before two minutes were up, a police officer made the next group of cars wait ten seconds at the tunnel entrance. The result? The tunnel now handled 1,320 vehicles per hour. (I will explain why shortly.)

On streets with traffic signals, engineers set progressions with a certain speed in mind that will enable the driver to hit a line of constant greens. To drive faster than this only ensures that the driver will be forced to come to a stop at the next red light. Each stop requires deceleration and, more important, acceleration, which costs the driver in time and fuel. A queue of drivers stopped at a light is a gathering of "start-up lost time," as engineers call it (in an appropriately forlorn echo of Proust). The first cars in a queue squander an average of two seconds each, two seconds that would not have been lost had the car sailed through at the "saturation-flow" rate. The first driver at a light that turns from red to green, because he must react to the change, make sure that the intersection is empty, and accelerate from a standstill, generates the most "lost time." The light is green, but for a moment the intersection is empty. The second driver creates a bit less lost time, the third driver less still, and so on (assuming everyone is reacting as soon as they can, which is not a given). SUVs, because they are longer (on average, 14 percent longer than cars), and take longer to accelerate, can create up to 20 percent more lost time.

Some of the start-up lost time could be "found" if drivers approached at a slower, more uniform speed that did not require them to come to a stop. (If they came *too* slowly, however, time would also be lost, as green signal time would be wasted on an empty intersection.) Much of the time being lost these days is "clearance lost time," the time between signals when the intersection is momentarily empty. This is because traffic engineers are increasingly lengthening the "all-red phase," meaning that when one direction gets the red, the competing direction has to wait nearly two seconds before getting a green. They do this because more people cannot seem to stop on red.

Now picture a highway during stop-and-go traffic. Like those drivers stopped at the light, each time we stop and start in a jam we are generating lost time. Unsure of what the drivers ahead are doing, we move in an unsteady way. We are distracted for a moment and do not accelerate. Or we overreact to brake lights, stopping harder than we need to and losing more time. Drivers talking on cell phones may lose still more time through delayed reactions and slower speeds. The closer the vehicles are packed together, the more they affect one another. Everything becomes more unstable. "All of the excess ability for the system to take in any sort of disturbance is gone," says Coifman. He uses the metaphor of five croquet balls. "If you put them a foot apart and tap one lightly, nothing happens to the other four. If you put them all up against one another and tap one lightly, the far one then moves out. When you get closer to capacity on the roadway, if there's any one little tweak, it impacts a lot of the cars."

When the first in a group of closely spaced cars slows or stops, a "shock wave" is triggered that moves backward. The first car slows or stops, and the next one slows or stops a little farther back. This wave, whose speed usually seems to register at about 12 miles per hour, could theoretically go on for as long as there was a string of sufficiently dense traffic. Even a single car on a two-lane highway, by simply changing its speed with little rhyme or reason (as people so often seem to do, in what I like to call "speed-attention-deficit disorder"), can itself pump these waves back down a stream of following vehicles. Furthermore, even if that car's average speed is fairly high, the fluctuations wreak progressive havoc. This was the secret behind the Holland Tunnel experiment: With cars limited to "platoons" of forty-four vehicles each, the shock waves that were triggered were confined to each group. The platoons were like croquet balls spaced apart.

Many times we find ourselves stuck in traffic that seems to have no visible cause. Or we make it through a jam and begin to speed up, seeming to make progress, only to quickly drive into another jam. "Phantom jams," these have been called, to the annoyance of some. "Phantom jams are in reality nonexistent," thunders Michael Schreckenberg, a German physics professor at the University of Duisburg-Essen so noted for his traffic studies that he has acquired the epithet "jam professor" in the German media. There is always a reason for a jam, he says, even if it is not apparent. What seems to be a local disturbance might just be a wave pumped up from downstream in what is in reality a big, wide moving jam. It is wrong, says Schreckenberg, to simply call the whole thing stop-and-go traffic: "Stop-and-go is the dynamic *within* a jam."

We fall for the phantom-jam illusion because traffic happens in both time and space. You may be driving into a space where a jam has been. Or you may not be driving into a jam—instead, the jam might be driving into you. "In my bucket analogy," says Coifman, "the driver would be a water molecule. If the water level's rising, then the jam's coming to us." We are also driving into history—or, perhaps more accurately, we are being driven back into history. By the time we actually arrive where something triggered the shock wave, in all likelihood the event will be only a memory. It may have been an accident, now cleared. "The queue's going to persist for a while as it's dissipating," says Coifman. "It's that water sitting in the bucket. In this case you've enlarged the hole in the bucket, but it does not disappear instantaneously."

Or the hiccup in heavy traffic that passes through you might be the echo of someone who, forward in space and backward in time, did something as simple as change lanes. The car that changes lanes moves, eating up capacity in the new lane and causing the driver behind to slow; it also frees up capacity in the lane it has left, which triggers a bit of acceleration in that lane. These actions ripple backward in a kind of seesaw effect. This is why, if you pick one car in the neighboring lane as your benchmark, you will often find yourself passing that car and being passed by that car continuously. This is equilibrium asserting itself, the accordion of traffic flow stretching and compressing, the lingering chain reaction of everyone who thought they could get a better deal.

Since it takes so long for traffic to resume flowing freely once it has plunged past the critical density, it would seem the best way to avoid the ill effects of a jam would be not to drive into it, or let it drive into you, in the first place. This is the thought that occurred one afternoon a few years ago to Bill Beatty, a self-described "amateur traffic physicist" who works in the physics laboratory at the University of Washington. Beatty was on State Highway 202, returning from a state fair. The road, a "little four-lane," was thronged with traffic from the fair. The traffic was "completely periodic," as he describes it. "You'd drive real fast and then almost get to sixty and then you'd slow down and come to a stop, for almost two minutes," he says.

So Beatty decided to try an experiment: He would drive only 35 miles per hour. Rather than let the waves drive into him, he would "eat the waves," or subdue the wildly varying oscillations of stop-and-go traffic. Instead of tailgating and constantly braking, he would try to drive at a uniform speed, leaving a large gap between himself and the car ahead. When he looked in his rearview mirror, he saw a revelation in the pattern of headlights: Those behind him looked to be in a regular pattern, while the other lane had clusters of clumped stop-and-go vehicles. He had "damped" the wave, leveled off the extremes. "It cuts off the mountains and puts them in the valley," he says of his technique. "So instead of getting to drive at sixty miles per hour briefly, you're forced to drive at thirty-five miles per hour. But you don't have to stop, either."

Without analyzing the total traffic flow of the highway, it would be hard to know for sure what good Beatty's experiment did. People may have just merged in front of him, pushing him back (if he wanted to keep the same following distance), while those behind him who thought he was going too slow may have jumped into the next lane, causing additional disturbance. But even if Beatty's technique did little more than take a tightly congested traffic jam and stretch it backward, so that a car spent the same amount of time traveling a section of road, it would still save fuel and reduce the risk of rear-end accidents—two added benefits for the same price. Only how do you get everyone to cooperate? How do you prevent people, as so often seems to happen, from simply consuming the space you have left open? How, in essence, can we simulate ant-trail behavior on the highway?

One way is the "variable speed limit" system now being used on any number of roads, from England's M25 "controlled motorway" to sections of the German autobahn to the Western Ring Road in Melbourne, Australia. These systems link loop detectors in the road to changeable speed-limit signs. When the system notices that traffic has slowed, it sends an alert upstream. The approaching drivers are given a mandatory speed limit (enforced by license-plate cameras) that should, in theory, lessen the effects of a shock wave. Even though many drivers suspected it was the lowering of speeds to 40 kilometers per hour that was *causing* the congestion, a study of the M25 found that drivers spent less time in stop-and-go traffic, which not only helped lower the crash rate by 20 percent (itself good for traffic flow) but cut vehicle emissions by nearly 10 percent. As drivers adjusted to the system, their trip times declined. Again, slower can be faster.

Smart highways also require smart drivers. The sad truth is that the way we drive is responsible for a good part of our traffic problems. We accelerate too slowly or brake too quickly, or the opposite; since we do not leave enough space between vehicles, the effects are often magnified as they move back up the line. Traffic is what is known as a nonlinear system, meaning most simply a system whose output cannot be reliably predicted from its input. When the first car in a long platoon comes to a stop, one cannot exactly predict how quickly or how far back each car behind it will stop (if they come to a stop at all). And the farther back, the harder it is to predict.

A driver's overreaction (or underreaction) may amplify a shock wave that snaps, like the crack of a whip, several cars back, helping to cause a collision in the space that the originating driver has since left. One study examined a crash on a Minneapolis highway involving a platoon of seven vehicles that had been forced to come to a sudden stop. The seventh car in the group crashed into the sixth. Since we normally assume that cars keeping an adequate following distance should be able to stop in all conditions, that should be the end of it.

But the researchers, examining the braking trajectories of the vehicles in the platoon, found that the *third* car arguably bore a considerable responsibility for the crash. How so? Because the third car was overly slow to react, it "consumed" a larger portion of the "shared resource" of braking distance allocated among the cars. This left the cars farther down the line with progressively less time and space in which to stop—to the point where the seventh car, even though it reacted faster than the third, was following too closely to the sixth car to stop under the amplified conditions. Had the third car's reaction been faster, the crash might have been prevented. For these sorts of reasons, the researchers pointed out, people who tailgate—that is, do not follow at the "socially optimal" distance—increase their risk not only of striking the vehicle they're following but of being struck *by the car following them.*

What if drivers' reaction times could be predicted with mathematical precision? The ultimate answer may be to combine smart highways with smart cars. It's probably no accident that whenever one hears of a smart technology, it refers to something that has been taken out of human control. L. Craig Davis, a retired physicist who worked for many years in the research laboratories of the Ford Motor Company, is one of a number of people who have run simulations showing how equipping cars with adaptive cruise control (ACC), already found on many high-end models, can improve traffic flow by keeping the distance between cars at varying speeds mathematically perfect. This would not kill traffic waves entirely, says Davis. Even if a line of stopped cars could be coordinated to begin accelerating at the same time, he says, "if you wanted to get them up to speed with a normal distance between them at sixty miles per hour, you would still have this wave effect."

Remarkably, the simulations show that if just one in ten drivers had ACC, a jam could be made much less worse; with as few as two in ten drivers, the jam *could be avoided altogether.* In one experiment, Davis located the precise moment the jam was avoided, just as one additional manual car was given ACC. This putative straw that broke the camel's back brings to mind the example of the locusts. When the locusts reached critical density—one more locust—they began to behave entirely differently.

Just one problem has arisen in Davis's simulations. Since the simulated vehicles with ACC like to keep very tight gaps between themselves, it may be difficult for a non-ACC car entering from an on-ramp to find a safe space between them. Also, like human drivers, ACC cars may not feel obliged to yield to entering drivers. These problems can surely be solved scientifically, but in the meantime, as we suffer the effects of our failure to always act cooperatively on the highway, we can draw one comforting lesson: Even machines sometimes have trouble merging.

## Chapter Five

# Why Women Cause More Congestion Than Men (and Other Secrets of Traffic)

### Who Are All These People? The Psychology of Commuting

> You're not stuck in a traffic jam. You are the traffic jam.
>
> —advertisement in Germany

One of the curious laws of traffic is that most people, the world over, spend roughly the same amount of time each day getting to where they need to go. Whether the setting is an African village or an American city, the daily round-trip commute clocks in at about 1.1 hours.

In the 1970s, Yacov Zahavi, an Israeli economist working for the World Bank, introduced a theory he called the "travel-time budget." He suggested that people were willing to devote a certain part of each day to moving around. Interestingly, Zahavi found that this time was "practically the same" in all kinds of different locations. The small English city of Kingston-upon-Hull's physical area was only 4.4 percent the size of London; nevertheless, Zahavi found, car drivers in both places averaged three-quarters of an hour each day. The only difference was that London drivers made fewer, longer trips, while Kingston-upon-Hull drivers made more frequent, shorter trips. In any case, the time spent driving was about the same.

The noted Italian physicist Cesare Marchetti has taken this idea one step further and pointed out that throughout history, well before the car, humans have sought to keep their commute at about one hour. This "cave instinct," as he calls it, reflects a balance between our desires for mobility (the more territory, the more resources one can acquire, the more mates one can meet, etc.) and domesticity (we tend to feel safer and more comfortable at home than on the road). Even prisoners with life sentences, he notes, get an hour "out in the yard." When walking was our only commuting option, an average walking speed of 5 kilometers per hour meant that the daily commute to and from the cave would allow one to cover an area of roughly 7 square miles (or 20 square kilometers). This, remarks Marchetti, is *exactly* the mean area of Greek villages to this day. Moreover, Marchetti notes, none of the ancient city walls, from Rome to Persepolis, encompassed a space wider than 5 kilometers in diameter—in other words, just the right size so that one could walk from the edge of town to the center and back in one hour. Today, the old core of a pedestrian city like Venice still has a diameter of 5 kilometers.

The growth of cities was marked, like tree rings, by advances in the ways we had to get from one place to another. The Berlin of 1800, Marchetti points out, was a walkable size. But as horse trams came along, then electric trams, then subways, and, finally, the car, the city kept growing, by roughly an amount proportional to the speed increase of the new commuting technology—but always such that the center of the city was, roughly, thirty minutes away for most people.

The "one-hour rule" found in ancient Rome still exists in modern America (and most other places), even if we have swapped sandals for cars or subways. "The thing to recognize is that half the U.S. population still gets to work in almost twenty minutes, or under twenty minutes," says Alan Pisarski, the country's leading authority in the field of "travel behavior." For decades, Pisarski has been compiling numbers for the U.S. Census on how we get to work and how long that trip takes us. There seems to be some innate human limit for travel—which makes sense, after all, if one sleeps eight hours, works eight hours, spends a few hours eating (and not in the car), and crams in a hobby or a child's tap-dance recital. Not much time is left. Studies have shown that satisfaction with one's commute begins to drop off at around thirty minutes each way.

The enduring persistence of the one-hour rule was shown in a paper by urban planning researchers David Levinson and Ajay Kumar. Looking at the Washington, D.C., metropolitan area over a number of years from the 1950s to the 1980s, they found that average travel times—around thirty-two minutes each way—had hardly budged across the decades. What *had* changed were two other factors: distance and average travel speed. Both had gone up. They suggested that people were acting as "rational locators." Because they did not want to spend too long commuting, they had moved to more distant suburbs. They had longer distances to drive, but they could now travel on faster suburban roads, rather than crowded city streets, to get to where their jobs were located. (Those in the center city, meanwhile, were probably walking to work or taking the Metro, meaning their times had hardly changed as well.)

"Wait," I can hear you say, "I thought traffic was getting worse." For many people, it undoubtedly is. The Texas Transportation Institute estimates that total traffic delay in the United States went from 0.7 billion hours in 1982 to 3.7 billion hours in 2003. In the twenty-six largest urban areas, the delay grew almost 655 percent in those same years. The U.S. Census noted that in most large cities, it took longer to get to work in 2000 than it did in 1990. The authors of the "rational locator" study took another look at the issue and decided that perhaps travel times were *not* stable after all. Perhaps, they suggested, it was a "statistical artifact." Cities were growing larger every year, gobbling up new counties into their "metropolitan region," so maybe more-distant drivers who were not tallied in previous surveys were now being captured, jacking up the numbers. Or maybe the suburbs that they had moved to previously to escape congestion were now themselves getting congested. Perhaps the total outcome of all that rational location had itself become irrational.

But why exactly is it getting worse? Or, to ask a question I sometimes do when I encounter unexpectedly heavy congestion in the middle of the day, "Who are all these people?" There are obvious answers, the ones you yourself suspect, like the fact that we add new drivers faster than we keep adding new blacktop. To take a quite typical American example: In suburban Montgomery County, Maryland, just outside Washington, D.C., the population grew by some 7 percent between 1976 and 1985. The number of jobs grew too, by 20 percent. But vehicle registrations nearly doubled. The county, which hardly built any new roads at all during that period, was suddenly awash in cars. Studies show that when a household has more vehicles, it not only drives more as a total household, as one would expect, but *each person* puts on more miles, almost as if the presence of those extra vehicles prompts more driving.

Affluence breeds traffic. Or, as Alan Pisarski describes it, congestion is "people with the economic means to act on their social and economic interests getting in the way of other people with the means to act on theirs." The more money people have, the more cars they own, the more they drive (with the exception of a few Manhattan millionaires). The better the economy, the more miles traveled, the worse the traffic congestion. This is the interesting thing about studying traffic behavior: It reveals what Pisarski terms our "lines of desire." The U.S. Census is like a staid group portrait of the country. It shows us all in our homes, with our 2.3 bathrooms and 1.3 cats. But it does not really show us how we got there. The travel census is like a frantic, blurred snapshot of a nation in motion. It catches us on the move, in an unrehearsed moment, busily going about our daily lives in order to afford that house with 2.3 bathrooms. It may tell us more about ourselves than we know.

One striking thing the numbers seem to reveal is that women now make the largest contribution to congestion. (Another way to look at this is that they also suffer from it the most.) This seems like a controversial statement, and indeed one like it got a highway official booed at a conference. The statistic doesn't assign fault or suggest that women working is a bad thing; it does provide a fascinating example of how traffic patterns are not just anonymous flows in the models of engineers, but moving, breathing time lines of social change.

Many of us can remember or envision a time when the typical commute involved Dad driving to the office while Mom took care of the kids and ran errands around town. Or, because many American families had only one car, Dad was driven to the morning train and picked up again just in time for cocktail hour and Cronkite. This is a blinkered view, argues Sandra Rosenbloom, an urban planning professor at Arizona State University whose specialty is women's travel behavior. "That was just a middle-class model," she says. "Lower-class women always worked. Either alongside husbands in stores, or at home doing piecework. Women always worked."

Still, the *Leave It to Beaver* commute was not a total fiction, given that in 1950 women made up 28 percent of the workforce. Today, that figure is 48 percent. How could the roads *not* have gotten more crowded? "The rise in the number of cars, driver's licenses, miles traveled—it totally tracks women going into the labor force," says Rosenbloom. "It's not that men wouldn't have driven more, but you wouldn't see these astonishing increases in traffic congestion in all indices of travel if women weren't in the labor force, driving."

The rise in working women is only part of the story. After all, they still represent a minority of the workforce, and studies show that men still rack up more miles when they drive to work. But work is an increasingly small part of the picture. In the 1950s, studies revealed that about 40 percent of daily trips per capita were "work trips." Now the nationwide figure is roughly 16 percent. It's not that people are making fewer trips to work but that they're making so many other kinds of trips. What kinds of trips? Taking the kids to school or day care or soccer practice, eating out, picking up dry cleaning. In 1960, the average American drove 20.64 miles a day. By 2001, that figure was over 32 miles.

Who's making these trips? Mostly women. This is the kind of social reality that traffic patterns lay on the table: Even though women make up nearly half the workforce, and their commutes are growing increasingly close in time and distance to men's, they're still doing a larger share of the household activities that, back in the *Leave It to Beaver* days, they may have had the whole day to complete (and, as Rosenbloom points out, 85 percent of single parents are women). "If you look at trip rates by male versus female, and look at that by size of family," Pisarski says, "the women's trip rates vary tremendously by size of family. Men's trip rates look as if they didn't even know they *had* a family. The men's trip rates are almost independent of family size. What it obviously says is that the mother's the one doing all the hauling."

In fact, women make roughly double the number of what are called "serve-passenger" trips—that is, they're taking someone somewhere that they themselves do not need to be. All these trips are squeezed together to and from work in a process called "trip chaining." And because women, as a whole, leave later for work than men, they tend to travel right smack-dab in the peak hours of congestion (and even more so in the afternoon peak hours, which is partially why those tend to be worse). What's more, these kinds of trips are made on the kinds of local streets, with lots of signals and required turning movements, that are least equipped to handle heavy traffic flows.

Another way trip chaining has helped increase traffic congestion is that it has made carpooling virtually impossible. Who wants to share a ride with someone who is going to day care, picking up laundry, dropping by Blockbuster, stopping at Aunt Clarice's ("but just for a second")? Carpooling keeps dropping in the United States (save among some immigrant groups), but "fam-pools," car pools made up of family members (and almost 100 percent of fam-pools are *only* family members), keep rising. An estimated 83 percent of car pools are now fam-pools.

This raises the question of whether car-pool lanes are a good idea that has gone bad. If most people "carpooling" are simply toting their families around, taking no additional cars off the road and statistically driving more miles (thus creating more traffic), why should they get a break on the highway? Is a policy meant to reduce the number of drivers just acting as a "mommy lane," enabling drivers with children to do their trip chaining more quickly and thus encouraging more of it? (Some pregnant women have taken this to extremes, arguing that their unborn children are precocious car poolers).

That women suffer more from congestion, even if at the hands of other women, is demonstrated in the high-occupancy toll lanes (HOT lanes, a.k.a. Lexus lanes) in cities like Denver, where drivers pay more to travel on less congested roads. Rosenbloom notes that studies show that women pay to use the lanes more often than men do—despite making less money on average. "And they are not just high-income women," she says. "Even if you don't make very much money, you've got to get your kids from day care. Every minute they stay over, they penalize you. Or these women have second jobs they have to get to on time."

Women are not to be *blamed* for congestion, Rosenbloom argues. "The fault is the way families live today. The car is the way the two-worker families balance all the things they have to do." Where children might once have been cared for at home, they are now shuttled to day care. Where it was once the overwhelming norm for children to walk to school, today only about 15 percent do. Parents on the "school run" are thought to boost traffic on the roads by some 30 percent.

Parents' chauffeuring duties hardly end there, however, as the increasingly hyperscheduled "free time" of children, with its scores of games, lessons, and playdates, requires route planning and logistics that would turn a La Guardia air-traffic controller's hair gray. It's estimated that from

1981 to 1997, the amount of time children spent in organized sports in America *doubled*. All those games and all those practices, in increasingly far-flung suburbs, required rides. A new demographic entity, the so-called soccer mom, started hitting the roads big-time. "In the entire time I played baseball, my parents didn't watch me play ball once," recalls Pisarski, who is in his sixties. "I didn't feel slighted, because no other kid's parents were there either. Today you go to a game, and there's a hundred and fifty people and everybody gets a trophy."

Traffic, Pisarski emphasizes, is the expression of human purpose. Another huge way in which those purposes have changed is due to rising affluence. It's not just that American households have more cars, it is that they are finding new places to take them. And once you have shelled out for a car, the comparatively marginal cost of another trip is barely noticeable—in other words, there is little incentive *not* to drive.

Given that Americans increasingly spend much of what they make, it should come as little surprise that much of our increase in driving seems to stem from trips to the mall. From 1983 to 2001, the number of annual shopping trips per household almost *doubled*—and those trips are getting longer. Each year, the amount of driving we do for shopping would take us across the country once and almost all the way back again. Statistics now show that more people travel on Saturday at one p.m. than during the typical rush hours. The more money one has, the more choices one has, and so it's not surprising that nearly half of trips families make to supermarkets are not to those closest to their home. Pisarski notes that he, like many Americans, does not suffer for choice when it comes to food shopping, and his driving reflects this. "I go to Trader Joe's because I like their string beans. I go to Harris Teeter because their seafood is better than Giant. In effect, we are just more selective." Studies confirm that people shop at more grocery stores than they did a few decades ago.

You might think that the rise of larger, consolidated stores like Costco or Wal-Mart Supercenters, which offer one-stop shopping, might have actually helped cut down on the amount of shopping trips. But larger stores need to serve more people, which means, in effect, that they're farther away from more people. (A similar trend has also occurred with schools, which explains some of the decline in children's walking.) A study of Seattle grocery stores found that in 1940, the average store was .46 miles from a person's house, while in 1990, it was .79 miles. That small change in distance was basically the death knell for any thought of *not* driving to the store, for a half mile is as long as planners believe the average person is willing to walk. Even if the stores are bigger, moreover, we are going to them more frequently—the number of grocery trips per week almost doubled from the 1970s to the 1990s.

The reason we see so many people on the roads, getting in our way, is that so many of them are doing things that used to be done at home. This, too, is a function of affluence, but it's a complicated relationship. Do we drive to a restaurant for take-out food because we can afford it or because we are so busy trying to make money we have little choice? Either way, these sorts of social changes have their effects on traffic—often so fast that engineers can't keep up. When Starbucks began serving customers at drive-throughs a few years ago, the people who study traffic flows were caught flat-footed. Their models for what is called "trip generation"—basically, the additional traffic flow a new business will create—included numbers for "Fast Food Restaurants with Drive-Through Window," as well as for "Coffee/Bread/Sandwich Shop," but "Coffee Place with Drive-Through Window" was completely alien. For Starbucks, which will go so far as to put stores on opposing corners to capture different traffic flows and spare drivers the agony of having to make a left turn during rush hour, the drive-through represented a natural progression in its slow evolutionary insertion into the daily commute.

"Can you imagine, thirty years ago, saying nobody will make coffee at home?" Nancy McGuckin, a travel researcher in Washington, D.C., asked me on a break during an annual traffic conference. In her research, McGuckin (whom one colleague called "the queen of trip chaining") fingered coffee as a prime culprit in a dramatic new shift in traffic patterns. Men, it seemed, were suddenly doing more trip chaining. Sure, *some* were dropping off kids, but more were making a latte stop. She calls this the "Starbucks effect." The prime demographic, she says, is middle-aged men. "Who knew they needed 'me time'?" she asks. "We're used to women saying this: 'We're so busy, we need "me time."' But it was middle-aged men who were making that stop at Starbucks in the morning. I had some of them saying they were leaving their homes before it becomes chaotic with the backpacks and the school. [He wants] to get up and leave the house and go to Starbucks, where, by golly, there's somebody there who greets him by name, knows what his favorite drink is. It's like his time to prepare for the office environment. I don't think the psychology of that has been explored very well."

The same might be said for the psychology of commuting itself. It does not seem unreasonable to wonder why, if traffic is so bad, more people keep choosing to drive more miles. This question puzzles all kinds of people, from economists to psychologists to traffic engineers.

One important thing to consider, of course, is that for many Americans it is not so bad. They still get to work and back in that same roughly one-hour time frame. In relative terms, American commute times, Pisarski argues, should be "the envy of most places in the world." In cities like São Paulo, where the congestion is so bad "motorcycle medics" are needed to ferry patients between stalled queues of cars to the hospital, average daily travel times clock in at upward of two hours. The average car journey takes up to one-third longer in Europe than it does in the United States (which is perhaps why Europeans make fewer car trips). Driving to work alone, which is what nearly nine out of ten Americans do, is still, on average, about one and a half minutes faster than the average time for all other travel methods. One study that looked at the working poor found that those with a car were able to get around three times more quickly than those without one. Even people who do not *own* a car are more likely to commute via car than public transit.

Trying to crack the commuter psyche is rather bewildering work. On the one hand, people seem to hate commuting. When Princeton University psychology professor Daniel Kahneman and some colleagues surveyed a group of women about their experiences in a typical day and how they felt about them, commuting came in at the bottom. ("Intimate relations" and "relaxing with friends" were near the top.) On the other hand, Patricia Mokhtarian, a professor of civil engineering at the University of California, Davis, has found that when people were asked to name an "ideal" commute time, their mean response was not, as you might expect given its popularity in the aforementioned survey, "no commute" but sixteen minutes.

In another study, Mokhtarian and two colleagues located what they described as "an apparent paradox." When people were asked if they drive more than they would like to, the response was a unanimous yes.

When those same people were asked if they drive more than they need to, the response was a nearly unanimous yes. Why were people seemingly acting against their own interests? Why were they doing more of what they wanted to do less of? The researchers surmised that the driving people didn't want to do was, in fact, the driving they *needed* to do. Maybe it was the reasons they were driving that they wanted to eliminate, rather than the driving itself. Or maybe driving just seemed easier than figuring out alternatives.

A pair of Swiss economists have identified another kind of commuting paradox. They begin with the assumption that commuting, with its toll of time, stress, spilled coffee, and crash risk, is a "cost" that people rationally figure into their decisions about where to live in relation to where they work. If you have a long commute, that should be reflected in either a high-paying job or a nice house. The benefits that those things bring should offset the longer commute; in other words, a longer commute should not make you more unhappy. But that's precisely what the economists found in a study of German commuters; the researchers concluded that people making just the average twenty-three-minute commute would need a 19 percent salary raise to make the commute "worth it" from a rational perspective.

Commuters may, of course, have little choice. Housing might simply be too expensive close to where people work, so they're forced to live farther away from their jobs than they would like, out past the billboards that chide, "If You Lived Here, You'd Be Home by Now." The economist Robert H. Frank, comparing U.S. Census data between 1990 and 2000, found that commuting grew the most in counties in which income inequality had grown the most. He calls this the "Aspen effect," after the affluent Colorado city, which keeps expanding because the middle-class people who work in the town keep having to move farther away to find affordable housing. But there is a paradox here as well: Statistics show that commuting miles *rise*, not fall, with income. In other words, the people with the money to live close to the action seem to be doing more of the driving. Maybe those people are moving farther out in Aspen *because* they have more money, and they're choosing to buy bigger homes despite the commute.

This, however, is where things start to go wrong, according to many psychologists. A commuter who lives in the older suburb of Eagle Glen decides he wants to move to a new, bigger house in Fledgling Ridge. Getting the bigger house requires adding twenty minutes to his commute. This seems worth it because the bigger house provides such a boost to his quality of life. But gradually, that rosy glow fades. He quickly undergoes what psychologists call "hedonic adaptation." Suddenly, the newer, bigger house just seems normal. Everyone else has the same newer, bigger house. Meanwhile, the commuter has lost time (more of which cannot be made, unlike money). This means less time to do the things that are shown to actually bring happiness. He's locked into a longer commute, and studies show that the longer a commute is, the more prone it is to variability—to be longer or shorter than you expect. And some studies show that we are bothered more by *changes* to our commute time than by the actual time itself. As Harvard University psychologist Daniel Gilbert argues, "You can't adapt to commuting, because it's entirely unpredictable. Driving in traffic is a different kind of hell every day."

For a portrait of a driver in purgatory, consider urban bus drivers. Few drivers face as much traffic or are as affected by changes in their commuting schedule. The hassles they endure are legion, from the simpleton car drivers who accuse them—irony of ironies—of "causing" congestion to passengers yelling at them for being late. Despite the size of the buses they drive, they are struck by other vehicles at a higher rate than are passenger cars. And what happens to them? Studies of drivers in various countries have shown they have more stress-related hormones in their system than other people—including *themselves* before they started driving for a living. The worse the traffic, the higher the hormones. Medical ailments send more than half of them into early retirement. No wonder Ralph Kramden of *The Honeymooners* was always so grouchy!

The problem with the models used in the Swiss researchers' commuting paradox is that they rely, essentially, on asking people to turn their feelings into numbers. This is slippery stuff, prone to all kinds of biases. Psychologists have found, for example, that when college students were asked two questions, one concerning the number of dates they had in the previous month and the other concerning their general sense of life satisfaction, the results varied with the order in which the questions were asked. Ask about life satisfaction first, and this does not change the way they answer the dating question. Ask about dating first, and suddenly the students' idea of how happy they are seems to vary with how much they're dating. This has been called the "focusing illusion." Things become more important when we think about them. Ask a person how long their commute is, and then how happy they are, and they might give an answer that is different than if you had not first asked them about commuting. Maybe this betrays how unhappy their commute is really making them. Or maybe commuting is just not that important to their overall lives—until a researcher's question makes them think it is.

This is the murky, human side of traffic. Engineers can look at a section of highway and measure its capacity, or model how many cars will pass in an hour. That traffic flow, while it may mathematically seem like a discrete entity, is made up of people who all have their own reasons for going where they are going, for enduring that traffic. Some may have no choice; some may choose.

Moreover, Brian Taylor, a professor of urban planning at UCLA, observes that when we travel to work by car, there may be any number of parts to that journey. We may walk to our car, drive down our residential block, briefly cruise a larger arterial, then pop onto the highway for a spell before exiting onto another arterial, continue on to a smaller street, then drive up a parking-garage ramp, walk to the elevator, and finally walk to our desk. In the course of Taylor's hypothetical commute, the highway portion could be over half the distance traveled but less than half the *time* (and we perceive a minute of driving on a highway as shorter than a minute of walking to our car). Taylor notes that even if the speed on the congested highway were doubled, the total time saved would be less than 15 percent. For all these reasons, we cannot look at a jammed section of urban highway moving at fifteen miles an hour and assume that everyone is suffering, or suffering in the same way.

## The Parking Problem: Why We Are Inefficient Parkers and How This Causes Congestion

> Everyone in New York City knows there's gotta be way more cars than parking spaces. You see cars driving in New York all hours of the night. It's like Musical Chairs except everybody sat down around 1964.
>
> —Jerry Seinfeld

The next time you find yourself at a shopping mall or a store with a large parking lot where the store entrance more or less divides the lot in half by width, take a moment to observe how the cars are arrayed. Unless the lot is completely filled, you may be able to observe a common pattern. Chances are, the row that is dead opposite the store entrance will be the most filled, with cars stretching far out along the row. In each adjacent row, there are likely to be slightly fewer cars. This pattern will continue sequentially in each row so that if one were able to gaze down at the lot from above (as anyone can with Google Earth), the cluster of cars might look, depending on the lot's occupancy, like a giant Christmas tree or, perhaps, like a bell.

If you were to further study this bell-curve arrangement, you might conclude that the cars parked in the row closest to the store entrance but farthest out along the row are actually farther from the store entrance than many of the open spaces in the rows off to the side. Why is this so? Why don't parkers gravitate first toward the closer spaces? Perhaps parkers are not good geometricians. People may park in the row opposite the entrance, no matter how far away, because it will be easier to locate their car later. Parkers may find the center aisle, with its line of sight to the entrance, safer—even in open-air lots during the day. Or it may be that parkers optimistically sail to the closest row and, once having entered it, simply grab the first piece of what seems to be an increasingly scarce resource.

Whatever the case, something curious happens in parking lots. It seems that the people who actively look for the "best" parking place inevitably spend more total time getting to the store than those people who simply grab the first spot they see. This was the conclusion that Andrew Velkey, a psychology professor at Virginia's Christopher Newport University, came to after he studied the behavior of parkers at a Wal-Mart in Mississippi. Incoming cars were tracked; once they had acquired a parking spot, Velkey and his team measured the distance of the car to the entrance, as well as the time it took the driver to walk into the store. They observed two distinct strategies: "cycling" and "pick a row, closest space." They compared the results. "What was interesting," Velkey told me, "was although the individuals cycling were spending more time driving looking for a parking space, on average they were no closer to the door, time-wise or distance-wise, than people using 'pick a row, closest space.'" This is precisely what the pattern mentioned above had suggested: The best parking spaces, by distance or time, were not necessarily being chosen.

Were people just being lazy, or were they succumbing to cognitive biases? Bring a stopwatch on your next trip to the mall and see for yourself. Research has shown that people tend to underestimate the time it will take to get somewhere in a car and overestimate the time it will take to walk somewhere. The time spent cycling in the lot may have seemed less than it actually was, and the time spent walking greater than it actually was—and this could inform how they parked in the future.

In a previous study on a campus parking lot—a lot that was crowded but usually had some spaces in the back rows—Velkey polled students about how long they thought it typically took them to find a parking space. "They said four and a half minutes," Velkey told me. "In reality, when we watched them, it takes about thirty seconds. I said, 'Where did that extra four minutes come from?' " Velkey suggests that the psychological principle known as the "availability heuristic" was at work. Students were tending to remember the few times when it was very difficult to find a spot, instead of the everyday experience in which it was quite easy. They were remembering the things that stuck out in their memory.

In the Wal-Mart lot, there was something else interesting about the two groups of parkers. More women seemed to adopt the "cycling" strategy, while more men seemed to opt for the "pick a row, closest space" tactic. Velkey wondered if a "gender effect" existed in the way women and men perceived distance and travel time (previous studies have arrived at mixed conclusions on this). So he gathered a group of subjects and had them estimate the distance to an object at varying locations, and then asked them to estimate the time it would take them to walk there. Men seemed to underestimate how long it would take to walk, while women seemed to overestimate it—which might explain the differences in parking strategies. Both genders underestimated distances, an effect that grew larger as the distance did.

What had led Velkey, clipboard in hand, to parking lots? Interestingly, it was an offshoot of his prime research interest: the foraging behavior of animals, particularly how animals develop certain strategies in the face of constrained resources such as food or territory. He was studying this at the University of Montana, where wildlife abounds. It turned out there was an interesting example right outside the psych department window: the crowded parking lot. The value of the resource was clear—a faculty member had recently spent a day in jail after keying the car of someone who had stolen his parking spot. (Here we must remember the old dictum about what keeps a university running smoothly: "Beer for the students, parking for the faculty, and football for the alumni.")

In this lot, Velkey saw two kinds of behavior emerge: an active and a passive search strategy. Some people would drive around the lot looking for a space, while others would sit at the head of a row and wait for someone to leave. In terms of the avian foraging models Velkey usually studied, the active searchers were like condors, soaring and looking for prey; the passive searchers, meanwhile, were like barn owls, perched and lying in wait.

Most people were active searchers, spending about as much time looking as it would take them to drive to the next available lot, while the smaller group would wait for minutes on end for someone to exit. This group, Velkey noted, almost always got a spot in the lot, while others found one elsewhere. (In that study the "postacquisition" costs of walking from the car were not measured, so it is hard to say who came out ahead in terms of total time.) A set of "evolutionarily stable strategies" had taken hold: If everyone tried to be condors, they would all be endlessly circling; if everyone tried to be barn owls, they would all be hovering around the same spot. Depending on circumstances (e.g., whether classes were about to let out), one strategy or another might bring more "local" success than another, but, Velkey notes, eventually everyone gets a spot.

The way humans hunt for parking and the way animals hunt for food are not as different as you might think. Many scientists believe that animals' foraging habits can be explained by a model known as "optimal foraging"—animals seek to gather the most food with the least effort (thus leaving them with more time and energy to, say, reproduce). These strategies evolve in response to the myriad numbers of life-or-death decisions that are made in each successive generation: Does the hunter go after the easy, low-protein prey or the elusive, higher-protein prey? How long does one stay in a particular patch before moving on to a possibly more productive patch? Does one look for food in a group or on one's own?

For some optimal foraging in your own backyard, consider the bumblebee and the foxglove. Bees, it turns out, begin looking for nectar in the flowers arrayed on the bottom of the spike, slowly working their way up. Why? Because foxgloves add new flowers in an upward progression, so that those at the top contain less nectar. Bees also know to skip flowers they have already visited, and when a new bee lands on a foxglove that has already been visited by another bee, the odds are it will leave immediately. The chances of finding any missed nectar, it seems, are not worth the effort of looking.

Now, instead of bees, think of humans parking. The parkers in the Montana lot who followed the "perching" strategy had evolved a very specific optimal strategy: They knew that near the top of the hour, as classes emptied, spots would become available, but it was better to search for the exiting *driver* than the spot. New visitors to the lot, however, or visitors who arrived too late, would circle in vain before ultimately deciding not to expend any more of their energy in this "patch."

In our daily lives as parkers, we face these foraging questions. We must decide whether to act like condors or barn owls. And we're sometimes on the other end: It is not difficult to feel unnervingly like dying prey in the eyes of a stalking buzzard when you come out of a crowded shopping mall during the holidays and suddenly find yourself tailed by a creeping car. Is it faster to tail drivers to their cars and wait for them to load their merchandise or to look for an open space? Do we pass up less valuable spaces (i.e, "prey") for higher-value spaces that might be lurking around the corner? In some cases in the animal world, it is better to hunt for food in groups, but in other cases, going solo is the better option. You may have experienced this dilemma as you cruised the streets of a city (or the rows of a mall) looking for a parking spot, realizing with a sudden dread that the person ahead of you, taillights flashing hopefully in front of potential spaces (which turn out to house fire hydrants or compact cars), is doing *exactly the same thing*. It no longer makes sense to look in the same places, as the car ahead will consume the resource first—better to head elsewhere.

But neither animals nor humans always follow optimal strategies. One reason is that not enough information might be available—an issue that the parking industry is trying to address with technology that alerts people, via real-time signage or through cars' navigation systems, to available (paid) parking spaces. Another reason might be the cognitive illusions I have already mentioned. Urban planners have pointed out that people seem willing to walk about a half mile from a parking spot to a destination. But they seem more likely to do so when they're walking in the massive parking lot to a sports stadium, for example, than on downtown streets. There is an interesting explanation for this: Studies by geographers have shown that people tend to overestimate distances on routes that are "segmented," versus those where the destination is in sight. Thus a football stadium a half mile away in a big parking lot *seems closer* than a half-mile walk involving multiple turns in a city.

The Nobel Prize–winning economist Herbert Simon has suggested, in a seminal theory he called "satisficing" (a mix of *satisfying* and *suffice*), that because it is so hard for humans to *always* behave in the optimal way, we tend to make choices that leave us not with the "best" result but a result that is "good enough." To take the bell-curve parking patterns described earlier as an example, drivers may have entered the lot with a general goal of getting the "best" spot, that is, in the row closest to the entrance. Once they were in the row, however, the goal changed to getting the best spot in that row. This is good in that it helps them feel satisfied with the spot they acquire. But if their strategy to get the "best" spot left them worse off overall, it might not be so good. Simon called the human limitations in making decisions "bounded rationality." In Velkey's study, people who focused on finding the "best" parking spot, in terms of distance, failed to account for all the time they were losing while searching—and they didn't get closer anyway. We do not know if they were happy or not with their spot. When Velkey tried to conduct interviews, he was unsuccessful. Ironically, many people said "they didn't have time."

The ways in which we hunt for parking, whatever their biological basis, are one of those subtle, almost secret patterns of traffic. They matter more than you might think.

Parking occupies a strangely marginal place in the whole traffic equation. Engineers focus their energy on traffic-flow models, not parking models. We do not get morning "parking reports" on the radio. We tend to think of traffic as cars in motion; parking spaces seem more like real estate (indeed, they can be priced as high as houses, as the sale of quarter-million-dollar spots in New York and Boston has shown). But the simple, if often overlooked, fact is that without parking there would be no traffic. Every car on the road needs a place where it can begin and end, and mostly just sit there: Cars spend about 95 percent of their time parked.

Parking is the innocuous gateway drug to a full-blown traffic-abuse problem. One survey found that a third of cars entering lower Manhattan were headed to free or subsidized parking spots. If those spots were not free or subsidized, there would be fewer drivers during the morning rush hour. Ironically, near the Department of Transportation itself, the streets are filled with DOT vehicles bearing special parking permits. How much do they add to peak-hour congestion? (This brings to mind a headline from the satirical newspaper the *Onion:* URBAN PLANNER SITS IN TRAFFIC OF HIS OWN MAKING.)

When the city of Copenhagen was looking to reduce the number of cars entering the central city in favor of bicycles and other modes of transportation, it had a very crafty strategy, according to Steffen Rasmussen of the city's Traffic and Planning Office: Get rid of parking, but without anyone noticing. From 1994 to 2005, Copenhagen cut parking spaces in the city center from 14,000 to 11,500, replacing the spaces with things like parks and bicycle lanes. Over that same time, not accidentally, bicycle traffic rose by some 40 percent—a third of people commuting to work now go by bike—and Copenhagen has become one of the few places in the world where one can read, in a report, a sentence that would seem like a comical misprint almost anywhere else: "Cycle traffic is now so extensive that congestion on certain cycle tracks has become a problem, as has cycle parking space."

What you may not realize, when you find yourself driving on a crowded city street, is that many of your fellow drivers on that crowded street are simply cruising for parking. The problem is not so much the lack of street parking but the plentiful abundance of free or underpriced parking. This finding has sparked the fiery crusade of Donald Shoup, a bearded, bow-tied, and bicycling economist at the University of California, Los Angeles, and the author of a seven-hundred-page, cult-sensation tome titled *The High Cost of Free Parking.*

The mantra used by Shoup, and his growing legion of supporters (dubbed "Shoupistas"), is the "85 percent solution." In other words, cities should set prices on parking meters at a level high enough so that an area's spots are only 85 percent occupied at any time. The ideal price, says Shoup, is the "lowest price that will avoid shortages." Spaces with no meters at all, in a city like New York, are total anathema to Shoup. "People who want to store their car shouldn't store it on the most valuable land on the planet, for free," he told me in his office at UCLA, where a vintage parking meter sits atop his desk. "Something that is free is very misallocated." This is why people who want to see free Shakespeare in the Park performances in New York City have to begin waiting in line as early as the day before (or hire people to do it for them), why cafés that offer free Internet access soon find themselves having to limit the time patrons can spend at a table, and why it can be so hard to find a parking spot.

The reason people cruise is simple: They're hunting for a bargain. In most cities, there is a glaring gap between the cost of a metered parking spot and that of an off-street parking garage. Looking at twenty large U.S. cities, Shoup has found that, on average, garages cost *five times* more per hour than metered street spots. The reason garages can charge so much, of course, is that the streets charge so little. When free parking spaces are available, the discrepancy is even higher, particularly for a free spot that can be held for many hours. And so people are faced with a strong incentive to drive around looking for parking, rather than heading into the first available garage.

On the individual level, this makes sense. The problem, as is so often the case in traffic, is that the collective result of everyone's smart behavior begins to seem, on a larger scale, stupid. The amount of extra traffic congestion this collective parking search creates is shocking. When Shoup and his researchers tracked cars looking for parking near UCLA (they rode bikes, so other cars would not think *they* were looking for parking and throw off the results), they found that on an average day cars in one fifteen-block section drove some 3,600 miles—more than the width of the entire country—searching for a spot.

When engineers have tried to figure out how many cars in traffic are looking for parking, the results have ranged from 8 percent to 74 percent. Average cruising times clock in at anywhere from three minutes to thirteen minutes. What's so bad about three minutes? you might ask. As Shoup points out, small amounts can have big consequences. In a city where it takes three minutes to find street parking, and where each space turns over ten times per day, each of those spaces will generate thirty minutes of cruising per day. At 10 miles per hour, that means the average space generates five miles' worth of driving per day, which works out to a yearly sum that would get you halfway across the United States—not to mention a heap of pollution.

But it is not simply that cars are driving while looking for parking. They're driving in specific ways. There is the inevitable slowing to check out a prospective spot, the stopping to study whether a spot is valid, the actual jockeying into the spot, or what Shoup calls "parking foreplay," in which the person detects that a space is about to be vacated and stops to wait. This may seem a minor offense, but as I discussed earlier, one car stopped on a two-lane street creates a bottleneck that cuts traffic capacity *in half*.

This is worsened further by the inevitable delays and gaps caused by drivers battling to merge before they reach the stalled car. One person's small act is felt by many. The famed urbanist William H. Whyte once espied this phenomenon during a traffic study of Manhattan. In his "mind's eye," he observed, one particular street was always "jammed" with double-parked cars (a result of underpriced parking, in Shoup's view). But when he actually counted the number of double-parkers, he was shocked to only find "one or two" at any time. "It seemed odd that so few could do so much," he wrote. "But the number, we found, was not the critical factor. It was the amount of time a lane was out of action because of double parking. Just one vehicle per block was enough."

The more time one spends looking for parking, of course, the greater chance one has to get in a crash, which then creates even more congestion. Interestingly, parking itself, according to some studies, is responsible for almost one-fifth of all urban traffic collisions. While some engineers think curb parking should be done away with entirely for safety and traffic-flow reasons, others counter that the rows of parked cars actually make things safer for pedestrians, both as a physical barrier and a source of "friction," like street trees, that can drop traffic speeds by an estimated 8 miles per hour.

To return to the Wal-Mart study mentioned earlier, the massively capacious big-box lots might seem to have little to do with crowded city streets. But there is plenty of cruising in large, free lots. It is simply that the incentive to save money has been entirely replaced by the incentive to save distance (and, theoretically, time, even if that ends up not being the case). In fact, there is *always* parking at Wal-Marts, so much so that the company lets people in recreational vehicles treat it like a campground. As Shoup points out, at places like Wal-Mart, the planners who dictate what size the parking lot should design for "peak demand"—that is, Christmas Eve—thus guaranteeing that most of the year, the lot has an abundance of empty spaces. The estimated demand comes from the parking-generation models of traffic engineers, which are filled, Shoup notes, with strange irregularities, like the paradoxical fact that banks with drive-up windows are required to have more parking spaces than banks without drive-up windows.

Shoup argues that there is a circular logic at work in parking-generation models, one similar to that found in other kinds of traffic models. The demand for parking is treated as a foregone conclusion: Planners measure the number of people parked at a typical free parking lot in a location without much public transportation. The new Wal-Mart is built and, lo and behold, it attracts lots of cars. As Shoup writes, "The parking demand at new land uses with free parking then confirms the prediction that all the required spaces are 'needed.' " Planners seem to ignore the fact that they are helping to dictate demand by providing supply. There are lots of cars in lots because parking is free.

As Shoup reminds us, though, Wal-Mart's free parking, like the free curb parking in cities, is not really free; the term is an oxymoron. We pay for "free" parking all sorts of other ways—and not just as a surcharge on the goods we buy. Parking lots are not only the handmaidens of traffic congestion, they're temperature-boosting heat islands, as well as festering urban and suburban floodplains whose rapid storm-water runoff dumps motor oil and carcinogenic toxins like polycyclic aromatic hydrocarbons (from shiny black sealcoat) into the surrounding environment and overwhelmed sewer systems. They represent a depletion of energy and a shockingly inefficient use of land—in a study of one Indiana county, Bryan Pijanowski, a geographer at Purdue University, found that parking spaces outnumbered drivers by *three to one*. The whole parking equation is like a large-scale version of that person at the mall, circling to get a "better" spot to save time and energy, and not realizing how much time and energy they have wasted looking for a better spot.

Traffic patterns are the desire lines of our everyday lives. They show us who we are and where we are going. Examined more closely, this movement, like all desires, is not always rational or efficient. Traffic is a great river of opportunity, but often, as with the poor choices made with parking policy, we're just spinning our wheels. In the next chapter, we'll look at some more ways to get unstuck.

## Chapter Six

# Why More Roads Lead to More Traffic (and What to Do About It)

### The Selfish Commuter

> When a road is once built, it is a strange thing how it collects traffic.
>
> —Robert Louis Stevenson

In the summer of 2002, a labor dispute at the ports of Los Angeles and Long Beach halted the flow of goods for ten days. Ships backed up, containers of Nikes and Toyotas lay dormant, and five-axle trucks, the kind that carry the containers from the ships to their destination, suddenly had nothing to haul. The impact on I-710, the route most trucks take from the ports, was immediate: In the first seven days of the shutdown, there were nine thousand fewer trucks on the highway.

Frank Quon, deputy district director of operations for Caltrans, the state highway authority, noticed something peculiar happening that week. The *total* traffic flow dropped by only five thousand vehicles. "Nine thousand trucks disappeared off the system," Quon told me in his office in downtown Los Angeles. So why did the total flow drop by barely over half that? "Cars filled in the volume. Another four thousand cars just jumped in the mix."

Almost instantly, drivers just seemed to know that the 710, where speeds jumped an average 67 percent during the shutdown, was a good place to be. They may have heard it on the traffic report, or a friend may have told them. Or they took it one day, learned that it was uncongested, and decided to take it the next day as well. What is curious is that the 710 was not necessarily sucking drivers off more crowded routes. "If you look at the parallel routes, like the 110 freeway," Quon said, "the volumes remained essentially the same."

It was as if drivers had suddenly materialized out of nowhere to take advantage of a highway that was, by Southern California standards, almost too good to be true. And it was: By the following week, when the ports reopened, the traffic was even worse than before the shutdown as trucks scrambled to catch up on deliveries—truck traffic, as you might have guessed, jumped much more than the total traffic. Now those new cars were deciding to stay away from the 710.

Engineers like Quon call what happened on the 710 a case of "latent demand." "It's the demand that's there but because the system is so confined that demand doesn't materialize," Quon explained. "But when you create capacity, that latent demand comes back and fills it in." Basically, people who would have never taken the 710 because it was too crowded suddenly got on. We don't really know what they did before. Perhaps they used local streets. Perhaps they took public transportation. Perhaps they simply stayed home.

The point is that people are incredibly sensitive to changes in traffic conditions (sometimes *too* sensitive, as we shall soon see) and they seem capable of quickly adapting to even the most drastic changes in a road network. Engineers have a phrase: "It'll be all right by Friday." This rough rule of thumb means that even if on Monday something major happens that throws off the usual traffic patterns—a road is closed, a temporary detour set up—by the next Friday (or so) enough people should have reacted to the change in some way to bring the system back to something resembling normal. "When a change in a traffic pattern occurs, there's a state of flux for a period of time," Quon said. "We usually have everybody plan on expecting a two-week period. Things are going to keep balancing. Some days will be good, some days will be not so good, and then at the end of the two weeks, there will be an equilibrium in the system based on those changes."

The latent demand that the newly fast 710 highway in Los Angeles had unlocked is often described by another phrase, "induced travel," which is really just a twist on the same thing: There was a new incentive to drive on the highway. Imagine that instead of trucks disappearing from the 710, two new lanes were added. The result would be the same. Congestion would drop, but the highway would become more attractive to more people, and, when it was all said and done, traffic levels might be even higher than before. This is the "more roads create more traffic" argument you have no doubt heard before. It is actually an argument older than automobile traffic itself. In 1900, William Barclay Parsons, chief of New York City's subway system, wrote, "For New York there is no such thing as a solution to the rapid transit problem. By the time the railway is completed, areas that are now given over to rocks and goats will be covered with houses and there will be created for each new line a special traffic of its own. The instant that this line is finished there will arise a demand for other lines."

Over a century later, people are still arguing. There is a huge and enervating literature about this, which I heartily do not recommend. Do we build more roads because there are more people and more traffic, or does building those roads create a "special traffic all its own"? Actually, both of these things are true. What's in dispute are political and social arguments: Where and how should we live and work, how should we all get around, who should pay for it (and how much), what effect does this have on our environment?

But studies suggest that induced travel is real: When more lane-miles of roads are built, more miles are driven, *even more so* than might be expected by "natural" increases in demand, like population growth. In other words, the new lanes may immediately bring relief to those who wanted to use the highway before, but they will also encourage those same people to use the highway more—they may make those "rational locators" move farther out, for example—and they will bring new drivers onto the highway, because they suddenly find it a better deal. Walter Kulash, an engineer at Glatting Jackson, argues that road building, compared to other government services, suffers disproportionately from this feedback loop. "You build more roads and you generate more use of the roads. If you add mightily to the sewer capacity, do people go to the bathroom more?"

If you do not believe that new roads bring new drivers, consider what happens when roads are taken away. Surely all the traffic must simply divert to other roads, no? In the short term, perhaps, but over time the total level of traffic actually *drops*. In a study of what they called "disappearing traffic," a team of British researchers looked at a broad list of projects in England and elsewhere where roads had been taken away either for construction or by design. Predictably, traffic flows dropped at the affected area. Most of the time, though, the increase in traffic on alternative routes was nowhere near the traffic "lost" on the affected roads.

In the 1960s, as Jane Jacobs described in her classic book *The Death and Life of Great American Cities*, a small group of New Yorkers, including Jacobs herself, began a campaign to close the street cutting through Washington Square Park, in Greenwich Village. Parks were not great places for cars, they suggested. They also suggested *not* widening the nearby streets to accommodate the newly rerouted flow. The traffic people predicted mayhem. What happened was the reverse: Cars, having lost the best route through the park, decided to stop treating the neighborhood as a shortcut. Total car traffic dropped—and both the park and the neighborhood are doing just fine.

We have already seen how engineers' models are unable to fully anticipate how humans will act on "safer" roads, and it is no different for congestion. It makes sense, mathematically, that if a city takes out a road in its traffic network, traffic on other streets will have to rise to make up for the lost capacity. If you removed one pipe in a plumbing system, the other pipes would have to pick up the slack. But people are a lot more complex than water, and the models fail to capture this complexity. The traffic may rise, as engineers predict, but that in itself may discourage drivers from entering a more difficult traffic stream.

Or it may not. Los Angeles currently operates with a freeway system largely built in the 1950s and 1960s. Its engineers never imagined the levels of traffic the city now sees. As John Fisher, head of the city's DOT, put it, "They say, 'If you build it, they will come.' Because we didn't build it doesn't mean the people stopped coming. Freeways weren't built, but the traffic is still coming anyway. There's more and more traffic. The bottom line is that the L.A. area is going to be a magnet whether we build freeways or not. People are still going to want to come here."

This raises the question of how much more successful a city Los Angeles could be if it had built all the freeways it never did, if one could magically whisk from downtown to Santa Monica in a few minutes. Then again, how desirable would a place like Beverly Hills be if the freeway that had been planned for it, to "cure" L.A. traffic, was now running through it? Wouldn't the increased speed just attract even *more* people? Is traffic failing Los Angeles, or is it a symptom of a thriving Los Angeles? Brian Taylor, the planner at UCLA, argues that people often focus single-mindedly on congestion itself as an evil, which, leaving aside for a moment the vast, negative environmental impacts, misses the point: What great city has not been crowded? "If your firm needs access to post-production film editors or satellite-guidance engineers," Taylor notes, "you will reach them more quickly via the crowded freeways of L.A. than via less crowded roads elsewhere." Density, economists have argued, boosts productivity. Traffic engineers like to use the example of an empty restaurant versus a crowded restaurant: Wouldn't you rather eat at the crowded one, even if it means waiting in line?

Users of Match.com, a dating service, are said, in places like Washington, D.C., to specify that they would like to meet someone who lives no more than ten miles away, presumably to avoid the hassles of congestion. Some have seen this as a social problem: Traffic is literally *killing* romance! Cupid is thwarted by congestion! This, too, misses the point: People move to places like Washington, D.C., in fact, because there are so many other people nearby. This is why cities play host to speed-dating events. There is so much "romantic congestion" packed into one room that daters must speed through all the potential choices. In Idaho, you will not face traffic trouble in driving well beyond the ten-mile range to meet dates; actually, you will probably have little choice. In any case, as anyone who has been in a long-distance relationship knows, those intervening miles can be a good way of deciding if a potential mate is really worth it.

What about all that time wasted in traffic? Surely that is costing us—$108 billion in the United States in 2000, according to one estimate. But a number of economists, most notably Anthony Downs of the Brookings Institution, have pointed out the potential flaws in these estimates. The first is that people seem willing to accept much of the delay, instead of paying to eliminate it (which means the "real" loss is closer to $12 billion). Another problem is that some models measure the costs of congestion against a hypothetical ideal of a major city in which all commuters could move at free-flowing speeds during rush hours—a situation that has not been possible since Juvenal's Rome. Still another complication is that models judge the money people lose in traffic by a hypothetical wage rate, but this assumes that people would get paid for any time saved in traffic—or that they would somehow use the time saved in traveling to do something productive, not simply travel *more*. (As mentioned in the last chapter, many people seem to enjoy the time spent in their car.) Finally, no one really knows how much money we *make* because of our transportation system, so the losses due to congestion may be marginal. A useful comparison is the Internet. It imposes all kinds of costs on our productivity—YouTube videos, spam, fantasy football—but does anyone not think these are an acceptable cost for all the good we derive from it?

There is another way, a bit more subtle and complicated, that new roads can cause more traffic: the Braess paradox. This sounds like a good Robert Ludlum novel, but it actually comes from a classic 1968 paper by a German mathematician, Dietrich Braess. Put simply, the paradox he discovered says that adding a new road to a transportation network, rather than making things better, may actually slow things down for all its users (even if, unlike in the "latent demand" example, *no* new drivers have been induced onto the roads). Braess was actually tapping into the wisdom of a long line of people who had in some way thought about this problem, from the famous early-twentieth-century British economist Arthur Cecil Pigou to operations researchers in the 1950s like J. G. Wardrop.

You would need an advanced math degree to fully understand Braess and his ilk, but you can grasp the basic problem they were all getting at by thinking in simple traffic terms. First, imagine there are two roads running from one city to another. There is Sure Thing Street, a two-lane local street that always takes an hour. Then there is Take a Chance Highway, where the trip *can* be half an hour if it's not crowded, but otherwise also takes an hour. Since most people feel lucky, they get on Take a Chance Highway—and end up spending an hour. From the point of view of the individual driver, this behavior makes sense. After all, if the driver gets off the highway and goes to Sure Thing Street, he or she will not save time. The driver will save time only if others get off the highway—but why should they?

The drivers are locked into what is called a Nash equilibrium, a strategic concept from the annals of Cold War thinking. Popularized by the Nobel mathematician John Nash, it describes a state in which no one player of an experimental game can make himself better off by his own action alone. If you cannot improve your situation, why move to a different road? The irony is that when everyone does what is best for him- or herself, they're not doing what is best for everyone. On the other hand, if a traffic cop stood at the junction of the two roads and directed half the drivers to Sure Thing Street and half to Take a Chance Highway, the drivers on Sure Thing Street would get home no sooner, but the highway drivers would get home twice as fast. Overall, the total travel time would *drop*.

If all this puzzles you, Braess's finding really makes the head spin. To simplify greatly, imagine again the two hypothetical roads I mentioned, but this time imagine that halfway between the two cities, Take a Chance Highway (where the trip takes less than an hour by however many fewer drivers choose it) becomes like Sure Thing Street (always an hour), and vice versa. Since each two-part route is likely to take the same amount of time, drivers split between the two routes, putting us in one-hour equilibrium.

But now imagine that a bridge is built connecting the two roads, right at the halfway point where Take a Chance becomes Sure Thing, and vice versa. Now drivers who began on Take a Chance Highway and found that it was not so good take the bridge to the *other* Take a Chance Highway segment. Meanwhile, drivers who began on Sure Thing Street are not about to cross the bridge and move to the other Sure Thing Street when, instead, they could stick around as their road becomes Take a Chance Highway (who knows, they might get lucky).

The problem is that if everyone tries to do what they think is the best thing for themselves, the actual travel time for all drivers goes up! The new link, designed to reduce congestion, has made things worse. The reason lies in what computer scientist Tim Roughgarden has called "selfish routing." The way each person is moving through the network seems best to them ("user optimal"), but everyone's total behavior may be the least efficient for the traffic network ("system optimal").

This really brings us to the heart of traffic congestion. We are "selfish commuters" driving in a noncooperative network. When people drive to work in the morning, they do not pause to consider which route they could take to work, or at which time to take that route, so that their decision would be best for everyone else. They get on the same roads and wish that not so many others had also chosen to do the same thing.

As drivers, we are constantly creating what economists call, in the thorny language of economics, "uninternalized externalities." This means that you are not feeling the pain you are causing others. Two legal scholars at the University for California at Berkeley have estimated, for example, that every time a new driver hits the road in California, the total insurance cost for everyone else goes up by more than $2,000. We do not pay for the various unsavory emissions our cars create—to take just one case, the unpaid cost of Los Angeles' legendary haze is about 2.3 cents per mile. Nor do we pay for the noise we create, estimated by researchers at the University of California, Davis, to be between $5 billion and $10 billion per year. How can you estimate the cost of something like noise? Real estate provides a clue. Studies have shown that house prices decline measurably as traffic rates and speeds increase on the adjoining street, while, on the other hand, when traffic-calming projects are installed on streets, house prices often rise. One might argue that the lower price of a house on a high-traffic street already takes into account these costs, but what happens when a buyer purchases a house at a certain price and then traffic increases on that street, lowering its value? Living near a major road also exposes people to more hydrocarbons and particulates of car exhaust, and any number of studies have reported links between proximity to traffic and conditions like asthma and coronary problems.

There are other kinds of costs, more difficult to measure, that you as a driver put on the people you drive by. When the urban planner Donald Appleyard surveyed San Francisco in the 1970s, he found that on streets with more road traffic, people had fewer friends and spent less time outside. In the same way that traffic has been blamed for habitat fragmentation of the wild, cutting off species from foraging areas or reducing the tendency of birds to breed, high traffic helps starve social interaction on human streets (maybe *this* is how congestion hurts romance). Somewhat paradoxically, Appleyard found that people who lived on the streets with less traffic (who made more money and were more likely to own their homes) actually *created more traffic themselves*, while the people who lived on the high-traffic streets were less able to afford cars. The rich, in effect, were taxing the poor.

The most basic externality, however, is congestion itself. Your presence in the traffic stream helps add time to others' commutes, just as others' presences add time to yours. But no one driver is gaining more than those others are collectively losing. In economics, a "public good" is something that a person can consume without reducing someone else's ability to consume that same thing or exclude them from doing so—sunlight, for example. An empty road late at night might be thought of as a public good, but a road with any kind of congestion on it quickly becomes "subtractable"—the more people who use it, the worse it performs.

This is the famous "tragedy of the commons," as described by Garret Hardin, in which a pasture open to all is quickly filled up by herders who want to graze as many cattle as possible. Every time a herder adds a cow, he gains. The pasture eventually begins to suffer from overgrazing, but a herder *still* adds animals because he alone benefits from his gain, even if the returns are diminishing (and they ultimately vanish), while everyone shares the costs of that new animal. (Overfishing is another such oft-invoked "tragedy.")

The "tragedy of the highway" is seen as every car joins the peak-hour freeway. As each car gets on, things get worse for everyone, but as there is still a gain for each driver (getting to work, getting home) that exceeds the gain from not driving, and as the loss is shared by all, people keep joining the freeway.

## A Few Mickey Mouse Solutions to the Traffic Problem

98% OF U.S. COMMUTERS FAVOR
PUBLIC TRANSPORTATION FOR OTHERS

—headline in the *Onion*

So how can traffic congestion, this age-old dilemma, be solved? "Build more roads!" is a typical answer. "But more roads bring more traffic!" is the typical response. "Then build even more roads!" "But that will bring even more traffic!" Looking beyond that hall of mirrors, it's worth pointing out a few things. The most obvious problem with building more roads to alleviate congestion is that we, in the United States at least, cannot afford them. Talk to just about any traffic engineer and they will repeat what the numbers already tell us: We do not have enough money to maintain the *current* roads, much less build new ones. What about all those fuel taxes? Drivers in the United States pay one-half the fuel taxes of drivers in Canada, one-fourth that of the Japanese, and *one-tenth* of the English. Adjusted for inflation, the fuel tax brings in less revenue than it did in the 1960s.

But even if we could afford to build more roads, that might not be the best way to spend the money. For one, as the transportation scholar Martin Wachs has pointed out, "Well over 90 percent of our roads are uncongested for well over 90 percent of the time." Many congested roads are congested for only a few hours a day, which brings up the Wal-Mart parking lot problem of the previous section. Do you build a parking lot that will be below capacity for 364 days of the year so that it can accommodate every shopper on Christmas Eve? On the one hand, it might be a socially negative thing that some people have to get on the roads at five a.m. in Los Angeles to make it to work on time, or that both directions of the highway are crowded at many hours of the day. On the other hand, this is a good thing. It means the road network is being used efficiently. Empty roads may be fun to drive on, but they're also wasteful.

Adding more lanes to a road is not always the traffic-busting silver bullet you might think it is. Imagine that you're at the extremely crowded intersection of two three-lane roads. Why can't they make it bigger? you ask. Look at all those people who want to turn left—why can't they add another left-turn lane? The problem, as two Canadian researchers have pointed out, is that adding more lanes is a process of diminishing returns.

The bigger intersections grow, the less efficient they become. Adding a second left-turn lane, for instance, means that, for safety reasons, "permissive" (or on the green) left turns can no longer be allowed. Only "protected" left turns (on the green arrow) will be allowed. As fewer cars can now turn left on the green signal (through gaps in oncoming traffic), the arrow phase will have to be longer. This means most other movements have to be halted. More lanes also mean more "friction," as engineers call it; a car wanting to turn left, for example, will find it harder going—and have a greater impact on the total traffic flow—when it has to cross three lanes instead of one. Given that bigger intersections take longer to cross, the clearance phase—that dead zone engineers introduce to make sure everyone has gotten through, including pedestrians—needs to become longer as well, further increasing delay. The result is that where an intersection with a single-lane approach would handle an average of

625 vehicles per hour, the next lane allows only 483 vehicles per hour, the third 463, and the fourth just 385. The more you spend on new lanes, the smaller the return—and the faster it becomes recongested.

Another problem is that most traffic jams are what engineers call "nonrecurring congestion." This means a highway that normally functions fine is congested, perhaps because of construction or weather but, most often, because of crashes. Rather than build more lanes, the best congestion solution here is for people to get in fewer crashes—which, as described in Chapter 3, would happen if drivers simply paid more attention to their driving.

The actual crash, which may or may not close a lane, is only part of the problem, of course. The highway's capacity drops an estimated 12.7 percent because of the line that forms—often on *both* sides of the highway—to take a look. This is where human psychology fails us. Not only do we have a morbid curiosity to rubberneck, but we feel we should not miss out on what others have had a chance to see. The economist Thomas Schelling points out that when each driver slows to look at an accident scene for ten seconds, it does not seem egregious because they have already waited ten minutes. But that ten minutes arose from everyone else's ten seconds. Because no individual suffers from the losses he inflicts on others, everyone is slowed. "It is a bad bargain," concludes Schelling. The ubiquity of cell phone cameras is making things worse, as "digi-neckers" slow things even more to take photos of incidents. To top it off, drivers looking at crashes quite often get into crashes themselves. A study by researchers at Virginia Commonwealth University found that the second-leading cause of distraction-related crashes (behind fatigue) was "looking at crashes, other roadside incidents, traffic, or other vehicles."

What this means is that, at times, we have a perfect self-generating traffic jam: People slowing to look at crashes get into crashes, which causes other people to get into crashes, and so on. If traffic were a cooperative network and we could agree not to slow and look, Schelling notes, everyone could save time. Since that will never happen, traffic engineers have instead countered with antirubbernecking screens, which can be unfurled at crash scenes to block prying eyes. In theory these should help matters, but they have severe limitations. Just getting a screen to a crash site, past the traffic that has already developed, is hard enough. Then picture emergency responders, who probably have more pressing matters to attend to, trying to erect—in strong winds or snow—a giant wall of fabric, as if imitating the artist Christo. Plus, ironically, there is the interest in the screen itself. Janet Kennedy, a researcher at England's Transport Research Laboratory, told me the screens had been tried on construction projects on the M25 motorway. "To start with it didn't have much effect because people just looked at the screen anyway," she said. "But already we're finding people have stopped looking at the screen. They're used to it." That's fine for construction sites, which the same people drive past each day. Unfortunately, this suggests that for crashes, the events that generate the most rubbernecking, the screens are of little help—the crash would be cleared long before drivers became accustomed to seeing the same screen.

But what about the congestion that's "recurring," that happens on the same roads every day? If money was available, we could build more lanes. Only this still does not get us past the pasture problem: Create a bigger pasture, and people will bring even more cows. Traffic congestion is a kind of two-way trap. Because driving is a bargain (drivers are not picking up the full tab for the consequences of their driving), it attracts many people to roads that are not fully funded; this not only makes them crowded, it makes it hard to find revenue to build new ones.

When Costco discounts televisions during its Christmas shopping promotions, pricing them so low that stores do not make a profit, what happens? There are huge lines at the door at five a.m. When cities provide roads that are priced so low that they lose money on them, what happens? There are huge lines on the highway at five a.m. Pricing changes behavior. This is hardly a revelation, but it's always striking to see it in action. At a Pizza Hut in Beijing, I watched with some wonder as patrons at the salad bar carefully arranged towering piles of salad on their plates, then carefully walked away with mounds of teetering greens. Why did they do this? There was a flat fee for one visit, so patrons made sure they got their money's worth. They traveled as efficiently as they could. What if the fee was good for unlimited visits to the salad bar? People would have made multiple trips, carrying smaller portions of salad. The traffic flow back and forth to the bar would have gone up.

In traffic, the basic model has been a state-subsidized, all-you-can-eat salad bar. Take as many trips on the roads as you like, whenever you want, for whatever reason. It may be a good deal for society—a loss leader, like Costco's cheap televisions—but it's such a good deal that everyone does it. Recently, however, as we have been running out of money and space for new roads, the thinking has turned from "How can we get more people on the roads?" to "How can we get fewer?" The answer, of course, is congestion pricing. As an idea, it's hardly new. The idea of taxing people for the "externalities," like congestion, that they create goes all the way back to economists like Arthur Pigou, who talked about the problems road users create for other road users in his 1920 book, *The Economics of Welfare*.

Later, the Nobel Prize–winning economist William Vickrey led a long, lonely crusade to get people to accept the idea that urban roads are a scarce resource and should be priced accordingly. After all, as Vickrey pointed out in 1963, hotels charge more for in-season rooms, railways and airlines charge more for peak travel periods, and telephone companies charge more during the times when more people are likely to call—why should roads not cost more when more people want to use them? (Vickrey was a bit ahead of his time: Told in the early 1960s that there was no way to track where people drove, or how much they drove, Vickrey, the story goes, built a cheap radio transmitter and installed it in his car, displaying the results to friends.)

Congestion charging, in cities like London and Stockholm, has been shown to work because it forces people to make a decision about—and gives them a precise benchmark against which to measure—whether a given trip is "worth it." We may have been paying before, in *time*—which hardly helps fund the roads—but the human mind handles time differently than money. We seem less sensitive to the value of time, even if, unlike money, time can never be regained. It is easier for people to rationalize its loss. The problem with the crowded highway is that everyone suffers the same loss of time, even if some people's use of the highway might be worth more to them—to take an extreme example, think of a woman about to give birth on the way to the hospital, stuck in a traffic jam alongside someone who simply "needed to get out of the house." They may each feel that their trip is valid, but is that really how a scarce resource should be distributed?

When people are forced, by means of how much it will cost them, to think about when, where, and how they are going places, interesting things begin to happen. You might assume that a rush-hour highway is filled with people driving to work who have no other way to get there—and no other time they can travel—but studies suggest that this is not the case. When researchers have exhaustively tracked the license plates of every car traveling on rush-hour highways and matched the results to other days, they have typically found that only about 50 percent are the same people each day. Sometimes people's patterns emerge when you look deeper into what would seem to be random behavior. In what the English traffic researcher Richard Clegg calls the "See you next Wednesday effect," research has found that when people use a rush hour on Wednesday of one week, they're more likely to be on that same highway on the next Wednesday than on another day.

Not everyone is so rigid in their habits. In 2003, a group of drivers in Seattle were outfitted with electronic devices that would tell researchers where and when they had driven. Baseline data was collected on these people's typical habits. Then the drivers were informed that they would be given a hypothetical cash account. They would automatically be charged more for driving in the most crowded places at the most crowded times. Matthew Kitchen, director of the Puget Sound Regional Council, the group that sponsored the program (called Traffic Choices), said he was struck by how differently people acted day to day even *before* they were charged tolls.

Once the tolls kicked in, things really began to change: People left sooner, took different routes, took buses, "collapsed" trips into shorter bundles. "The reality which is emerging is that I think people are very intelligent agents, working on their own behalf," he said. "They understand the unique trade-off they face between time and money. The range of response is extremely broad. For instance, my willingness to pay to save ten minutes today might be very different than tomorrow."

How much did the charging affect driving? The total "tours," as they are called in transportation-planning lingo, dropped by 13 percent. That may not seem like much, but in the world of bottlenecks, small changes can have big effects (a 5 percent drop in traffic, it is said, can increase speeds by 50 percent, even if that only means going from 5 to 10 miles per hour). With traffic jams, Kitchen noted, "Once you start falling off the cliff, you fall pretty fast and pretty hard. That's why between 5 and 10 percent less traffic restores what are really credible speeds on the network. You don't have to hit people over the head with something that is punitive. You can achieve reasonable results with incentives that result in fairly modest behavioral response."

By getting just some people to change their behavior, congestion pricing can help reverse a long-standing vicious cycle of traffic, one that removes the incentives to take public transportation. The more people who choose to drive to work, the worse the traffic. This raises the time buses must spend in traffic, which raises the cost for bus companies, who raise the fares for bus commuters—who are being penalized despite their own efforts to reduce total traffic. As the bus becomes less of a good deal, more people defect to cars, making things worse for the bus riders, who have even less incentive to ride the bus.

It doesn't take much to set this avalanche in motion. The historian Philip Bagwell notes that in 1959, only 7 percent of the total traffic entering London was via private car. But if just 1 percent of the people taking public transportation shifted to cars, the percentage of car journeys would rise 12 percent, and the number of cars in the traffic stream would jump by 5 percent. Which is exactly what happened, and London soon had "traffic thrombosis." Everything engineers did to ease the flow just seemed to make it worse.

Congestion pricing reverses the cycle. Driving becomes more expensive, so traffic is reduced. The fees raised by pricing go into buses, which benefit in time and in money from the reduced traffic. This makes buses cheaper, and thus more popular. Small changes in traffic levels make all kinds of other things possible. In London, a familiar lament was the decline of Trafalgar Square, the city's symbolic heart, home to Nelson's Column and countless demonstrations through the years. But on most days it merely seemed the elaborate centerpiece of a busy traffic circle, a noisy and noxious holding pen for pigeon-feeding tourists. Then came a plan to close the street between the square and the National Gallery, uniting the two entities into a grand civic space. This was deemed, from a traffic point of view, impossible. As Malcolm Murray-Clark, the director of London's congestion-pricing program, told me over tea in his office, congestion pricing changed all that. By removing the "background levels" of traffic from London, as he put it, planners had the wiggle room to remove the Trafalgar road without catastrophic consequences. "Eighteen percent of the traffic through Trafalgar Square did not have a destination in central London," he said. "It was just a through trip. Those were the first to go, if you like."

Congestion pricing is really just another spin on making the system optimal, or, to put it another way, saving people from their own instincts: How do you persuade everyone not to go to the same place at the same time? Cities like London are, in effect, learning from Disneyland. That may seem like a stretch, but consider that Disney theme parks open each day to a flood of people, many wanting to first go on the most popular attractions. Cities "open" each day with people all wanting to go to the same "attractions" at once. Disney executives are as much in the traffic business as the entertainment business: moving people around, from ride to ride (and through the shops and restaurants), in the most efficient manner and with the least customer grumbling. They hire talented engineers, like Bruce Laval, to manage these flows and queues.

Laval, now retired, joined the company's industrial engineering department in 1971. His master's thesis was on traffic signal coordination, and his first task at Disney was to figure out a way to reduce the wait times on its popular monorail. "Management wanted to put together justification to buy a sixth monorail train," he told me. "They figured they needed more capacity to move more people." But Laval ran simulations that came to a counterintuitive solution: Disney could move people faster by *removing* a train, not adding one. The reason was that each train had a buffer zone, for safety, in front of it; as it neared another train, it slowed or stopped. Reducing the number of trains meant they all moved faster (one of those "slower is faster" effects that show up so often in networks).

Early on, Disney realized that as the park grew in popularity, managing the queues of people would prove difficult, particularly on the marquee attractions like Space Mountain. What could you do? Disney could take the approach of our traffic networks, which is simply to let an inefficient kind of equilibrium take hold. Let people wait, and if the line is too long, they may decide on their own not to get in line (or get on the highway), and thus be diverted to other rides (roads). The queue will regulate itself. You can also make the line not seem as long, through various psychological tricks (like posting longer wait times than are really the case or having the queue itself wind through mini-attractions). But that still means people are waiting in lines (i.e., in traffic) and not being as productive as they might be, rather than shopping and eating (i.e., working or spending time at home). Disney could, and sometimes did, add capacity to its rides. But that, too, had limitations. "It costs a lot of money to add capacity," Laval said. "If you eliminate wait times during your peak days, you're over capacity for the other ninety-five percent of the year. You don't design a church for Easter Sunday."

So Disney tried a form of congestion pricing. It issued ticket books in which the tickets' values reflected the capacity of the rides. Popular rides like Space Mountain required E tickets, which were more expensive than A tickets, good for tamer attractions like the Horseless Carriage on Main Street. The idea was not only to prevent people from simply lining up for the top attractions but to spread people out across the park, avoiding traffic jams at places like Space Mountain. "One way of increasing capacity is rerouting demand," Laval said. This was successful to a point, but the signals that prices send can work in different ways. At Disney World, Laval explained, where 80 percent of park entrants were first-time visitors (Disneyland has more repeat visitors), many of whom had no particular itinerary of which rides to go on first, the E tickets were like a big red flashing sign saying, "Ride me first." Everyone wanted to get their money's worth, so they immediately gravitated to the most expensive rides. The rides were not only expensive because they were popular, they were popular because they were expensive.

Phenomena like this shows up in traffic as well: The HOT lanes in Southern California charge more as more people enter them (in order to help keep them from becoming congested); yet sometimes people enter a tolled lane precisely *because* it is expensive—they think the toll must be so high because the untolled lanes are really jammed. (This sort of behavior subverts the normal economics principle of "price elasticity," in which the number of users should drop as the toll goes up.)

Disney finally hit upon the ultimate solution in 1999, when it introduced the FastPass, the system that gives the customer a ticket telling them when to show up at the ride. What FastPass essentially does is exploit the idea that networks function both in space and in time. Rather than waiting in line, the user waits in a "virtual queue," in time rather than space, and can in the meantime move on to other, less crowded rides (or buy stuff). People can take a chance on the stand-by line, or they can have an assured short wait if they can simply hold off until their assigned time. Obviously, FastPass could not literally work on the highway. Drivers do not want to pull up to a tollbooth and be told, "Come back at two-thirty p.m." But in principle, congestion pricing works the same way, by redirecting demand on the network in time.

Traffic can be made to flow better by redirecting demand in *space*, of course, if traffic engineers know what the demand and available supply are on a network at any given time—and if they can find a way to get that information to drivers. In the past, this has been a necessarily crude process, hindered by delays in getting and sending the information, and the ability to see the network at once with all its interacting flows. Surely you have had the experience of listening in vain to a rapidly spoken traffic report, hoping against hope to get the details on the jam you're sitting in (and by some law, you never can). And as we saw in Los Angeles, traffic information often comes too late for us to do anything about it, or is not even accurate.

Rather than surgical strikes at congestion, one can always try carpet bombing. Sam Schwartz (a.k.a. "Gridlock Sam"), New York City's former traffic commissioner, claims that by declaring "gridlock alert" days, he could "knock fifty thousand or sixty thousand cars out of traffic" by plastering the airwaves with dire warnings. "The Heisenberg principle exists in traffic. If you look at it and announce and tell people about it, it has an effect." When he wanted to reduce traffic on one parkway so construction crews could work on an overhead rail link, he rolled out more horror stories. "I was able to scare away forty percent of the vehicles from that corridor," he says. "We measured it. I was amazed at how effective we were. Sometimes when you hear on the radio, they talk about how terrible the traffic is—it's really me, like the Wizard of Oz behind the curtain."

But human psychology has a way of rearing its complicated head. One problem is that you can never quite know how people will react. In one study, researchers assembled a panel of drivers who regularly commuted on U.S. 101 in California's Silicon Valley. Following a multiple-vehicle crash that took over a half hour to clear, causing extensive delays, the researchers interviewed the commuters. They found that only half the drivers had heard about the incident, and that even the majority of those drivers simply headed to work at the normal time, the normal way. Many people simply seemed unconvinced that they could save any time by changing their plans.

We have all had these moments. Do I take the local streets when I see there is a crash ahead? Is it better to leave early on Sunday morning to go back to the city, or will everyone else have that same idea? Do I get in the right lane because it seems empty, or is there a reason no one else is in it? It boils down to how we make decisions when we do not have all the facts. We rely instead on heuristics, those little strategies and mental shortcuts we all have in our head: Well, this road is usually busy for only a few minutes, so I'll stay on it. Or: I bet that since the radio called for snow there will not be many people at the mall. We use our experience; we make predictions.

This recalls the famous "El Farol problem," sketched by the economist W. Brian Arthur, after a bar in Albuquerque, New Mexico. The hypothetical scenario imagines that one hundred people would like to go to the bar to listen to live music, but it seems too crowded if more than sixty show up. How does any one person decide whether or not to go? If they go one night and it's too crowded, do they return the next night, on the thought that people will have been discouraged—or will others have precisely the same thought? Arthur found, in a simulation, that the mean attendance did indeed hover around sixty, but that the attendance numbers for each night continued to oscillate up and down, for the full one hundred weeks of the trial. Which means that one's chances of going on the right night are essentially random, as people continue to try to adapt their behavior.

This kind of equilibrium problem happens frequently in traffic, even when people have some information. In 2006, for example, the Dan Ryan Expressway in Chicago was undergoing massive repairs. The first day that eight express lanes were closed, traffic moved surprisingly well. The recommended detours moved more slowly than the highway. This was reported on the news. You can guess what happened on Tuesday: More people flocked to the highway. We can surmise that the expressway's traffic went down on Wednesday, though it may equally likely have gone up.

What happens when we no longer have to guess? We are now just at the beginning stages of a revolution in traffic, as navigation devices, increasingly often equipped with real-time traffic information, enter the market.

The navigation part alone has important consequences for traffic. Studies have shown that drivers on unfamiliar roads are roughly 25 percent less efficient than they should be—that is, they are lost—and that their total mileage could be cut by 2 percent if they were always shown the best routes. Logistics software now helps cut delivery times and fuel emissions for UPS and other truck fleets simply by finding ways to avoid, when possible, time-consuming left turns in two-way traffic. But the biggest change will occur when each driver can always know which roads are crowded and what alternate routes would be best—not through guesses but through accurate real-time data.

In theory, this will help reduce inefficiency in the system. Drivers are told there has been a crash ahead, and their in-car device gives them another route that is estimated to save ten minutes. But nothing in traffic is ever so simple.

The first problem is that real-time data is not yet what its name promises. At Seattle's Inrix, for example, one of the key providers of traffic information, traffic-pattern data is gathered from a variety of sources, current and historical—from loops to probes on commercial vehicles to the schedule of conventions in Las Vegas, some five billion "data points"—and weighted according to its perceived accuracy and age. "So a thirteen-minute-old traffic speed estimate from a Caltrans sensor in the Los Angeles market would get a sub-five-percent weighting in our estimate of current conditions," explains Oliver Downs, Inrix's principal research scientist. Inrix estimates the current conditions every minute, but as Downs notes, "it's a 3.7-minute-old estimate of the current conditions." Customers, meanwhile, get a new feed every five minutes. "When we say 'real-time,' " Downs says, this means "less than five minutes." That might not seem like much, but, as Downs admits, "it's long relative to how quickly things can change on the roadway."

The other problem comes in how people will use that information, or what you should tell them to do based on the information. Michael Schreckenberg, the German physicist known as the "jam professor," has worked with officials in North Rhine–Westphalia in Germany to provide real-time information, as well as "predictive" traffic forecasts. Like Inrix, if less extensively, they have assembled some 360,000 "fundamental diagrams," or precise statistical models of the flow behavior of highway sections. They have a good idea of what happens on not only a "normal" day but on all the strange variations: weeks when a holiday falls on Wednesday, the first day there is ice on the road (most people, he notes, will not have yet put on winter tires), the first day of daylight savings time, when a normally light morning trip may occur in the dark.

This kind of information, along with the data gathered from various loops and sensors, can be used to make precise forecasts about what traffic will be like not only on "normal" days but when crashes or incidents occur. There is, however, a problem: Does the forecast itself change the way people will behave, thus changing the forecast? As the economist Tim Harford notes about Wall Street forecasting, if everyone knew today that a stock was going to rise tomorrow, everyone would buy the stock today—thus making it so expensive it could no longer rise tomorrow.

Shreckenberg calls this the "self-destroying prognosis." In his office at the University of Duisburg-Essen, he points to a highway map with its roads variously lit up in free-flowing green or clogged red. "The prognosis says that this road becomes worse in one hour," he says. "Many people look at that and say, 'Oh, don't use the A3.' Then they go somewhere else. The jam will not occur since everyone turned to another way. This is a problem." These sorts of oscillations could happen with even short lags in information, in what Shreckenberg calls the "ping-pong effect." Imagine there are two routes. Drivers are told that one is five minutes faster. Everyone shifts to that route. By the time the information is updated, the route that everyone got on is now five minutes *slower*. The other road now becomes faster, but it quickly succumbs to the same problem.

This raises a question: Has the information provided actually helped drivers or the system as a whole—or has it triggered the "selfish routing" mentioned before? Moshe Ben-Akiva, the director of MIT's Intelligent Transportation Systems program, has studied such travel behavior issues for decades. He calls traffic predictions a "chicken-and-egg problem." "The correct prediction must take into account how people are going to respond to the prediction," he says. "You cannot predict what will happen tomorrow without taking into account how people are going to respond to the prediction once the prediction is broadcast."

And so researchers create models that anticipate how people will behave, based on how they have behaved in the past. Shreckenberg, in Germany, wonders if this means, in essence, not giving drivers the whole picture. "You have to structure the information. What you want is for the people to do certain things. Telling them the whole truth is not the best way." This is something on the minds of the big commercial providers of traffic information. As Howard Hayes, president of NAVTEQ, said at the firm's headquarters in Chicago, "What happens if once this really good predictive traffic information becomes available, everyone starts getting shunted over to a different direction, which itself becomes jammed? Ideally you need something sophisticated, so that a certain number of people get shunted to one route and others to another."

Since the information is still so limited, and since so few people actually have access to it, we do not really know how it will all play out once everyone is able to know the traffic conditions on every road in a network. Most simulations have shown that more drivers having more real-time information—the closer to actual real-time, the better—can reduce travel times and congestion. Even drivers *without* real-time information can benefit, it is argued, because better-informed drivers will exit crowded roads, thus making those roads less crowded for uninformed drivers stuck in traffic. But as you might expect, studies suggest that the benefit for any one driver with access to real-time information drops as more people have it. This is, in essence, the death of the shortcut. The more people know the best routes at all times, the less chance of there being some gloriously underutilized road. This is good for all drivers (i.e., the "system") but less good, say, for the savvy taxi driver.

Real-time traffic and routing is most valuable, it has been suggested, during nonrecurring congestion. When a road that is normally not crowded is backed up because of a crash, it's useful to know of better options. During recurring congestion, however, those peak-hour jams that result from too many people going to the same place at once, the advantage shrivels once the tipping point of congestion has been passed. (It is most effective right on the brink, when alternative routes are on the verge of drying up.) In a traffic system that is always congested, any good alternative routes will have already been discovered by other drivers.

Another shortcoming of real-time routing is due to a curious fact about urban road networks. As a group of researchers observed after studying traffic patterns and road networks in the twenty largest cities in Germany, roads follow what's called a "power law"—in other words, a small minority of roads carry a huge majority of the traffic. In Dresden, for example, while 50 percent of the total road length carried hardly any traffic at all (0.2 percent), 80 percent of the total traffic ran on less than 10 percent of the roads. The reason is rather obvious: Most drivers tend to drive on the largest roads, because they are the fastest. Even though they may have slowed due to congestion, they are still fastest. Traffic engineers, having built the roads, are generally aware of this fact, and would rather have you stay on the road that was designed for heavy use, instead of engaging in widespread "rat runs" that play havoc with local roads.

Both the promise and the limits of real-time traffic and routing information were demonstrated to me one day as I drove on Interstate 95 in Connecticut, using real-time traffic information provided by TeleNav via a Motorola mobile phone. The phone had been cheerily giving directions, even offering an evolving estimated time of arrival. Suddenly, an alert sounded: Congestion ahead. I queried the system for the best alternate route. It quickly drew one up, then delivered the bad news: It would take *longer* than the route I was on. The road I was on, congested or not, was still the best.

Real-time traffic and routing information and congestion pricing are two sides of the same coin. One tells drivers how to avoid traffic congestion; the other *impels* drivers to avoid traffic congestion. When the roads are congested, real-time information does little good, except to tell drivers, like the people in line for Disney World's Space Mountain, how long they can expect to wait. This alone may be enough of a social good. But real-time congestion information, provided by the very cars generating that congestion, promises something else. It can be used to calculate the exact demand for any stretch of road at any time. With congestion pricing, the traffic on the roads will finally be made to act like the traffic in things, with market prices reflecting and shaping the supply and demand.

## Chapter Seven

# When Dangerous Roads Are Safer

### The Highway Conundrum: How Drivers Adapt to the Road They See

An overturned cart is a warning to oncoming drivers.

—Chinese proverb

Just before dawn on Sunday, September 3, 1967, there was an unusually festive air in the streets of Stockholm. Cars honked, passersby cheered, people gave flowers to police officers, pretty girls smiled from the curb. The streets were clogged with cars, many of which had been waiting for hours to participate in a historic traffic jam. People stole bicycles simply to be a part of traffic. At the moment the bells chimed for six o'clock, Swedes began driving on the right.

It had taken years of debate, and much preparation, to get to this point. Motions to switch from left-side driving had been raised in Parliament several times in previous decades, only to be shot down. The issue was put before Swedes in a 1955 referendum, but the measure was overwhelmingly defeated. Undeterred, backers of right-side driving finally got a measure approved by the government in 1963.

Proponents said that driving on the right, as was the practice in the rest of Scandinavia and the bulk of Europe, would lower the number of accidents in which foreigners were increasingly becoming involved. Most cars in use already had steering wheels on the left side. Those opposed, which was most of Sweden, grumbled about the huge costs of the changeover, and said that accident rates were bound to rise.

As "H-Day" (after *höger*, the Swedish word for "right") approached, the predictions of ensuing chaos and destruction grew dire. "What is going to happen here in September has cast many grotesque shadows all over Sweden," the *New York Times* observed darkly. This despite four years of preparation and an especially energetic blitz of public-service announcements in the final year before the changeover. There was even a pop song, titled "Håll dej till Höger, Svensson!" or "Let's All Drive on the Right, Svensson!" (after a stereotypically common Swedish surname).

And what happened when Swedes started driving on the other side of the road, many for the first time in their lives? The roads got safer. On the Monday after the change, the traffic commissioner reported a below-average number of accidents. True, this may have been anticipated, despite the gloomy predictions. For one, many Swedes, scared witless of the spectacle, undoubtedly chose not to drive, or drove less. For another, a special speed limit, which had already been in place for some months before the changeover, was enforced: 40 kilometers per hour in towns, 60 on open roads, 90 on highways. Lastly, the whole operation was run with Scandinavian efficiency and respect for the law. This was the country that gave the world Volvo, by God—how could it not be safe?

Remarkably, it was not just for a few days, or even weeks, after the changeover that Sweden's roads were safer. It took a year before the accident rate returned to what it had been the year before the changeover. This raises the question of whether the changeover actually achieved anything in the long run for safety, but in the short term, when one might have predicted an increase in accidents as an entire nation went through the learning curve of right-hand driving, Sweden actually became safer. Faced with roads that had overnight theoretically become more dangerous, Swedes were behaving differently. Studies of drivers showed they were less likely to overtake another when a car was approaching in the oncoming lane, while pedestrians were looking for longer gaps in the traffic before choosing to cross.

Had Sweden's roads actually become more dangerous? They were the same roads, after all, even if drivers were driving on a new side. What had changed was that the roads *felt* less safe to Swedish drivers, and they seemed to react with more caution.

Most people have probably had similar moments. Think about a roundabout, quite common in Europe but still rare to these shores. For many Americans they are frightening places, their intimidation factor perhaps best captured by the plight of the hapless Griswold clan in *National Lampoon's European Vacation*, who, having entered a London traffic circle, find that they cannot leave. They orbit endlessly, locked in a traffic purgatory, until night closes in, the family has fallen asleep, and the father is babbling uncontrollably. Whether this rings true or not, it must be pointed out that the much-maligned traffic circle is not the same thing as a roundabout. A traffic circle varies in a number of ways, most notably in that cars already in the circle must often yield to cars entering the circle. Traffic circles are also larger, and cars enter at a much higher speed, which makes for less efficient merging. They may also rely on traffic signals. In roundabouts, which are free of signals, cars entering must yield to those already in the circle. We have already seen that roundabouts can be more efficient, but it may surprise you to learn that modern roundabouts are also much safer than a conventional intersection with traffic lights.

The first reason has to do with their design. Intersections are crash magnets—in the United States, 50 percent of all road crashes occur at intersections. At a four-way intersection, there are a staggering fifty-six potential points of what engineers call "conflict," or the chance for you to run into someone—thirty-two of these are places where vehicles can hit vehicles, and twenty-four are spots where vehicles can hit pedestrians.

Roundabouts sharply drop the total number of potential conflicts to sixteen, and, thanks to their central islands (which create what engineers call "deflection"), they eliminate entirely the two most dangerous moves in an intersection: crossing directly through the intersection, often at high speed (the average speed in most roundabouts is *half* that of conventional intersections, which increases safety for surrounding pedestrians), and making a left turn. This little action involves finding a suitable gap in oncoming traffic—often as one's view is blocked by an oncoming car waiting to make its own left turn—and then, as your attention may still be divided, making sure you do not hit a pedestrian in the crosswalk you are entering as you whisk through your turn. One study that looked at twenty-four intersections that had been converted from signals and stop signs to roundabouts found that total crashes dropped nearly 40 percent, while injury crashes dropped 76 percent and fatal crashes by about 90 percent.

There is a paradox here: The system that many of us would feel is more dangerous is actually safer, while the system we think is safer is actually more dangerous. This points to a second, more subtle factor in why roundabouts are safer. Intersections of any kind are complex environments for the driver, requiring high amounts of mental workload to process things like signs, other cars, and turning movements. Drivers approaching an intersection with a green light may feel there is little left for them to do; they have the green light. But traffic lights have pernicious effects in and of themselves, as Kenneth Todd, a retired engineer in Washington, D.C., has pointed out. The desire to "catch" a green makes drivers speed up at precisely the moment they should be looking for vehicles making oncoming turns or entering the main road from a right turn on red. The high placement of traffic lights also puts drivers' eyes upward, away from the street and things like the brake lights of the slowing cars they are about to hit. Then there are the color-blind drivers who cannot make out the red versus green, and the moments when sunlight washes out the light for everyone.

With a roundabout, only a fool would blindly sail into the scrum at full speed. Drivers must adjust their speed, scan for openings, negotiate the merge. This requires more workload, which increases stress, which heightens the feeling of danger. This is not in itself a bad thing, because intersections are, after all, dangerous places. The system that makes us more aware of this is actually the safer one.

Once, on a driving trip in rural Spain, I decided to take a shortcut. On the map, it looked like a good idea. The road turned out to be a climbing, twisting, broken-asphalt nightmare of blind hairpin turns. There were few guardrails, just vertigo-inducing drops into distant gulleys. The few signs there were told me what I already knew: PELIGRO. Danger. And how did I drive? Incredibly slowly, with both hands locked on the wheel, eyes boring straight ahead. I honked ahead of every blind curve. My wife, who fears both heights and head-on collisions, never trusted me with a Spanish map again.

Was the road dangerous or safe? On the one hand, it was incredibly dangerous. The "sight distances," as road engineers call the span required for one to see a problem and safely react to it (based on a certain travel speed), were terrible. The lanes were narrow and not always marked. There was only the occasional warning sign. Had there been a collision, there was little to keep me from tumbling off the edge of the road. And so I drove as if my life depended on it. Now picture another road in Spain, the nice four-lane highway we took from the airport down to Extremadura. There was little traffic, no police, and I was eager to get to our hotel. I drove at a healthy pace, because it felt safe: a smooth, flat road with gentle curves and plenty of visibility. The sun was shining; signs alerted me to every possible danger. And what happened? Grown briefly tired from the monotony of the highway (drivers have a greater chance of becoming drowsy on roads with less traffic and on divided highways free of junctions) and the glare of the sun, I just about fell asleep and ran off the road. Was this road dangerous or safe?

Of the two roads, the highway was of course the more objectively safe. It is well known that limited-access highways are among the safest roads we travel. There is little chance of a head-on collision, cars move at relatively the same speeds, medians divide opposing traffic streams, curves are tamed and banked with superelevation to correct drivers' mistakes, there are no bikes or pedestrians to scan for, and even if I had started to nod off I would have been snapped back to attention with a "sonic nap alert pattern," or what you might call a rumble strip. At the worst extreme, a guardrail may have kept me from running off the road or across the median, and if it was one of the high-tension cable guardrails, like the Brifen wire-rope safety fence, increasingly showing up from England to Oklahoma, it might have even kept me from bouncing back into traffic.

Those rumble strips are an element of what has been called the "forgiving road." The idea is that roads should be designed with the thought that people will make a mistake. "When that happens it shouldn't carry a death sentence," as John Dawson, the head of the European Road Assessment Programme, explained it to me. "You wouldn't allow it in a factory, you wouldn't allow it in the air, you wouldn't allow it with products. We do allow it on the roads."

This struck me as a good and fair idea, and yet something nagged at the back of my brain: I couldn't help but think that of the two roads, it was the safer one on which I had almost met my end. Lulled by safety, I'd acted more dangerously. This may seem like a simple, even intuitive idea, but it is actually an incredibly controversial one—in fact, heretical to some. For years, economists, psychologists, road-safety experts, and others have presented variations on this theory, under banners ranging from "the Peltzman effect" and "risk homeostasis," to "risk compensation" and the "offset hypothesis." What they are all saying, to crudely lump all of them together, is that we change our behavior in response to perceived risk (an idea I will explore more fully in Chapter 9), without even being aware that we are doing so.

As my experience with the two roads in Spain suggested, the question is a lot more subtle and complicated than merely "Is this a dangerous or safe road?" Roads are also what we make of them. This fact is on the minds of engineers with the Federal Highway Administration's Turner-Fairbank Highway Research Center, located in Langley, Virginia, just next to the Central Intelligence Agency.

The first thing to think about is, What is a road telling you, and how? The mountain road in Spain did not need speed-limit signs, because it was plainly evident that going fast was not a good idea. This is an extreme version of what has been called a "self-explaining road," one that announces its own level of risk to drivers, without the need for excessive advice. But, you protest, would it not have been better for that mountain road to have signs warning of the curves or reflector posts guiding the way? Perhaps, but consider the results of a study in Finland that found that adding reflector posts to a curved road resulted in higher speeds and more accidents than when there were no posts. Other studies have found that drivers tend to go faster when a curve is marked with an advisory speed limit than when it is not.

The truth is that the road itself tells us far more than signs do. "If you build a road that's wide, has a lot of sight distance, has a large median, large shoulders, and the driver feels safe, they're going to go fast," says Tom Granda, a psychologist employed by the Federal Highway Administration (FHWA). "It doesn't matter what speed limit or sign you have. In fact, the engineers who built that road seduced the driver to go that fast."

But those same means of seduction—the wide roads, the generous lane widths, the capacious sight distances, the large medians and shoulders—are the same things that are theoretically meant to ensure the driver's safety. This is akin to giving a lot of low-fat ice cream and cookies to someone trying to lose weight. The driver, like the would-be dieter, is wont to "consume" the supposed health benefits. Consider a key concept in traffic safety engineering: the "design speed" of roads. This is a confusing concept, not least because engineers are often not so good at explaining their concepts to nonengineers. The so-called Green Book, the bible of U.S. highway engineers, defines "design speed" as the following: "The maximum safe speed that can be maintained over a specified section of highway when conditions are so favorable that the design features of the highway govern." Got that? No? don't worry—it confuses traffic people too. An easier way to understand design speed is to think of the speed that most people—what engineers refer to as the "85th percentile" of drivers—generally like to travel (thus leaving out the suicidal speeders and stubborn slowpokes). As we saw in the previous chapters, leaving it up to drivers to figure out a safe speed is itself risky business.

Even more confusingly, sometimes this speed matches the speed limit, and sometimes it does not. Once engineers figure out the 85th percentile speed, they try to bring, where possible, the various features of the highway (e.g., the shoulders, the curves, the "clear zones" on the side of the road) into line with that speed. So does this mean that everyone then travels at the "safe" design speed? Not exactly. As Ray Krammes, the technical director of the FHWA's Office for Safety Research and Development, explained to me, drivers routinely exceed the design speed. "We know we can drive faster than the design speed," he said. "We're doing it every day. We set a design speed of sixty and people are driving seventy. If it's a seventy-miles-per-hour design, there are a number of people out there pushing seventy-five or eighty miles per hour." Drivers, in effect, are every day loading twenty-one people on an elevator that has a capacity of twenty and hoping that there's *just* that extra margin of safety left.

As we have seen, traffic engineers face a peculiar and rather daunting task: dealing with humans. When structural engineers build a bridge, no one has to think about how the stress factors and loads of the bridge will affect the behavior of the wind or water. The wind or water will not take a safer bridge as an invitation to blow or flow harder. It's a different story when engineers design a road. "When the engineers build something," Granda says, "the question everybody should ask is, What effect will it have on the driver? How will the driver react, not only today, but after the driver sees that sign or lane marking over a period of time? Will they adapt to it?"

To try to answer these questions, Granda, who works in the Human

Centered Systems Laboratory at FHWA, spends his days running drivers on test roads in the agency's driving simulator. "It is hard to know how human beings will react," he notes. "We can decide to do something, and we think we know how they're going to react. You don't really know." As Bill Prosser, a veteran highway designer for the agency, described it to me, "there are three things out there that affect the way a highway operates: the design, the vehicle, and the driver. We as design engineers can only control one of those. We can't control the driver, whether they're good, bad, or indifferent."

The best thing engineers can do, the thinking has gone, is make it easy. "You can't violate driver expectation," says Granda. Tests of what researchers call "expectancy" routinely show that it takes drivers longer to respond to something they do not expect than something they do expect. Think of the mental models described in Chaper 1: People were faster to respond when character traits corresponded to names in a way they expected ("strong John" versus "strong Jane"). Similar things happen in traffic. It takes us longer to process the fact that a car is approaching in *our* lane on a two-lane highway, instead of, as we would expect, in the other lane. A driver in Maine will brake faster for a moose than for a penguin. As David Shinar, a traffic researcher in Israel, has described it, "That 'second look' that we colloquially say we take when 'we can't believe our eyes' may be a very real and time-consuming effort."

This is expressed on the highway in all kinds of subtle ways. Highway engineers have long known that a set of curves, seemingly a dangerous road segment, is less dangerous than a curve that comes after a long stretch of straight highway. A similar principle exists in baseball: A batter can more easily hit a curveball if he sees nothing but curveballs than when he is thrown a curveball after a steady diet of fastballs. So engineers strive for what they call "design consistency," which basically means: Tell drivers what to expect, and then give it to them.

The flip side of this is that too *much* expectancy can be boring. You might feel, for instance, that interchanges, where the on-ramps and off-ramps swirl into the highway, are the most dangerous areas on the highway. They are certainly the most stressful, and they are home to the most crashes. But that's not where most people lose their lives. "In terms of fatalities," says Michael Trentacoste, the director of the Turner-Fairbank center, "the highest number is 'single-vehicle run-off road.' " I thought back to my near accident in Spain. "If you look at Wyoming," he continues, "they have a tremendous amount of single-vehicle run-off-the-road accidents. A few years ago they had the highest percentage of run-off-the-road [accidents] on the interstate. You've got long stretches, a lot of night-time driving, people falling asleep."

This is why road designers will often introduce subtle curvatures, even when it is not warranted by the landscape. One rough rule of thumb for highways is that drivers should not drive for more than a minute without having a bit of curve. But highway curves, most of which can be driven much like any other section, are often not enough to keep a tired driver awake. Which is why engineers, starting in the 1980s, began to turn to roadside rumble strips. The results were striking. After they were installed on the Pennsylvania Turnpike, run-off-road crashes dropped 70 percent in the period studied.

Those rumble strips would hardly lull drivers into falling asleep, knowing they'll be startled awake if they drifted off the road. But does something about the highway itself help drivers fall asleep in the first place? The line between safety and danger is not always well defined, nor is it always easy to locate.

When the U.S. Interstate Highway System was first built, engineers could not know what to expect once everyone got on the highway at the same time. "We never did have a cookbook when we started building the Interstate," the FHWA's Prosser told me. Engineers are still learning what works and what does not. Exiting on the left on interstate highways, a fixture in "the early days," has been phased out wherever possible—partially because its rarity makes us slower to react. Another fixture, the cloverleaf interchange, so named because its four looping ramps look like a clover from above, has also fallen out of favor. "When we started building interstates they were pretty much the interchange of choice," said Prosser. Cloverleafs were originally a brilliant, space-saving solution to a major problem: how to get traffic to flow across to two interconnecting roads without stopping. This made them useful for joining two intersecting highways (they are also quite good at preventing people from entering the freeway in the wrong direction of travel, an act that is said to be responsible for 350 deaths per year in the United States alone).

But they have one big drawback: The on-ramp loop enters the highway just beyond where cars are exiting via the off-ramp loop. The two streams must mix. Engineers call this the "weaving section," a mysterious, traffic-tossed tempest full of what engineers call "turbulence" and "friction," in which people coming onto and getting off the highway end up in each other's way. Drivers at different speeds, scanning for directional signs, have to probe openings (i.e. make "gap acceptance" decisions) and sometimes get across several lanes—often quite suddenly. Interchanges, as it happens, are where most crashes on freeways occur—according to studies, the shorter the weave section, the higher the crash rate. With light traffic, the cloverleaf presents less of a problem, but when "weaving volume" on the two loops tops the magic number of one thousand vehicles per hour (hardly a rarity these days), things begin to break down. Because of the curious nonlinear dynamics of traffic, when traffic volume doubles, the length of weaving section required to keep it moving smoothly *triples*. Over time, engineers have responded by moving the weaving section out of the main highway flow and onto special "collector" lanes, which, where possible, seems to be safer and more efficient.

Highways are continuing to evolve. Recently, as traffic volumes have grown, and with new highway building increasingly unaffordable or undesirable, some agencies have begun adding new lanes to highways by either eliminating the shoulder lane or making the existing lanes narrower. In theory, this is riskier because on narrow lanes there is a greater chance of one car drifting into another. There is literally less room for error. On the other hand, wider lanes, which are presumably safer, have been shown to increase speed and may encourage drivers to drive less cautiously. Indeed, some reports have even suggested that lanes wider than the typical U.S. twelve-foot standard may actually be *less* safe. So far, studies that have looked into the narrowing of highway lanes have come to mixed conclusions on whether the new layouts are more or less safe. In some cases, the difference was not statistically significant. This suggests that the way drivers behave is as important as the way a road is designed. As Ezra Hauer, a Canadian engineer and traffic-safety expert, once put it, "Drivers adapt to the road they see."

There is a simple mantra you can carry about with you in traffic: When a situation feels dangerous to you, it's probably more safe than you know; when a situation feels safe, that is precisely when you should feel on guard. Most crashes, after all, happen on dry roads, on clear, sunny days, to sober drivers.

## The Trouble with Traffic Signs—and How Getting Rid of Them Can Make Things Better for Everyone

Try to remember the last time you saw, while driving, a "School Zone" or "Children at Play" sign. Chances are you will not remember, but if you can, now try to recall what you did when you saw it. Did you suddenly slow? Did you scan for children? If you're like most people, you did nothing. You may not have understood what it was asking you to do, which is rather common—in one study, subjects who were shown a sign warning, WATCH FOR FALLEN ROCKS, were split equally between those who said they would look for rocks falling at the moment and speed up and those who said they would slow down and look for rocks already on the road. Perhaps signs should simply say, WATCH FOR ALL ROCKS, EVERYWHERE.

More likely, the reason you did nothing when you saw the sign is that there were no children playing. If there were children playing, you probably saw them before you saw the sign. "Children at Play" signs have not been shown to reduce speeds or accidents, and most traffic departments will not put them up. Yet why do we seem to see so many? City governments usually post them to assuage complaints by neighborhood residents that people are speeding down their streets. They may have even been put up after a child was hit or killed by a driver, in which case it would probably be more effective to erect a sign saying just that.

Similarly, drivers routinely see signs warning of deer crossings (in the United States) or elephant crossings (in Sri Lanka) or camel crossings (in Tunisia). It is difficult to say what's going on in the mind of a driver when he or she sees a deer or elephant or camel crossing sign, but studies have shown that most drivers do not change their speed at all. A Colorado trial featured a special animated deer sign (no, it wasn't Bambi). Researchers presumed that the animated sign would draw more attention and heighten driver awareness. For a few weeks, it was turned away from the road, then turned back. There were actually *more* deer killed when the sign was activated than when it was not, even though fewer deer had crossed. The researchers then went so far as to place a deer carcass next to the animated sign—only then did drivers finally slow.

Traffic engineers have tried putting signs up only during migratory seasons or using special flashing signs equipped with sensors to detect the presence of deer, but these so-called dynamic signs are not only costly but prone to false alarms and maintenance issues, not to mention being riddled with buckshot, particularly in parts of rural America. (Maybe in the off-season deer hunters practice on deer signs.) Researchers in Wyoming who put up a special deer-sensing, flashing system were able to get some drivers to slow down when they included a deer decoy, but they walked away with the opinion that "these reductions in vehicle speed would most likely not reduce the probability of a deer-vehicle collision." Maybe deer should simply be dressed in head-to-toe blaze orange outfits, like the people hunting them!

Perhaps the most absurd warning-sign case involved moose advisories in Newfoundland. One foggy stretch of road was home to not only many car-moose collisions but many collisions between cars and cars stopping to take pictures of moose. And so signs were erected that featured full-size, reflective silhouettes of moose. Unfortunately, tourists found these pretty interesting too, and as they slowed or stopped to take photos, the moose signs themselves became crash hot spots. The next logical step? Create new signs that read CAUTION: MOOSE SIGNS AHEAD.

Many traffic signs have become like placebos, giving false comfort to the afflicted, or simple boilerplate to ward off lawsuits, the roadway version of the Kellogg's Pop-Tarts box that says, "Warning: Pastry Filling May Be Hot When Heated." Engineers insist that they are necessary to protect municipalities from liability lawsuits.

But what is a sign actually telling a driver? As Carl Andersen of the FHWA pointed out during my visit, the same sign can mean two different things in two different places. Take the chevron warning sign, the one that looks like a mathematical "greater or less than" symbol. "You drive in Vermont and you see a chevron sign, you better start braking for that curve," Andersen said. "You see that chevron in Connecticut, you better ignore it. They pick different rates of curvature to put these chevron signs up to provide that kind of warning. So even though there's guidelines to do it consistently, there's enough leeway in there that they do it at different times." Nor does a sign always mean the same thing: "Bridge Freezes Before Roadway" does not tell the driver whether the bridge is frozen, and in July it tells the driver absolutely nothing. Should a "65 MPH" speed-limit sign say something else when it's raining? Engineers have created costly dynamic signs in response to all of these issues, but the real question may be, At what point must common sense do the work of a sign?

If "Slow: Children" and "Deer Crossing" signs do not seem to have noticeable effects, it hardly seems impertinent to ask, Do traffic signs work, and are they really needed at all? This question has been raised by Hans Monderman, a pioneer who was, until his death in January 2008, perhaps the world's best-known traffic engineer. It's probably no accident that he became famous by turning his back on decades of received wisdom in his profession and created traffic plans—like entire major intersections without lights or signs—that were radical even by the standards of his native Holland. "The Netherlands is different," noted Kerstin Lemke, a researcher at Germany's Federal Highway Research Institute, as if discussing the openness toward sex and drugs in Amsterdam. "They've got things on the motorway we would never do." Then again, the Netherlands has a better traffic-safety record than Germany, so maybe they're on to something.

If people have heard of Monderman, they tend to recall something about "the guy in the Netherlands who hated traffic signs." But there is, in fact, one traffic sign that Monderman loved. It stands at the border of the small village of Makkinga, in Friesland. It announces a 30 kilometers per hour speed limit. Then it says, WELKOM. Finally, it says: VERKEERSBORDVRIJ!! In English this means, roughly, "Free of traffic signs."

A traffic sign announcing the lack of traffic signs is a good joke, but it's also a perfect symbol of Monderman's philosophy. The sign itself is superfluous, for a driver can see that there are no traffic signs in Makkinga. After all, Monderman pointed out, what do traffic signs actually tell us? One day, driving through Friesland in his Volvo, Monderman gestured toward a sign, just before a bridge, that showed a symbol of a bridge. "Do you really think that no one would perceive there is a bridge over there?" he asked. "Why explain it? How foolish are we in always telling people how to behave. When you treat people like idiots, they'll behave like that."

Monderman's work was far more complex than a simple dislike of traffic signs. It revolved around a central theory that said there are two kinds of space: The "traffic world" and the "social world." The traffic world is best exemplified by the highway. This world is impersonal, standardized, meant only for cars. It is all about speed and efficiency and homogeneity. Monderman, a great fan of the German autobahn, happened to like this world. The social world, on the other hand, is seen in a place like a small Dutch village. These are places where the car is meant to be a guest, not the sole inhabitant. The street has other uses beyond being a means for people to drive quickly from one place to another. Behavior is governed by local customs and interpersonal contact more than abstract rules. Monderman liked this world too, but he did not want it to have anything in common with the German autobahn.

Yet the traffic engineers, argued Monderman, with their standardized signs and markings, have forced the traffic world upon the social world. "When you built a street in the past in our villages, you could read the street in the village as a good book," he said. "It was as readable as a book. Here is the entrance to the village, over there is a school, maybe you can shop in that shop over there. There's a big farmyard and perhaps there's a tractor coming out. Then the traffic engineers came and they changed it into an absolute uniform piece of space." Drivers, he maintained, are no longer taking cues from the social life of the village; they're working off the signs, which have become such a part of our world that "we don't see them anymore." Suddenly, the village's main road is just another segment of the highway passing through, with only a few small signs to tell anyone otherwise. This may be why speeding tickets are so common at the entrances to small towns all over the world. Rather than the simple greed of the local municipality, it is also that the road through the village so often feels the same as the road outside the village—the same width, the same shoulders. The speed limit has suddenly been cut in half, but the driver feels as if he or she is still driving the same road. That speeding ticket is cognitive dissonance.

In the mid-1980s, Monderman had an epiphany that is still reverberating throughout the world. He was called in to rework the main street of a village called Oudehaske. Villagers, as they do the world over, were complaining about cars speeding through the village, on a wide asphalt road with steady traffic volumes. Before Oudehaske, Monderman's response, like that of any good Dutch traffic engineer, had been to deploy the arsenal of what is known as "traffic calming."

Traffic calming is, essentially, the art of getting drivers to slow down. You have traveled down a street on which traffic-calming measures have been applied, even if you were not aware of the taxonomy of devices. The most famous is the speed bump, the steep, jarring obstruction that dates to the dawn of the car itself. With the exception of places like Mexico City, speed bumps are mostly restricted to school parking lots and the like. What you see on streets nowadays is the "speed hump," a wider, more gently sloping creature that, among other things, helps cities avoid lawsuits from car owners with ruined suspensions. There are a veritable Audubon guide's worth of different hump styles, from "parabolic" to "sinusoidal" to the popular English import known as the "Watts profile." A *really* wide hump with a flat plateau is called a "speed table." Apart from these myriad undulations, there are also "chicanes," which sound like French cigarettes but are really little S-shaped artificial curves that drivers must slow to navigate. "Neck-downs" (a.k.a. "bulb-outs," "nubs," or "knuckles"), meanwhile, are small extensions added to curbs to make intersections narrower, meant to induce drivers to slow and, at the very least, give pedestrians a shorter—and thus safer—distance to cross.

The list goes on—which should give you an idea of how hard it is to calm traffic—with any number of "diagonal diverters," "median chokers," and "forced-turn islands" (also called "pork chops," for their shape). If you want to sound smart around your friends, just remember that engineers refer to bumps and the like as "vertical deflection," while anything that relies on squeezing and narrowing is "horizontal deflection."

Traffic-calming devices have been shown to slow speeds and reduce the volume of through traffic. But as with any medicine, the right drug—and the right dosage—must be administered. Many people think that stop signs are a good way to calm speeds in neighborhoods. One problem is that the power of these signs diminishes with use: The more stop signs, the more likely drivers are to violate them. Studies have also shown that stop signs do little if anything to reduce speed—drivers simply go faster at the midblock location to make up time. This issue plagues speed humps too, which is why engineers advise placing them no more than three hundred feet apart, so drivers do not have time to speed. As with any drug, there are side effects: Slowing and accelerating for humps increases noise and emissions, while studies have suggested that speed humps on one block can lead to higher speeds or more traffic on another. People opposed to traffic-calming measures have argued that they delay emergency responders, but researchers in Portland, Oregon, found that they added ten seconds at most to these trips—no more than any other random delay. Would you want to live on a neighborhood street that made the rare fire-truck visit ten seconds faster but was also a safe haven for faster, noisier, and more dangerous traffic every day?

As it happens, many of these traffic-calming innovations were first popularized in the Netherlands. In the beginning, they were almost impromptu acts, a kind of radical street theater directed against the growing encroachment of cars in the city. Joost Váhl, a progressive engineer working for the city of Delft in the late 1960s, was one of their key architects. Sitting one afternoon in his tidy house in Culemborg, Váhl recalled a series of outlandish stunts that ranged from a "dial-a-bump" service (citizens could call and request "bumps" in front of their homes), to the staging of a bicycle accident ("we wanted to know if car drivers would stop and help or pass us by"), to putting up false construction sites on city streets ("we found out that when streets are broken up for repair, everything is functioning perfectly with half of the space"). These tactics, which were really investigations into how to get cars and people to coexist in cities, eventually made their way into genuine social institutions. The most famous of these were the *woonerven*—the word translates roughly into "living yards"—which began to spring up in European cities in the early 1970s.

For decades, planners had said that people and traffic should be segregated, with cars on speedy urban motorways and pedestrians shuttling around on elevated networks of bridges and walkways. Many saw this as a capitulation of the city to the car, while as early an observer as Charles Dickens understood the futility in trying to get pedestrians to ascend pedestrian bridges when people preferred to simply cross at street level. ("Most people would prefer to face the danger of the street," he wrote, "rather than the fatigue of getting upstairs.")

The *woonerven* reversed this idea, suggesting that it was people who lived in cities and that cars were merely guests. Neighborhood streets were "rooms" to be driven through, at no higher than walking speeds of 5 to 10 miles per hour, with drivers being mindful of the furniture and decor—not just speed humps but benches, flowerpots, and nice cobblestones—and, more important, the residents. Even today, *woonerven* plans seem radical, with children's sandboxes sitting cheek-by-jowl to the street and trees planted in the middle of traffic. The reports that trickled in, however, talked about how children were playing outside longer, often without supervision. In time, the *woonerven* got their own traffic signs (a small icon of a house with a child standing next to it). These were marks of the concept's success, but in the eyes of Monderman, those signs also rather defeated the purpose: Drive carefully near the *woonerven*, the sign implies, but drive less carefully everywhere else.

By the time Monderman had been called to rework the village of Oudehaske, the political winds of traffic planning had shifted, and suddenly things like speed bumps were out of favor. In any case, Monderman did not have the budget for traffic-calming infrastructure. At a loss, he suggested that the road simply be made more "villagelike." Maybe if the road looked more like a village road and less like the highway leading out of town, people would act accordingly. The village, coincidentally, had called in some consultants to redesign the village itself. Why not extend the treatment to the road? Working with the consultants, Monderman offered a design. "I thought, this must go wrong. There were no flowerpots, no chicanes. It was just a simple road in a village, nothing more." A month after the project was finished, Monderman took a radar gun and measured the speed of cars passing through the village. In the past, with his chicanes and flowerpots, he would have been lucky to get a 10 percent drop in speed. This time, the speed had dropped so much that he could not get a reading. "The gun only functioned at thirty kilometers per hour," he recalled.

What had happened? Monderman, in essence, had created confusion by blending the car, bike, and pedestrian realms. What had been a wide road with clearly marked delineations was suddenly something more complex. "The width of the road is six meters," Monderman told me as we stood on the sidewalk in Oudehaske. "That makes it impossible for two cars to pass each other together with a bicycle. So you're forced to interact with other people, negotiate your behavior." What adds to the complexity is that the road, now made of small paving blocks to give it a "village feel," is two-tone: The center segment is red, and two small "gutter" strips running alongside are gray. Even though the strips are slightly curved to channel water, they are perfectly usable. "So when you look at the street it looks like a residential street of five meters," Monderman explained. "But it has all the possibilities of a six-meter street. You can use it for all the traffic." There is also, noticeably, a quite low curb. "The height of the curb is very low because both of the parts are parts of the one scheme," he said. "We have the feeling we belong to one another. When you isolate people from each other by a high curb, 'This is my space, this is mine,' drivers drive faster. When you have the feeling that at this moment a child could drop in front of my car, you slow down."

Monderman's experiments were seminal steps in what would become known as "psychological traffic calming." Rather than hit people over the head with speed bumps they would resent and signs they would ignore, better results could be achieved if drivers were not actually aware that they were slowing down, or why. "Mental speed bumps" is the delightful phrase used by David Engwicht, a gregarious traveling Australian traffic activist who for years had been tinkering, on a less official basis, with ideas similar to Monderman's—even though neither knew each other at the time.

Instead of speed bumps, which tell drivers to drive as fast as they can before they hit the next speed bump, Engwicht argues that intrigue and uncertainty—the things that active cities are filled with—are the best remedies for traffic problems. Put a child's bike on the side of the road instead of a speed bump; hang a weird sculpture instead of a speed-limit sign. One of Engwicht's signature tactics is to set up a "Street Reclaiming Chair," a bright throne of sorts, in the middle of a local street and then, wearing a large colorful crown, chat with passing drivers who, not surprisingly, have slowed. The Danish Road Safety Council got at this idea in a different way in a film a few years ago that showed a mock new traffic-calming scheme: topless Danish models standing on the side of the road holding speed-limit signs. In this case, the "flashing" signs worked quite well.

More than twenty-five years after the Oudehaske incident, the speed through the village is the same—and no one has had to take off their shirts. "That experience changed my whole idea about how to change behavior," Monderman told me. "It proved that when you use the context of the village as a source of information, people are absolutely willing to change their behavior." Monderman was, in essence, thinking like an architect in a realm that had been handed over entirely to engineers. In constructing a building, engineers are essential to making it function, but it is architects we call upon to determine how the building will be used, to organize the space. "Each user of a house knows that a kitchen is used differently from the bathroom," Monderman said. "You don't have to explain." Why not make the difference between a village road and the rural highway that flows into it as legible?

Monderman continued to toil away in relative obscurity, his non-traditional techniques tolerated in small doses. Then came a request to do something about the traffic situation at the Laweiplein, a four-way crossing in the city of Drachten. The traffic volume was relatively high—twenty thousand cars a day, plus many scores of cyclists and pedestrians—and congestion was a growing problem. "The traffic lights were so slow," Monderman recalled. But the challenge, as he saw it, was not just moving traffic through as quickly as possible; the Laweiplein "was also the heart of the village. It was exactly the place meant for people. But it was a horrible place, all poles and paint and fences."

Simply replacing the four-way signalized intersection with a roundabout was only half a solution. "Roundabouts work for traffic wonderfully, but in a more city-building type of way they destroy any quality of space," Monderman said. "It's a circular pattern, and most cities have a grid. It doesn't fit in the space; it's telling the wrong story." What Monderman wanted was a traditional village square that just so happened to contain a roundabout: a "squareabout." After seven years of design and construction, the new Laweiplein was unveiled. It was the intersection heard around the world. Seeing it for the first time, one is immediately struck by how clean and open the space looks. Then one begins to realize why. There are no signs, no traffic lights, no zebra-striped poles, no raised curbs, none of the ugly and cheap roadside junk we have come to think is part of our "natural" world. There are simply four roads coming into a small circle at the center of a large square. The space is dominated not by the roads but by sidewalks and a series of fountains whose water gushes higher as more traffic enters the crossing.

As one looks longer, it becomes clear how well it all flows. No one ever seems to come to a stop, neither cars nor cyclists. "Sometimes a car has to slow down, you think he's stopping—no, he's creeping and is going on again. You actually see all the brains of people working together in a much more organic, fluid way," said Monderman. Then he demonstrated one of his favorite tricks. He began to walk into the roundabout, continuing our conversation. He walked backward. He closed his eyes. It may have just been unnatural Dutch patience at work, but cars, already on the lookout for other cars and cyclists, seemed to regard him as just another obstacle to interact with, and so they steered around him, slowly. "What is nice," he noted, "is that even in the strongest traffic-oriented type of crossing, behavior can be steered by the context."

This seemed a kind of group enactment of the traffic experiments Ian Walker had conducted on the roads in Bath. People were taking stock of one another, making decisions, and acting accordingly in the moment. Ben Hamilton-Baillie, an English transportation planner who has allied with Monderman in a movement known as Shared Space, talks about seeing scores of little moments in Drachten like the one in which a Dutch mother on a bike, carrying a kid, merges in front of a big truck with little more than the smallest flicker of eye contact and the slightest lift of a finger. To many people, this might seem scary, perhaps even slightly insane. And maybe just *Dutch*.

Hamilton-Baillie suggests that there is something crucial in the fact that above 20 miles per hour, humans begin to lose eye contact. "As social creatures it is incredibly important for humans to exchange rapid messages about status and other traits," he says. "I've spent a lot of time watching the junction. What are the rules? There's clearly a hierarchy. If you were a confident young businesswoman in a suit you sailed straight through; if you were a hesitant tourist you waited. Your position in the hierarchy could apparently be established in a microsecond." But all this has to happen at *human speeds*. The faster we drive, the less we see. Hamilton-Baillie suggests that it is more than coincidental that as drivers get above 20 miles per hour, we lose eye contact with pedestrians, while our chances of dying as pedestrians if hit by a car also begin to soar dramatically. As humans with an evolutionary history, we are presumably not meant to move faster than we can run, which tops out at around 20 miles per hour. In the modern world, Hamilton-Baillie adds, this may explain why being struck by a car becomes so much more exponentially deadly above that speed.

Monderman insisted that what he was doing was not anarchy. Instead, he said, he was replacing the traffic world with the social world. "I always say to people: I don't care if you wear a raincoat or a Volkswagen Golf, you're a human being, and I address you as a human being. I want you to behave as a human being. I don't care what kind of vehicle you drive." People, his argument goes, know what a roundabout looks like, and they know what its rules are, so why should they be told again? If they're unsure about what to do or feel insecure, they can do what people do in any situation where they're unsure or insecure, be it a cocktail party or the first day of school: Learn by watching others, and proceed cautiously.

This gets to the heart of a controversy about how to make traffic safer. Not everyone acts cautiously. People *do* drive like idiots. As I argued in the first chapter, traffic makes it hard for us to be human. Drivers, insulated in their anonymous cocoons and holding a three-thousand-pound advantage, kill hundreds of pedestrians every day around the world. Would it not be better to segregate people and cars and bikes to the greatest extent possible? Would it not be better to have as many signs, lights, guardrails, signals, bollards, and zebra crossings as possible?

Hamilton-Baillie does not agree that drivers are incapable of understanding social norms and conventions and need to be under the constant control of mechanical devices and signs. "You can quite quickly instill in children a sense of what's appropriate behavior: when you can talk loudly and quietly, or how to join a conversation; when you can fart and when you can't," he explained one night in a restaurant in the Dutch city of Groningen. "What you get by transferring the control systems to cultural or social norms [is that] you then empower other people to tackle the issue themselves. If someone was misbehaving in here, there would come a point at which someone would say, 'C'mon, mate, get out of here.' " But anyone driving the roads today can see that many people do not obey social conventions, or even laws. "Of course there will be people who ignore those conventions," he said. "Such behavior will exist even in a legislated context. But you don't control teenage joyriding through legislation."

Most of our daily life is governed by social conventions. In the elegant Tiffany store on Fifth Avenue in New York, there aren't any "No Spitting" signs, but there are probably few people who choose to expectorate there (and not simply because a security guard would toss them out). To return to the queues of the last chapter, when one enters a McDonald's there are no signs that say, "Do Not Cut in Line." But chances are people do not (of course, in some places they may, but this is a point I'm saving for Chapter 8). I can hear you protesting: People violate social conventions every day. They talk on their cell phones when signs ask them not to. And traffic is dangerous. How could you take the "Yield" sign away from a roundabout and not cause chaos? How would people figure out how to negotiate the crossing without traffic signals? If anything, we need more signals and signs!

We have a strange, almost fetishistic belief in the power of signals. If a visitor from a planet without cars were to visit Earth, he might be truly perplexed by the strange daubs of paint on the street, the arrows blinking in the air. Do you remember the children's game Red Light, Green Light? The person acting as the stoplight would stand with his back to the other players and announce, "Green light." The players would move forward. Then he would say, "Red light" and spin around. If you didn't stop before he saw you, you were "out." What makes the game work is that children do not always stop in time. Nor do adults in real life, which is even more complicated, because we have things like yellow lights—do I stop or do I go? A line on the street or a light in the air may keep cities from getting sued (as long as it doesn't malfunction), but it does nothing to prevent a driver from misbehaving, perhaps even killing someone. Traffic signals assign priority; they do not provide safety. The high number of people killed by drivers running a red light—the sort of thing a roundabout with a nice big fountain in the middle tends to cure—is proof enough of this.

Or consider, for a moment, the urban pedestrian "Walk" signal. Surely this seemingly enlightened bit of design must be vital to the safety of people on foot? Yes, except that at most intersections it happens to accompany the invitation for drivers to make a turn. The result is that every year, many pedestrians, correctly believing themselves to have the right-of-way, are killed while walking in the crosswalk by perfectly sober drivers who have paid slavish attention only to their own green light. (Or they may have had their view obscured by their car's roof pillar, a problem particularly in left turns, when the pillar looms in the center of the driver's vision.) Things are even worse where right turns are permitted on red; for drivers, rights on red may be the only "cultural advantage" of Los Angeles, as Woody Allen joked, but studies have shown that they are a distinct disadvantage for the health of pedestrians. The sad fact is that more urban pedestrians are killed while legally crossing in crosswalks than while jaywalking. Granted, the number of people who use the crosswalk is higher, but this does not diminish the point that more pedestrians are killed in New York City while obeying the law than while not.

Careful jaywalking, particularly on one-way streets, can be safer than confident crossing at the crosswalk (where the pedestrian may have to worry about streams of traffic from different directions). A similar phenomenon seems to occur at the crosswalks one finds at places without traffic signals. Confusingly, there are two types; they seem different but are legally the same: "marked" versus "unmarked." Marked crosswalks are easy to identify: two lines across the pavement. In most jurisdictions in the United States and elsewhere, unmarked crosswalks exist at any place, like intersections, where there are connecting sidewalks on either side of the street. Even though there may be no visible crosswalk line connecting the sidewalks, legally, there is: Drivers must yield to crossing pedestrians, even at intersections that are "uncontrolled" (i.e., there are no stop signs). One might think that marked crosswalks, which send clear signals to all, would be preferable. But marked crosswalks are actually no safer than unmarked crosswalks, and in some cases are actually more dangerous, particularly when pedestrians, like the hero of the old video game Frogger, must navigate several lanes.

Studies do show that motorists are more likely to yield to pedestrians in marked crosswalks than at unmarked crosswalks. But as University of California, Berkeley, researchers David Ragland and Meghan Fehlig Mitman found, that does not necessarily make things safer. When they compared the way pedestrians crossed at both kinds of crosswalks on roads with considerable traffic volumes, they found that people at unmarked crosswalks tended to look both ways more often, waited more often for gaps in traffic, and crossed the road more quickly. Researchers suspect that both drivers and pedestrians are more aware that drivers should yield to pedestrians in marked crosswalks (even though 35 percent of drivers polled did *not* know this). But neither are aware of this fact when it comes to unmarked crosswalks. Not knowing traffic safety laws, it turns out, is actually a good thing for pedestrians. Because they do not know whether cars are supposed to stop—or if they will—they act more cautiously. Marked crosswalks, by contrast, may give pedestrians an unrealistic picture of their own safety.

If signs and symbols do not always achieve their intended results, *removing* road markings can have surprising effects. White lines on the road are commonly thought to be a fundamental element of a safe road. Indeed, on high-speed roads they are essential. Drivers are able to travel at high speeds without crashing into one another or running off the road only if they have a consistent sense of their lane position. Think of the nervous moment, as you're approaching a toll station, when all the lines disappear and the road opens into a vast alluvial fan (not to mention the equally disturbing confusion on exiting as everyone jockeys for position).

But what about on roads with 30-mile-per-hour speed limits? Don't we still need lines to keep people in their own lanes and to prevent them from smashing into one another? A study in England's Wiltshire County looked at two similar roads, one that had a center line, and a narrower one on which the line had been removed. Drivers actually did a better job of staying in their own lane on the road with no center line. Even though the road with no center line was narrower than the road with a line, vehicles still managed to stay farther away from oncoming vehicles (by 40 percent) than on the road with a line. They also tended to slow down in the face of oncoming traffic. What was going on? Apparently, the drivers were not using the road markings but were using their brains—and the results, far from chaos, seemed to indicate more order. What white lines do is enable drivers to drive faster and, intentionally or not, closer together. Similarly, several studies in different countries have found that drivers tend to give cyclists more space as they pass when they are on a street *without* a bicycle lane. The white marking seems to work as a subliminal signal to drivers that they need to act less cautiously—that it's the edge of the lane, and not the cyclist, they need to worry about. (This suggests that *no* bicycle lanes are better for cyclists than insufficiently wide bicycle lanes.)

Hans Monderman was quite aware that by removing signals and markings, he made people feel more at risk in the Laweiplein. This was a good thing. "We feel it is unsafe," residents told him. "I think that's wonderful," he told me. "Otherwise I would have changed it immediately." There would even be a benefit to some crashes, he added: "I hope that some small accidents happen, as part of the learning process of society." Monderman was delighted when his son had his first minor mishap in a car. In fact, he said, he would have paid for him to have it: "He knows that he is vulnerable, with his own responsibility for his behavior. Having an accident should be part of the driving course. I think that these small accidents help in avoiding severe accidents."

But here is the funny thing. Since the Laweiplein was converted into a "squareabout," the number of crashes, according to a preliminary study by the local technical college, has dropped. In 2005, there were none at all. Well, that's because everyone is moving more slowly, right? Perhaps. But there are a few other interesting facts. Since the conversion, the average time to cross the intersection has dropped by 40 percent, even as traffic has increased. The time buses had to wait to get through has been cut by as much as half. All traffic, the college found, seems to move at a constant flow, and even at peak hours, the movement is steady, if slow—and any traffic engineer will tell you how important it is for drivers to be able to sense progress. The report noted something else interesting: More cyclists were using hand signals when moving in the roundabout; this, the report claimed, was unusual behavior in the Netherlands. More drivers were using their signals, as well. The responsibility for getting through the intersection was now up to the users, and they responded by communicating among themselves. The result was that the system was safer, even though the majority of users, polled in local surveys, felt that the system was more dangerous!

In changing the design in Drachten, Monderman was really asking, What is this street for? What are cities for? Monderman had said he would never widen the streets leading into the Drachten crossing. People were coming for the city, not for the traffic. "Cities are never roads" is how Joost Váhl describes it. Freed from a sense that they are in a city or a village, and instead are simply on a road, drivers respond in kind. They take their information not from local context but from standardized signs. "When you removed all the things that made people know where they were, what they were a part of, then you had to explain things," Monderman said.

There can be a power in not explaining things. In Culemborg, Váhl and I, joined by Hamilton-Baillie, pedaled out to a crossing on the outskirts of town where a long, straight highway comes into the village. Marking the crossing are two yellow lights that rise out of the ground. They are actually lanterns, of the sort hung around the canals in the Dutch city of Utrecht, turned upside down. They're not standard traffic devices. Váhl installed them in an effort to get drivers to slow down as they careened in off the rural highway. "It's making clear that there is something strange," he told me. "It's not common that there are lights like this." But doesn't the strange become familiar quickly? That's why

Váhl placed them so close together. It looks as if two cars may not make it through. But, as Váhl explained, with a hint of whimsy, "It is four meters and twenty in between the yellow things. It makes it possible that you don't hit the mirror of the other car." With practice the drivers may get used to this as well—but how can they be sure that the approaching driver is a local? Best to slow down.

What if, instead of the strange lanterns, approaching drivers were faced with a speed-limit sign? First, they might not even look at it. Second, they might worry about getting a ticket, but perhaps experience has taught them there is usually no cop there. Third, a speed-limit sign just announces a number. It says nothing about the fact that one is now in a village, where children or bicyclists might be present. Nor does it communicate risk. Forcing drivers to slow down, in order to save their own skins, just might be the best way to help save others' skins.

All this crazy stuff might be fine for provincial Dutch cities and English villages, with their relatively low traffic volumes and speeds. And in the Netherlands, where 27 percent of daily local trips are made on bicycles, drivers are much more experienced in interacting with cyclists. This sort of thing simply would not work in a large city in another country, you might think. Or would it?

Kensington High Street, the main commercial thoroughfare in one of London's poshest neighborhoods, is worth taking a look at, as I did one day with Peter Weeden, a senior engineer with the Traffic Section of the Royal Borough of Kensington and Chelsea. By the 1990s, Weeden recalled, the street was in a sorry state, and merchants were concerned about losing business to a large new shopping development being planned nearby. There was little aesthetic coordination, with the streets and sidewalks a jumble of different materials. "There was lots of clutter and street signs," Weeden explained. "They were put up with the best of intentions, but always on a very piecemeal basis. Someone comes along and puts up a sign for speed humps, someone else comes along and puts up another. Over time you end up with a forest of signs, most of which, it turns out, are not actually required."

The borough wanted the street to look better, but not at the expense of traffic flow or safety. "As well as being a shopping and residential high street, it's also one of the main arterial routes in and out of West London," said Weeden. Some 2,500 cars course down the street in a busy hour, while as many as 3,000 pedestrians spill out of the main tube station. Where the usual approach would have been to dig into the trusty traffic engineer's "toolkit," this time the Kensington planners began by throwing out everything that had been done before. "What we did was to actually strip out ninety-five percent of the signs in Kensington High Street," said Weeden.

They wanted to see what was really necessary and what was simply there because some engineer assumed it had to be. The guardrails lining both sides of the street, a not uncommon sight in London, were also removed in an effort to reduce visual clutter. "There is a very strong case for taking out guard railing," noted Weeden. "Wheelchair users don't like it; there are vision problems. Cyclists don't like it; they can get trapped between the vehicle and the rail if they get cut off. And the segregation between travel modes has been found to increase vehicles' speeds—you think you're going to own that space." The plan was not without critics—including the city's department of traffic engineering. "Transport for London thought we were taking unacceptable risks," Weeden said. But the Kensington engineers were not just casually saying, "Let's rip out all the traffic signs." They began by altering only a small test section, then waited to see what would happen.

Walking down the street, I noticed, as with Drachten, how much more clean and pleasant it looked without all the traffic markings, railings, and signs. It felt more like a city street should, and not like a slalom course for cars or a veal pen for pedestrians. The sidewalk felt connected to the street. There were several traffic lights, and while some pedestrians did cross at the light, there was no marked zebra crossing. Most people crossed elsewhere, in any case. No longer steered toward the crosswalk by the railing, they crossed where they chose to, navigating their way through the slow but steady flow of cars, buses, and bikes, pausing halfway on a center island.

Having tossed away the bulk of the safety improvements put in over the years for cars and for pedestrians, what happened? Chaos and destruction? Quite the reverse. Pedestrian KSIs ("killed or seriously injured") dropped 60 percent, with a similar decline for minor injuries. Weeden and his colleagues were as surprised as anyone. "The scheme itself never set out to be an accident-reduction scheme," he told me. "It was really just for aesthetic reasons, to encourage people to shop there. As a by-product we found that accident rates had dropped."

By making the street look better, they also made it safer. Perhaps this is not an accident. Cities are meant to be places for mixing with others, for improvised encounters, for observing details at a human scale. (Hamilton-Baillie says that London taxi drivers he interviewed reported liking the new scheme without quite knowing why, though they did cite the presence of "pretty girls" as a positive.) "This world of standardized, regulated kit—traffic islands, bollards, road markings, safety barriers, signs, signals—it's all a world completely separated to whatever happens behind it," said Hamilton-Baillie. "It's a world that we have been taught, and created policy, to say is an alien world. You've got to press a button to get permission to cross it." Drivers, absolved from their social responsibility by the mandates of the traffic world, accordingly act in antisocial ways. Pedestrians, tired of being steered far out of their way to cross the street or being inordinately delayed by cars in the many cases where they are the majority, rebel against the safety measures that have supposedly been erected for their benefit. The safety measures cause drivers and pedestrians to act in more dangerous ways.

A favorite example for Hamilton-Baillie of how things can be different is Seven Dials in London, the small circular junction in the Covent Garden district where seven streets converge. At a small plaza in the center, marked by a sundial, it's not uncommon to find people eating their lunch or to see them strolling across the roundabout, even as cars navigate their way slowly around the space. There are no guardrails protecting the pedestrians sitting in the center from the road. There are no speed bumps on the approaches. There are no signs warning, PEOPLE EATING LUNCH AHEAD. Rather, the uncertainty of the space and its human-scaled geometry dictate the behavior. There is an element of mystery and surprise, one that Charles Dickens remarked upon over a century before in *Sketches by Boz:* "The stranger who finds himself in the Dials for the first time . . . at the entrance of Seven obscure passages, uncertain which to take, will see enough around him to keep his curiosity awake for no inconsiderable time."

That awakened curiosity is still present today, and for drivers and pedestrians it translates into a need to pay attention. Even as a pedestrian navigating the Dials, I found myself confused. Which of the seven streets led to the Tube? If only there was a sign to point the way. Instead I paused, looked around, and decided to take the road that had the most people on it. This was the social world, and I was relying on human instincts. My choice was correct, and I found the Tube.

## Forgiving Roads or Permissive Roads? The Fatal Flaws of Traffic Engineering

One of Hans Monderman's many interesting ideas about traffic was that it is a network not only in space but in time. What this means is that the farther we drive, the faster we expect to be able to go. "When I start at home, I drive very slowly," he told me. "All my neighbors know me, they are part of my world, and I part of theirs, and it's absolutely unacceptable that I speed in my own street. But after a few minutes, I'm a bit more anonymous, and the more anonymous I get, the more my foot goes down and I'm speeding more and more." At the beginning of his trip, he was in the social world, and at the end of it, perhaps arriving in another village, he was as well. But what about the in-between? This was when he appreciated the traffic world, with all its signs and markings and safety measures and speeds. "When you want nice villages," he noted, "you need freeways."

But there is a problem with that in-between. Sometimes the roads on which people drive fast, as if they are the restricted-access highways of the traffic world, still have elements of the social world. People live near them, do their shopping on them, perhaps even have to cross them on foot. "I always say the road in-between is the most dangerous road," Monderman remarked. "It's not a highway, but it's not a residential street. All these roads have the biggest accident problem. The road is often telling you this is a traffic system: We have organized everything around you for all your needs. But the same road is cutting as a knife through the social world. The traffic world and the social world are shouting at each other."

One finds a striking example of this situation not in the Netherlands but in Orlando, Florida. Dan Burden is a widely acclaimed traffic guru who now works with the Orlando transportation planning firm Glatting

Jackson. We were cruising down East Colonial Drive, which is the Orlando stretch of U.S. Highway 50, heading for Baldwin Park, a New Urbanist community built on a former naval base that Burden was eager to show me. Burden, famously known for his elaborate walruslike mustache, was newly clean-shaven ("It's for charity," he explained). As we drove, Burden gave a running commentary on the nature of the street, which bears a dubious distinction: One analysis found it to be the twelfth-deadliest road in America. (The deadliest road, according to another survey, is U.S. 19, also in Florida, a few hours away.)

In the beginning, we were in the urban section of East Colonial Drive, which runs through the heart of north Orlando. It looked a bit like Los Angeles, a mixture of strip malls with a smattering of people on the sidewalks. Buildings were not set back very far, and the road was lined with concrete utility poles and other obstacles. As we passed a speed-limit sign, I did a double take. It read, 40 MPH. That struck me as strange. We were driving in what seemed to be a place that would be posted for 35 at the most. This is not uncommon in Florida, according to Burden. "If you looked on a city-by-city basis, county by county, you're going to find our high speeds are seven to fifteen higher than they will be in most states."

Continuing on Colonial, we entered the historically newer sections of town, and the road began to change subtly. The lanes became wider, the speed limit was raised to 45, and the sidewalks, when they existed at all, were dozens of feet from the road. "Notice how far back the sidewalk is," Burden exclaimed. "What is it, fifty feet? It's so far back it's like another world. There's no trees, and they've pushed the clear zone as far back as they could." Pulling into the parking lot of a Circle K convenience store, we saw a small white memorial posted in the swath of grass between the road and the gas pumps. Florida, somewhat controversially, is one of the few states that allows family members to place memorials on the site of fatal crashes. (The states that don't cite reasons ranging from the perceived safety risks of the memorials themselves to highway aesthetics.) It wasn't the first memorial I had seen. But I hadn't seen any in the more downtown part of Colonial Drive. Had I just not looked carefully enough, or was something else going on?

Colonial Drive is a tale of two roads. The first section of the road, with its narrow lanes, many crosswalks, thicker congestion, and bountiful collection of utility poles, parked cars, and other hazards, is the kind of road conventional traffic engineering has judged to be more dangerous. More people packed more tightly together, more chances for things to go wrong. The newer section of Colonial, with its wider lanes, its generous clear zones (i.e., roadsides without obstacles), its less-congested feel, and its fewer pedestrians, would be judged to be safer.

But when Eric Dumbaugh, an assistant professor of urban planning at Texas A&M University, did an in-depth analysis of five years' worth of crash statistics on East Colonial Drive, his results were surprising. He looked at two sections: what he terms a "livable" section, with the narrower lanes and lack of clear zones, and a section with wider lanes and more generous clear zones. In many respects, the two sections were similar, and thus ideal for comparison: They had the same average daily traffic, the same number of lanes, and the speed limits were similar (40 miles per hour versus 45). They had similarly sized painted medians in between the opposing streams of traffic, and the lengths of roadway were the same. They'd even had the same number of crashes at intersections, and the age of the at-fault drivers in those crashes was the same.

When Dumbaugh looked at the number of midblock crashes, precisely those types that should be reduced by the safety features of the road with wider lanes and wider clear zones, he found that the livable section was safer in every meaningful way. On the livable section, there had not been a fatality in five years (and hence there were no white memorial markers). On the comparison section, there had been six fatalities, three of them pedestrians. The livable section, which offered a driver many more chances to hit a "fixed object," had fewer of these crash types than the section designed to avoid those crashes. What about cars crashing into other cars? Surely the livable section, with all its drivers slowing to look for parking or coming out of parking spots, with all those cars packed tightly together, must have had more crashes. But across the board, from rear-end crashes to head-on crashes to turn-related crashes to sideswipe crashes, the numbers were higher in the section that the conventional wisdom would have deemed safer.

Why might this be so? Without a detailed reconstruction of each crash, it is impossible to be certain. But there are plausible hypotheses. Speed is a prime suspect. The wider lanes and lack of any roadside obstacles in the comparison section make 45 miles per hour seem optional, and some drivers are hitting near-highway speeds as other drivers are slowing to enter Wal-Mart or coming out of Wendy's. The painted median down the middle, known colloquially as a "suicide lane," allows people to make turns wherever they like. But these turns are across several lanes of oncoming high-speed traffic, and as we saw in Chapter 3, choosing safe gaps is not often an easy task for humans.

For pedestrians, a seemingly trivial variance in a car's speed can be the difference between life and death. A Florida study found that a pedestrian struck by a car moving 36 to 45 miles per hour was almost twice as likely to be killed than one struck by a car moving 31 to 35 miles per hour, and almost *four times* as likely as one struck by a car moving 26 to 30 miles per hour. In the livable section, pedestrians have an ample number of crosswalks, placed closely together. In the newer section, there are few crosswalks, and the ones that do exist are found at large intersections with multiple lanes of turning traffic. The "curb radii," or the curves, are long and gentle, enticing drivers to take them quickly, and do nothing to remind drivers about the pedestrians that may be legally crossing with the signal around that bend. In the livable section, drivers must slow to take tight turns, and parked cars buffer pedestrians from cars that veer off the road—not to mention that parked cars themselves cut speeds by some 10 percent.

Dumbaugh's research challenges a school of thought that has long held an almost unassailable authority in traffic engineering: "passive safety." This line of thinking, which emerged in the United States in the 1960s, says that rather than trying to prevent crashes, highway engineers (as well as car makers) should try to reduce the consequences of crashes, or, as one highway manual put it, "to compensate for the driving errors [the driver] will eventually make." Engineers running cars on "proving ground" test roadways found that once they departed the roadway, cars came to a stop an average of thirty feet off the road—so this became the standard minimum "clear zone," that section of legally required nothingness beyond the edge marking and before any obstacle. At General Motors, a "crash-proof highway" was designed with *one-hundred-foot* clear zones. Its engineer was so impressed with the performance that he declared, "What we must do is operate the ninety percent or more of our surface streets just as we do our freeways . . . [converting] the surface highway and street network to freeway and proving ground road and roadside conditions."

In many cases, like on East Colonial Drive, that is exactly what happened. The traffic world was brought to the social world. The design is well in line with the stated current engineering guidance: "The wider the clear zone, the safer it is." But far from ensuring safety, the road was home to more crashes than the section of the street that looked more like a traditional city street, even though the traffic was similar. What went wrong?

Part of the problem may be the in-betweenness of the newer sections of East Colonial. Walter Kulash, another noted traffic engineer with Glatting Jackson, says traffic engineers are not always to blame. Roads like Highway 50, he told me, are being used in ways engineers never intended. Designed as arteries to ferry people from one city cluster to another, they have instead become the "Main Streets" for suburban sprawl, lined with busy shopping centers and strip malls. "The engineers had nothing to do with that development, fronted by parking, for miles along the arterials, like you saw on Colonial Drive," Kulash said. "That is highly injurious to the function of the highway. The fact that fifty thousand travelers a day are bundled together, thereby making that irresistible to commerce, you might say, Okay, who's responsible for that? But you can hardly say a majority of blame ought to go to highway engineers."

From a strict engineering perspective, the "proving ground" approach makes sense. As Phil Jones, a traffic engineer based in the Midlands of England, argues, engineers are taught to work in "failure" mode. To design a bridge on a highway, engineers calculate the loads the bridge will need to carry, find out at which point the bridge would fail, and then make it more safe than that, for redundancy. But what happens when the factors involved are not just loads and stresses but the more infinitely complex range of humans behind the wheel?

In designing the approach path to a T-intersection, engineers use the factor of driver reaction time to determine what the appropriate sight distance should be—that is, the point at which the driver should have a clear view of the intersection. The sight distance is typically made longer than needed, to accommodate drivers with the slowest reaction times (e.g., the elderly). As with the highway bridge, the road design has a safety cushion to help it withstand extremes. So far, so good. But designing the road for slow reaction times, Jones explains, creates "very long sight distances, so someone who's younger and more able and can react faster than that will consume that benefit. What the safety model doesn't recognize is that yes, the elderly person will react more slowly, but they're not the ones driving fast in the first place. You're giving license to people to drive more quickly." This may be why, as studies have shown, railroad crossings where the sight distance is restricted—that is, you can see less of the track and the oncoming train—do not have higher crash rates than those with better views. Drivers approached the tracks more quickly when they felt it was safer.

What is meant to be the "forgiving road," argues Dumbaugh, becomes the "permissive road." Safety features meant to reduce the consequences of driver error encourage drivers to drive in a way *requiring* those generous safety provisions. Sometimes, passive-safety engineering makes things more dangerous. Dumbaugh studied a Florida road on which a number of cars had crashed into trees and poles. Simple, right? Just get rid of the obstacles and make the clear zones bigger. Looking carefully at the crash records, however, Dumbaugh found that the majority of crashes happened at intersections and driveways, as cars were turning. Were the obstacles the problem or was it, as Dumbaugh suggested, that drivers were unable to complete the turn because they were traveling too fast as they entered the turn, at the high speed the road design was telling them was "safe"?

In both Drachten and London, choices were made to remove traffic-safety infrastructure like signs and barriers. These choices were influenced by aesthetics, but they had the perverse outcome of making things safer. The problem with applying typical highway-engineering solutions to cities, villages, and the other places people live is that the same things that often signify "livability" are, in the eyes of a traffic engineer, "hazards."

Take the case of trees. In my Brooklyn neighborhood, they add to the desirability of a street. They raise property values. They may protect pedestrians from wayward cars. Yet they're also a common bane of traffic engineers, who have been—perhaps with the best of intentions—removing them from roadsides for decades. While many people have indeed died from colliding with trees, there is nothing inherently dangerous about a tree. What matters is the context. In his research Dumbaugh looked at a section of a road in Florida that travels through Stetson University. It's lined with mature trees, a few feet from the road. In four years, Dumbaugh found, there was not a single crash. What's more, he observed, most cars traveled at or even below the speed limit of 30 miles per hour (which many studies—and probably your own experience—have shown is rarely the case in cities). The hazards *were* the safety device. Drivers left with little room for error seemed quite capable of not making errors, or at least driving at a speed that would help "forgive" their own error.

The tree-lined road goes against the typical engineering paradigm, which would have deemed the trees unsafe and in need of removal. With the trees (the potential source of system failure) removed, a typical pattern would have happened: Speeds would have increased. The risk to pedestrians (students at Stetson, mostly) would have gone up; perhaps a pedestrian would have been struck. The police would have been called in to set up speed traps. Eventually, vertical deflection—a.k.a. speed bumps—would have been installed to calm the traffic. Having made the road safer, new measures would have been needed to again make it safe.

The pursuit of a kind of absolute safety, above all other considerations of what makes places good environments, has not only made those streets and cities less attractive, it has, in many cases, made them less safe. The things that work best in the traffic world of the highway—consistency, uniformity, wide lanes, knowing what to expect ahead of time, the reduction of conflicts, the restriction of access, and the removal of obstacles—have little or no place in the social world.

## Chapter Eight

# How Traffic Explains the World: On Driving with a Local Accent

### "Good Brakes , Good Horn, Good Luck": Plunging into the Maelstrom of Delhi Traffic

> Opening his eyes, he would know the place by the rhythm of movements in the street long before he caught any characteristic detail.
>
> —Robert Musil, *The Man Without Qualities*

"What other city in the world is like Delhi?" demanded Qamar Ahmed, the city's joint commissioner of traffic, as we sat drinking *chai* in his office. Clad in a khaki uniform topped with bright epaulettes on each shoulder, Ahmed brusquely shifted his attention between me and any one of the three mobile phones on his desk that kept ringing. An air conditioner labored against the enveloping premonsoon heat. "Delhi has forty-eight modes of transport, each struggling to occupy the same space on the carriageway. *What* other city is like this?"

To exit the Indira Gandhi International Airport, typically at night, when the international flights arrive, and alight into one of the city's ubiquitous black-and-yellow Ambassador cabs is to enter a motorized maelstrom. As an anticongestion measure, trucks are allowed into Delhi only between ten p.m. and six a.m., and so the sparsely lit road is thronged with lorries. They lurch, belch smoke, and ceaselessly toot their pressure horns. This seems by invitation: The back of most trucks bears the brightly festooned legend "Horn Please," often accompanied by a request to "Use Dipper at Night" (this means "dim your lights"). "Horn Please" originally invited following drivers to honk if they wanted to pass the slower-moving, lane-hogging trucks on the narrower roads of the past, and I was told that it endures merely as a decorative tradition. Nevertheless, a cacophony of claxons filled the air.

By day, the mayhem is revealed as true chaos. Delhi's streets play host to a bewildering stream of zigzagging green-and-yellow auto-rickshaws, speeding cabs, weaving bicyclists, slow-moving oxen-drawn carts, multi-passengered motorcycles conveying helmetless children and sari-clad women who struggle to keep their clothing from getting tangled in the chain, and heaving buses, which are often forced out of the bus-only lane because it is filled with cyclists and pedestrians, who are themselves in the lane because there tends to be no sidewalk, or "footpath," as they say in Delhi. If there is a footpath, it is often occupied by people sleeping, eating, selling, buying, or simply sitting watching the traffic go by. Limbless beggars and young hawkers converge at each intersection, scratching at the windows as drivers study the countdown signals that tell them when the traffic lights will change. Endearingly, if hopelessly, the signals have been embellished with a single word: RELAX. In the roundabouts of New Delhi, the traffic whizzes and weaves defiantly past faded safety signs bearing blunt messages like OBEY TRAFFIC RULES, AVOID BLOOD POOL and DON'T DREAM OTHERWISE YOU'LL SCREAM. These signs are as morbidly whimsical as they are common, leading one to suspect that somewhere, lurking in Delhi's Public Works Department, is a desk-bound bureaucrat with the soul of a poet.

The most striking feature of Delhi traffic is the occasional presence of a cow or two, often lying idly in the median strip, feet away from traffic. The medians, it is said, provide a resting place that is not only dry but kept free from pesky flies by the buffeting winds of passing cars. I posed the question of cows to Maxwell Pereira, Delhi's former top traffic cop, who has of late been playing the Colonel Pinto character on Indian *Sesame Street.* "Let me correct a little misperception," he told me as we sat in his office in the Gurgaon district. "The presence of a cow in a congested urban area is no hazard. Much as I don't like the presence of a cow on the road when I am advocating smoother traffic and convenience, the presence of a cow also forces a person to slow down. The overall impact is to reduce the tendency to overspeed and to rashly and negligently drive." Cows, in effect, act as the "mental speed bumps" that Australian traffic activist David Engwicht described in Chapter 7. They provide "intrigue and uncertainty," as Engwicht put it, and the average Delhi driver would certainly rather be late for work than hit a cow.

I heard that particularly Indian phrase—"rash and negligent driving"—often while in Delhi, but after a few days I started to lose sight of how that could differ from the norm. Delhi drivers have a chronic tendency to stray between lanes, most alarmingly those flowing in the opposite direction. The only signal used with regularity is the horn. Instead of working brake lights (or indeed any lights), many trucks have the phrase KEEP DISTANCE painted on the back, a subtle reminder to the driver behind: *I may stop at any moment.* Some taxis, on the other hand, bear the inscription KEEP DISTANCE. POWER BRAKE. This means: *I may come to a stop faster than you expect.*

Many vehicles lack side rearview mirrors, or keep them folded in. Auto-rickshaw wallahs actually mount their side-view mirrors on the inside, presumably to keep them from getting clipped off—or from clipping others. When changing lanes, drivers seem to rely not on the mirrors but rather that the person behind them will honk if there is danger. (It is not uncommon, meanwhile, to see scores of bus passengers leaning out the windows and advising the driver about whether he can merge, or trying to guide traffic themselves.) As a result of this collective early warning system, the sound of horns, on a road like Janpath in New Delhi, is as constant as birdcalls. When I asked one taxi driver, who went by the moniker J.P., how he coped with Delhi traffic, his answer was quick: "Good brakes, good horn, good luck."

After spending some time in the city, one vacillates between thinking Delhi drivers (and pedestrians) are either the best or worst in the world—the best because they're so adept at maneuvering in tight spaces and tricky situations, or the worst because they put themselves there to begin with. "That is why we have a negative connotation to the phrase 'defensive driving' in India," said Pereira, who still speaks in the flowery but formal vernacular of Indian officialdom. "Defensive driving is defending yourself from all the vagaries, including the negligence contribution on the part of the other road user." Pereira advised me not to try Delhi traffic firsthand: "The Indian driver relies more on his reflexes, absolutely. Your reflexes would not be geared to expect the unexpected."

Conversely, when Pereira finds himself in the United States visiting relatives, his passengers, who may fail to appreciate the lingering aftereffects of Delhi traffic, are often perturbed by his driving style. "When I see a vehicle approaching from a side road, I tense up. Internally, I'm used to a condition in India where I'm not sure if when they are coming from the side road they will step into my path," he said, adding that in the States, "you expect that he will never; here I will not expect that he will never. The halt-and-proceed thing is not there."

Arguably, drivers anywhere should always try to expect the unexpected, but this is taken to a kind of high art in Delhi, where the unexpected perversely becomes the expected. There are nearly 110 million traffic violations *per day* in Delhi, I was told by Rohit Baluja as we sat in his office in the Okhla Industrial Area, eating lunch out of the small metal pails known as tiffins.

The dapper and successful owner of a shoe company, Baluja founded the Institute of Road Traffic Education in an effort to improve the conditions of Indian roads, on which an estimated 100,000 people die every year—one out of every ten road deaths in the world. He launched IRTE after a succession of business trips to Germany, where he was astounded by the well-defined and relatively orderly traffic system. "As soon as I returned to Delhi it felt as if everybody here is stealing your right-of-way, and that nobody understands there is something called a right-of-way," he said. In 2002, a group of English police studying Delhi traffic told Baluja that whereas in the United Kingdom one can predict with 90 percent certainty the behavior of the average road user, in Delhi they felt that no more than 10 percent compliance could be anticipated. They called it anarchy on the roads. "We have started living in indiscipline, so we don't feel there is an indiscipline," Baluja told me.

The estimate of daily traffic violations was obtained by IRTE researchers who followed and filmed random vehicles on the streets of Delhi in a camera-and-radar-equipped SUV they called the Interceptor. I was shown a sample of this footage by Amandeep Singh Bedi, a researcher at IRTE, and all the "vagaries" that Pereira had been discussing came to light. In one clip, a driver is rear-ended when he stops his car suddenly in the middle of a busy road. Why did he stop? So he could buckle his seat belt and not be *challaned*, or fined, by a traffic cop posted on the side of the road. In another, a bus illegally halts far from the marked curbside bus stop, making harried passengers weave through several traffic streams simply to board the bus. It soon becomes clear that one reason the number of violations is so high is that many drivers are forced to violate the rules in reaction to another driver violating the rules: The bus lane is filled with pedestrians or bicycles (who, in fairness, have nowhere else to go), so the bus cannot travel in the bus lane; thus begins a cascade of violations across the traffic stream.

Not everything can be strictly blamed on the driver. Lane markings are often missing, shattered wrecks sit in the middle of busy roads, foliage obscures traffic lights, and sometimes traffic signs in Delhi are no more than small, barely legible hand-lettered placards taped to utility poles; a "No U-Turn" sign may look more like a suburban garage-sale announcement. These are created by an artist with the Delhi Traffic Police. "Sometimes there is a gap in my request [for a new sign] and their installation," Ahmed admitted to me with a sigh. "To fill up this gap we make these signs."

Things are even worse in the countryside. "Our highways are built by consultants from across the world," Baluja said. "They have got no idea of mixed traffic conditions. Highways have been built cutting across villages. Villagers cross still, but underpasses were not made for them." And so what is meant to be a restricted-access highway becomes, unintentionally, a small village road, with animals crossing, vendors selling fruit and newspapers on the median strip, and bus passengers queuing up for buses that have stopped directly on the carriageway. Openings are cut into guardrails, or the guardrails themselves are stolen for scrap. In vain, localities do things like erect stop signs on high-speed national highways—taking "expect the unexpected" to a new level.

On one of my last days in Delhi, I witnessed an episode that seemed to contain the exasperating essence of the Delhi traffic experience. One afternoon, as the temperature swelled to over one hundred degrees, the air pregnant with the weight of the rainy season, I saw a funeral procession on the famously bustling Chandni Chowk in Old Delhi. A group of men were bearing aloft a body draped in white fabric and marigold garlands, jostling through the traffic of cycle-rickshaws, pedestrians, scooters, and carts heaped high with produce. A thought occurred to me then: The living may indeed fear for their lives on Delhi roads, but even the dead have to fight for space.

## Why New Yorkers Jaywalk (and Why They Don't in Copenhagen): Traffic as Culture

One of the first things that strikes a visitor to a new country is the traffic. This happens in part simply because foreign traffic, like a foreign currency or language, represents a different standard. The cars look odd (who makes *that*?), the road widths may feel unusual, the traffic may drive on the other side of the road, the speed limits may be higher or lower than one is used to, and one may struggle, as one does with showerheads at the hotel, with traffic signs that look somewhat familiar but still escape interpretation: A particular symbol might refer to rocks falling or sheep crossing the road—or both, at the same time. I was once in the back of a London taxi when I saw a red-and-white traffic sign that declared, CHANGED PRIORITIES AHEAD. Whose priorities, I thought with a panic—mine? All of ours?

Most of the standard stuff is fairly simple, requiring only slight adjustments to adapt. The more difficult thing to crack is the *traffic culture*. This is how people drive, how people cross the street, how power relations are made manifest in those interactions, what sorts of patterns emerge from the traffic. Traffic is a sort of secret window onto the inner heart of a place, a form of cultural expression as vital as language, dress, or music. It's the reason a horn in Rome does not mean the same thing as a horn in Stockholm, why flashing your headlights at another driver is understood one way on the German autobahn and quite another way on the 405 in Los Angeles, why people jaywalk constantly in New York and hardly at all in Copenhagen. These are the impressions that stick with us. "Greek drivers are crazy," the visitor to Athens will observe, safely back in Kabul.

But what explains this traffic culture? Where does it come from? Why did I find the traffic in Delhi so strange? Why does Belgium, a country for all intents and purposes quite similar to the neighboring Netherlands, have comparatively riskier roads? Is it the quality of the roads, the kinds of cars driven, the education of the drivers, the laws on the books, the mind-set of the people? The answer is complicated. It may be a bit of all of these things. There does, however, seem to be one overarching, "rule of thumb" way to measure the traffic culture of a country, its degrees of order or chaos, safety or danger; we will return to this in the next section.

The first thing to recognize is that traffic culture is *relative*. One reason Delhi traffic feels intense to outsiders is simple population density: The metropolitan area of Delhi packs five times the people into the same space as New York City, a place that already feels pretty crowded. More people, more traffic, more interactions. Another reason Delhi seems so chaotic (to me, at least) is the staggering array of vehicles, all moving at different speeds and in different ways. The forty-eight modes of transport I referred to earlier are a far cry from those of my hometown, New York City, which has roughly five: cars, trucks, bicycles, pedestrians, and motorcycles or scooters (with a few horse-drawn carriages and cycle-rickshaws thrown in for tourists). Many places in the United States are essentially down to two modes: cars and trucks.

Geetam Tiwari, a professor at the Indian Institute of Technology in Delhi, has posited that what may look like anarchy in the eyes of conventional traffic engineering (and Western drivers) actually has a logic all its own. Far from breaking down into gridlock, she suggests, the "self-optimized" system of Delhi can actually move more people at the busiest times than the standard models would imply. When traffic is moving briskly on two- and three-lane roads, bicycles tend to form an impromptu bike lane in the curb lane; the more bikes, the wider the lane. But when traffic begins to get congested, when the flows approach 2,000 cars per lane per hour and 6,000 bikes per lane per hour, the system undergoes a change. The bicyclists (and motorcyclists) start to "integrate," filling in the "longitudinal gaps" between cars and buses. Cars slow dramatically, bikes less so. The slowly moving queues grow not only lengthwise but laterally, squeezing out extra capacity from the roads.

In so-called homogenous traffic flows, where every vehicle is roughly the same size and same type, lane discipline makes sense: You cannot fit two cars into one lane. It is also easy to figure out the maximum capacity of a road and to try to predict driver behavior through relatively simple traffic models like the previously discussed "car following." But in heterogeneous traffic flows, like Delhi's, where nonmotorized traffic can make up as much as two-thirds of the traffic stream, those formal models are of little use—having bicycles or scooters queue one per lane at a traffic light, for example, would create massive traffic jams.

It can be unnerving to sit at a Delhi intersection in the back of an auto-rickshaw and feel humanity press to within inches, or to see bicycles slowly thread between teeming lorries. When the traffic compresses in this way, the number of what engineers call conflicts increases—there are, to put it simply, more chances for someone to try to occupy the same space at the same time as someone else. In conventional traffic-engineering thought, the more conflict, the less safe the system. But again, Delhi challenges preconceptions. In a study of various locations around Delhi, Tiwari and a group of researchers found that the sites that had a low conflict rate tended to have a high fatality rate, and vice versa. In other words, the seeming chaos functioned as a kind of safety device. More conflicts meant lower speeds, which meant fewer chances for fatal crashes. The higher the speeds, the better the car and truck traffic flowed, the worse it was for the bicycles and pedestrians. Even when the roads were crowded, however, they were hardly ideal for cyclists. Studies show that 62 percent of the cycle fatalities during peak hours were because of collisions with trucks and buses, which tend to use the same lane as the cyclists. Self-organization clearly has its limits.

The second point is that traffic culture can be more important than laws or infrastructure in determining the feel of a place. In China, which is undergoing the fastest motorization in history, the power of traffic culture was made clear to me one afternoon as I sat studying an intersection in the Jingan neighborhood of Shanghai, from the God's-eye perspective of my thirteenth-floor hotel room. At first glance, the intersection, ringed by office buildings and well marked with signs and signals, was unremarkable. But then I took a closer look.

Traffic engineers note that signalized four-way intersections have over fifty total points of conflict, or places where the turning movements and crisscrossing flows might interfere. At the intersection of Shimen Yilu and Weihai Lu, that number seemed hopelessly low. As groups of cars hurtled toward other groups of cars, I fully expected to see a collision. Instead, time seemed to slow, space compressed like an accordion, and in that small cluster the various parties worked a way through. Then the accordion expanded again, the space opened up, and the speed increased as all the parties went on their way. It seemed to be orchestrated by some giant invisible hand.

But the sheer range of ways for things to go wrong was staggering. Cars moving down Weihai Lu will use the oncoming left-turn lane to pass cars moving in the same direction. Bikes coming down Shimen Yilu and wanting to turn left onto Weihai Lu will park themselves in the middle of the big intersection, waiting to find an opening in three lanes of oncoming traffic. A pedestrian escapes one right-turning car only to be almost hit by a left-turning bicycle, who in turn narrowly avoids being struck by a vehicle that has crossed the yellow line to get around another car. There is no left-turn arrow, so when Shimen Yilu northbound gets the green, all four lanes of cars begin to move. But the cars turning left must navigate the two-way stream of bike and moped traffic before plunging farther into the wide, crowded zebra-striped pedestrian crosswalk. Cars pay little heed to the pedestrians crossing; even if there are huge massings, the cars will still push through, sometimes stranding pedestrians between two streams of probing cars. The two-way bike traffic does not look to necessarily follow any rule of thumb regarding being on the right or left, and on Weihai Lu, it's not uncommon to see bikes almost have head-on collisions.

In theory, this intersection could have been anywhere, from Houston to Hamburg. But what went on within that intersection was something else entirely. Crossings continued after the lights had changed, pedestrians seemed to cross as if they had given up on life, and drivers seemed to be doing their best to oblige that wish.

In a study a few years ago, a group of researchers examined a number of intersections in Tokyo and a number of comparable intersections in Beijing. Physically, the intersections were essentially the same. But those in Tokyo handled up to *twice* as many vehicles in an hour. What was the difference? The researchers had several ideas. One was that Tokyo had more new and higher-quality vehicles, which could start and stop more quickly. Another was that by contrast with Tokyo, Beijing had many more bicycles. In 2000, bicycles still accounted for 38 percent of all daily trips in the city, with cars at 23 percent, according to the Beijing Transportation Research Center (the gap has since been closing). Bicycles, the researchers noted, were often not separate from the main traffic flow, and so weaving bikes caused "lateral disturbance."

The most important difference had nothing to do with the quality or composition of Beijing's traffic flow; it concerned the behavior of its participants. In Tokyo, signal compliance by cars and pedestrians was, like Japanese culture itself, rigorously formal and polite. In Beijing, the researchers observed, drivers (and cyclists and pedestrians) were much more likely to violate traffic signals. People not only entered the intersection after the light had changed, the researchers found, but *before*. This impression was confirmed to me by Scott Kronick, a longtime Beijing resident who heads Ogilvy Public Relations' Chinese division. "Driving in China is total offense—you go for it. You'll see people on the green light trying to take left-hand turns before the traffic goes through."

One of the more outlandish transportation proposals made by the Red Guards during China's Cultural Revolution—along with banning private vehicles and demanding that rickshaw passengers pedal the rickshaws—was to change the meaning of traffic lights: Red would mean "go," green would mean "stop." To look at Chinese cities today, you might not realize that the proposal never took hold.

At first, the traffic disorder seems a bit surprising, given the strictness of the Chinese government in other areas of life (e.g., blocking Web sites). Then again, jostling traffic is not going to bring down a regime. The British playwright Kenneth Tynan observed in his *Diaries*, after seeing the wreckage of a car crash in Turkey, "Bad driving—i.e. fast and reckless driving—tends to exist in inverse ratio to democratic institutions. In an authoritarian state, the only place where the little man achieves equality with the big is in heavy traffic. Only there can he actually *overtake*." As amateur sociology, this is pretty good stuff. And people in China—drivers, pedestrians, cyclists—did at times seem to be going out of their way to assert their presence, to claim some ownership of the road.

This became clear one afternoon as I went cycling with Jonathan Landreth, the Beijing correspondent for *Hollywood Reporter* and a regular cyclist. Even within the bike lane, things were more complex than they seemed. Simply by having a mountain bike with gears, I was able to ride much faster than the typical Chinese commuter on their heavy Flying Pigeon, who years ago would have commanded the entire street. But I was still not top of the food chain in the bike lane—faster still are the electric-powered bicycles, one of which almost hit me head-on. Then there are the motorized three-wheeled vehicles commissioned to transport Beijing's handicapped—and, it seemed, to add to their ranks.

"Those guys use the bike lane too," Landreth told me, "and they get really annoyed when you're in the way."

I was given another theory on Chinese traffic behavior by Liu Shinan, a columnist at the *China Daily*, a government-owned newspaper. I happened to be in China at a time when several vigorous campaigns were under way, in part to improve traffic before the 2008 Beijing Olympics. In Shanghai, officials were threatening to post photographs of jaywalkers in their place of business. Liu thought the tactic might work. "We Chinese attach importance to face," he told me as we sat in the newspaper's canteen. "When they jaywalk they don't care too much about it, because all the people around them are strangers. They don't think they have lost face. But if you published a photo in my unit here, I would feel very embarrassed." What was happening in Shanghai was, in essence, a version of the eBay-style reputation-management system discussed earlier in this book. But why were such measures deemed necessary? The roots of Beijing's traffic lawlessness, Liu suggested to me, lie in history. "After the Cultural Revolution, which lasted for ten years, it was a chaotic society," he said. "People didn't show any respect to any law, because Chairman Mao encouraged the people to revolt, to question authority."

So were these countless infractions little acts of everyday rebellion? Were drivers still paying heed to Mao's praise of "lawlessness" as a social good? Or can the roots of China's disorganized traffic be traced even further back? It has long been argued, for example, that Confucian ethics, which emphasize personal relationships and the cultivation of private virtues, contribute to a diminished sense of public morality and civic culture. In his 1935 best-seller *My Country and My People*, Lin Yutang wrote that the lack of "personal rights" had led to an individualistic, deep-seated indifference toward the public good. "We are great enough to elaborate a perfect system of official impeachment and civil service and traffic regulations and library reading-room rules," Lin Yutang observed, "but we are also great enough to break all systems, to ignore them, circumvent them, play with them, and become superior to them." In opposition to the Socratic tradition of the West, Confucianism emphasizes personal ethics and virtues over the "rule of law." As the legal scholar Albert H. Y. Chen writes, "in situations where there were disputes, people were encouraged to compromise and give concessions rather than to assert their self-interest or rights by litigation." Indeed, one can find echoes of this on the streets of China today. In the span of a few weeks, I saw several instances where minor traffic collisions had occurred. When this happens in the United States, drivers generally exchange insurance information and move on; in Beijing, the parties involved were engaged in heated negotiation, often surrounded by a crowd that had enthusiastically joined the proceedings.

In China, things were happening in traffic faster than the government could keep pace. A few decades ago, a city such as Beijing did not have much in the way of cars, or even commutes. Privately owned vehicles were illegal, and many workers lived and worked in the same unit, known as the *danwei*. In 1949, Beijing had 2,300 automobiles. In 2003 it had 2 million—and this number is rapidly growing, with the capital adding upward of 1,000 new cars a day. A sweeping new Road Safety Act, the country's first, was passed in 2004 to cope with the radically changing traffic dynamics, but it has not been without controversy, particularly when it comes to assigning fault in a crash. Zhang Dexing, with the Beijing Transportation Research Center, told me of a well-known case in 2004 that involved a husband and wife, new arrivals to the city, who were illegally walking on the highway. A driver struck the two, killing the wife. Although the pedestrians' presence on the highway was illegal, the driver was still found partially at fault and was forced to pay the husband several hundred thousand *renminbi* (nearly U.S. $20,000).

One key to understanding traffic culture is that laws themselves can explain only so much. As important, if not more so, are the cultural norms, or the accepted behavior of a place. Indeed, laws are often just norms that have been codified. Take the example of the laws that say that in the United States, one must drive on the right side of the road, while in the United Kingdom, one must drive on the left side of the road. These emerged not from careful scientific study or lengthy legislative debate about the relative safety of each approach but from cultural norms that existed long before the car.

As the historian Peter Kincaid describes it, the reason why you drive on the right or left today has to do with two things. The first is that most people are right-handed. The second is that different countries were using different forms of transportation at the time that formalized rules of the road began to emerge. The way in which the first consideration interacted with the second consideration explains how we drive today. Thus a samurai in Japan, who kept his scabbard on his left side and would draw with his right arm, wanted to be on the left as he passed potential enemies on the road. So Japan today drives on the left. In England, horse-drawn carts were generally piloted by drivers mounted in the seat. The mostly right-handed drivers would "naturally" sit to the right, holding the reins in the left hand and the whip in the right. The driver could better judge oncoming traffic by traveling on the left. So England drives on the left. But in many other countries, including the United States, a driver often walked along the left side of his horse team or rode the left horse in a team (the left-rear horse if there were more than two), so that he could use his right arm for better control. This meant it was better to stay to the right, so he could judge oncoming traffic and talk to other drivers. The result is that many countries today drive on the right.

Even when laws are ostensibly the same, norms help explain why traffic can feel so different in different places. Driving on the Italian *autostrada* for the first time, for example, can be a shock to the uninitiated. Left-lane driving is reserved for passing, and for many drivers in the left lane, their entire trip is one epic overtaking, a process known in Italy as *il sorpasso*, a phrase freighted with additional meanings in social mobility. Get in the way of someone in the midst of a *sorpasso* and they will soon drive so close that you can feel, on the back of your neck, the heat of their headlights, which they're flashing furiously. This is less a matter of aggressiveness than incredulousness at your violation of the standard.

"The law in most European countries is to drive as far to the right as is practical," explained Per Garder, a Swedish professor of traffic engineering who now teaches at the University of Maine. "But in America that's just on paper—the person who comes from behind almost always yields to the person in front, while in Italy it's the person behind. You are supposed to move away and let them pass. As an American driver it is difficult to remember, especially if you're going above the speed limit yourself—why shouldn't you be allowed to be in the passing lane?" In the United States, a rather hazy norm (and a confusing array of laws) says that the left lane is reserved for the fastest traffic, but this is not as rigidly ingrained as it is in Italy. In fact, in the United States one is likely to see the occasional reaction (passive-aggressive braking, refusal to move, etc.)

to Italian-style tailgating. Americans, perhaps out of some sense that equality or fairness or individual rights have been violated, seem to take these acts more personally. In Italy, which has a historically weak central government and overall civic culture, the citizenry relies less on the state for articulating concepts like fairness and equality. This, at least, was the theory presented to me in Rome by Giuseppe Cesaro, an official with the Automobile Club d'Italia. "In American movies, they always say, 'I pay taxes. I have my rights.' In Italy no one's going to say this. You pay taxes? Then you are a fool."

Norms may be cultural, but traffic can also create its own culture. Consider the case of jaywalking in New York City and Copenhagen. In both places, jaywalking, or crossing against the light, is technically prohibited. In both places, people have been ticketed for doing it. But the visitor to either city today will witness a shocking study in contrast. In New York City, where the term *jaywalking* was popularized, originally referring to those hapless bumpkins, or country "jays," who came to the city with little notion of how to perambulate properly in big-city traffic, *waiting* for the signal is now the sign of a novice from the sticks. By contrast, the average Copenhagen resident seems to have a biological aversion to crossing against the light. Early on a freezing Sunday morning in January, not a car in sight, and they'll refuse to jaywalk—this in a city with the largest anarchist commune in the world! They'll stop, draw in a breath, perhaps tilt their head a bit skyward to catch a snowflake. They'll gaze at shop windows, or look lost in thought. Then the signal will change, and they'll move on, almost reluctantly.

It is tempting to chalk up the differences purely to culture. In New York City, a melting pot of clashing traditions and a hotbed of ruthless and obnoxious individualism, jaywalking is a way to distinguish yourself from the crowd and get ahead, a test of urban moxie. "Pedestrians look at cars, not lights," Michael King, a traffic engineer in New York City, told me. Jaywalking also helps relieve overcrowded clusters at intersections. In Copenhagen, which historically has had a more homogenous, consensus-seeking population, jaywalking is an act of bad taste, an unnecessary departure from the harmony that sustains communities. Waiting for the light to change, like waiting for spring, seems a test of the stoic and wintry Scandinavian soul. In the 1930s, the Danish-Norwegian novelist Aksel Sandemose famously described a set of "laws" (called the *Jantelagen*) inspired by the small Danish town in which he was raised.

They all basically had the same theme: Do not think you are better than anyone else. The "Jante laws" are a still popular shorthand toward explaining the relative social cohesion and egalitarian nature of Scandinavian societies, and it's not hard to imagine them applied to traffic. Jaywalking, like speeding or excessive lane changing (which one rarely sees on Danish roads), is just a form of ostentatious narcissism that disrupts communal village life.

When I offered these theories to the celebrated urban planner Jan Gehl as we sat in his office in Copenhagen, he brushed them aside and countered with a rival theory: "I think the whole philosophy of the city means you have good-quality sidewalks and frequent intersections. You know you only have to wait for a short while and then it gets green." By contrast, his firm had recently completed a study of London. "We found it was completely complicated to get across any street. We found that only twenty-five percent of the people actually did what the traffic planners suggested to do," he said. The more you make things difficult for pedestrians, Gehl argued, the more you downgrade their status in the traffic system, "the more they start to take the law into their own hands." I thought back to New York City, where the lights on Fifth Avenue seem purposely timed so that walkers have to pause at every intersection. Was it New York's traffic system, and not New Yorkers themselves, that made the city the jaywalking capital of the United States?

There is an iron law in traffic engineering: The longer pedestrians have to wait for a signal to cross, the more likely they are to cross against the signal. The jaywalking tipping point seems to be about thirty seconds (the same time, it turns out, after which cars waiting to make a left turn against traffic begin to accept shorter, more dangerous gaps). The idea that waiting time might be the real explanation behind jaywalking was brought home to me one afternoon in London as I looked at brightly colored computer maps of pedestrian crossings with Jake Desyllas, an urban planner who heads Intelligent Space. On certain streets in London, he pointed out, the proportion of people who crossed only during the "green man" would be 75 percent, but on a neighboring street, the number would be drastically lower. It was not that the culture of people waiting to cross the street changed as they walked one block, but rather that one street-crossing design paid more heed to pedestrians than the other. Not surprisingly, the places where it took pedestrians longer to get across had more informal crossings. At one of the worst spots in London, the crossing to the Angel tube station across the A1 Street in Islington, Desyllas found that pedestrians who make it to the center island can wait as long as sixty-two seconds for a "Walk" signal. The city is virtually compelling pedestrians to jaywalk.

As if traffic were not complicated enough, there is the additional problem that it regularly throws together people with different norms. Because each is convinced they are right—and traffic laws often disprove neither—they're that much more primed to "go off" at the other's perceived misdeeds (e.g., late merging, left-lane tailgating). Traffic also tosses together those with local knowledge and lesser-educated outside users, the pros with the amateurs. Any time-starved city dweller who has been stuck walking behind a group of slow-moving tourists has come across this phenomenon; proposals have been made for pedestrian "express lanes" in New York's Times Square or London's Oxford Street for this reason. Or take the local driver trapped behind someone looking for an unfamiliar address. The banal boulevard that one driver has seen a million times and wants to hurry through will be a fascinating spectacle for another driver, worthy of slow appreciation. In Florida, two bumper stickers embody this struggle: I BRAKE FOR BEACHES and SOME OF US AREN'T ON VACATION.

What's striking is how quickly the local norms can be picked up. Years of driver training or habit can be washed away like dirt from a windshield. David Shinar, an expert in the psychology of traffic at Israel's Ben-Gurion University, argues this point: "If you take an Israeli driver and transplant him to Savannah, Georgia, I guarantee that within two months he will be driving like the people there, like everyone around him. And if you transport someone from the American Midwest to Tel Aviv, within days he will be driving like an Israeli—because if he doesn't, he'll get nowhere." And so, like the visitor to England who begins to appreciate lukewarm beer, astute drivers will echo local inflections like the "Pittsburgh left," that act of driving practiced primarily in the Steel City (but also Beijing) in which the change of a traffic light to green is an "unofficial" signal for a left-turning driver to quickly bolt across the oncoming traffic. New arrivals to Los Angeles soon become versed in the "California roll," a.k.a. the "sushi stop," which involves never quite coming to a complete halt at a stop sign.

Traffic is like a language. It generally works best if everyone knows and obeys the rules of grammar, though slang can be brutally effective. If you're absolutely unfamiliar with it, it will seem confusing, chaotic, and fast. Learn a few words, and patterns begin to emerge. Become more fluent, and suddenly it all begins to make sense. Rome presents an interesting example here. As I mentioned in the Prologue, Rome has been grappling with traffic problems since it became Rome. As Caesar tried to ban carts, so did Mussolini, the "Twentieth-Century Caesar," try to regulate the city to his whims. Il Duce, as one story goes, grew so impatient with the chaos on the Via Corso that he attempted, in vain, to force pedestrians to walk in only one direction on each side of the street. Appropriately for a city whose history is steeped in mythology, the Roman driver has assumed an almost mythological status.

Roman driving is distinguished by space and pace. The narrowness of most streets, coupled with the quick acceleration of small, manually shifted cars, enhances the feeling of speed. Drivers focus on entering the smallest gaps possible. As Cesaro, the official with the Automobile Club d'Italia, explained one afternoon in his office on the Via Nazionale, Roman traffic behavior is "simply a need—there are so many cars on the tight road. We are always side by side. Sometimes we start talking to each other. The traffic lights change two or three times. Sometimes we become friends." Stuck at those lights, the car driver will notice a steady stream of scooters slowly filtering to the front of the queue, like the grains in a snow globe settling on the bottom. "They should follow rules like cars," said Paolo Borgogne, also of the ACI, of Rome's legions of scooters, "but for some reason it is believed they don't need to. . . . Traffic lights, for instance, they consider furniture on the corner of the road." But things are changing: Whereas for years scooter drivers required no license, a *patentino*, or "small driver's license," is now mandatory.

As with Delhi, however, it's not difficult to imagine that Roman traffic jams would be worse if scooters (which make up one-fifth of the traffic) always acted like cars. And the legendarily "crazy" Roman traffic might just be a matter of interpretation. Max Hall, a physics teacher in Massachusetts who often rides his collection of classic Vespas and Lambrettas in Rome, says that he finds it *safer* to ride in Rome than in Boston. Not only are American drivers unfamiliar with scooters, he maintains, but they resent being passed by them: "In Rome car and truck drivers 'know' they are expected *not* to make sudden moves in traffic for fear of surprising, and hurting, two-wheeler drivers. And two-wheeler drivers drive, by and large, expecting *not* to be cut off." In this regard, Rome is safer than other Italian cities where fewer riders wear helmets and studies have shown that scooters are much more likely to have collisions with cars. Reaching for the language of physics, Hall says, "The poetic and beautiful result is that four-wheelers behave like fixed objects, by moving very little relative to each other, even at significant speeds, while two-wheeler traffic moves 'through' the relatively static field of larger vehicles."

Thinking that the key to truly understanding Roman traffic might lie in physics, one afternoon I went to visit Andrea De Martino, a physicist with the Laboratory of Complex Systems at the University of Rome. In his office at La Sapienza, he drew diagrams on the chalkboard and spoke of "network optimalization" and "resource competition." Then he talked about Rome. "My girlfriend is not from Rome, she's not Italian," he said. "She tried to understand the logic behind the fact that a car can just cross the road even if it sees you coming. There is no logic." He contrasted this to driving in Germany, which he'd found to be "marvelous." This was not the first time I'd heard a Roman praise driving in some other, more "orderly," country. I asked him: If everyone likes it so much, why don't they drive that way here? He said: "I like the German system—*in Germany.*"

One could drive like a Roman in Frankfurt, or drive like a Frankfurter in Rome, only one might not do so well in either situation. But why is that? Where do these norms come from? The simplest answer may be that Romans drive the way they do because *other Romans do.*

This idea was expressed in a series of experiments by the psychologist Robert Cialdini. In one study, handbills were placed on the windshields of cars in a parking garage; the garage was sometimes clean and sometimes filled with litter. In various trials, a nearby "confederate" either littered or simply walked through the garage. They did this when the garage was filled with litter and when it was clean. The researchers found that the subjects, upon arriving at their cars, were less likely to litter when the garage was clean. They also found that subjects were more likely to litter when they observed someone else littering, but *only* if the garage was already dirty.

What was going on? Cialdini argues there are two different norms at work: an "injunctive norm," or the idea of what people should do (the "ought" norm), and a "descriptive norm," or what people actually do (the "is" norm). While injunctive norms can have an impact, it was the descriptive norm that was clearly guiding behavior here: People littered if it seemed like most other people did. If only one person was seen littering in a clean garage, people were *less* likely to litter—perhaps because the other's act was so clearly violating the injunctive norm. This is why so many public-service advertising campaigns fall on deaf ears, Cialdini and others have suggested. An advertisement about the many billions of dollars lost to tax cheating draws attention to the problem, but it also whispers: Look how many other people are doing it (and getting away with it). *Who* is violating a norm is also important: Studies of pedestrians have found that walkers are more likely to cross against the light when a "high-status" (i.e., well-dressed) person first does so; they're less likely to cross when that same person doesn't. "Low-status" violators prompt less imitative behavior either way.

Traffic is filled with injunctive norms, telling drivers what to do and what not to do. But the descriptive norm is often saying something else—and saying it louder. The most common example is the speed limit. The law on many U.S. highways is 65 miles per hour, but a norm has gradually emerged that says anything up to 10 miles per hour above that is legal fair game. Raise the speed limit, and the norm tends to shift; driving the speed limit starts to seem hazardous.

Some norms seem to hold more strongly than others. Leonard Evans, a trained physicist and traffic-safety researcher who worked for General Motors for more than thirty years, gives an example: "It's two a.m., some guy's just been speeding, to save time. He comes to this intersection. There's no traffic in sight anywhere. He sits stationary for thirty seconds. Objectively speaking, he is causing far more risk by his exceeding the speed limit than he would be if he stopped at the red light, looked this way and that way, and just went through it. But we have a robust social norm in the U.S. You just do not consciously and casually drive through a completely red light. Unfortunately, we don't have a robust norm against not going fast after it's turned green." Both acts are technically against the law, each bear similar penalties, but one act seems more illegal than the other. Perhaps in speeding the driver feels as if he's in control, while going through a red light, even carefully, puts one at the mercy of others. He may also speed because most other people do (whereas if everyone decided to cross through red lights, anarchy would ensue).

Most traffic laws around the world are remarkably similar. Many places have relatively similar roads and traffic markings. But the norms of each place are subtly different, and norms are powerful, curious things. Laws do not dictate how people should queue up in the United Kingdom or China—nor should they, most would argue—but try queuing up in either place and you will notice a striking difference. In the United Kingdom, queues are famously orderly, but in China, they often exist more in theory than reality—queue jumping, along with jaywalking, was another behavior targeted by the Chinese government for extinction before the 2008 Olympics.

Similarly, economists have long been puzzled by the fact that, in most places, restaurant patrons tip their server *after* they have already been served—which may boost the incentive for the server to give good service but hardly increases the incentive for the patron to tip well. Mysterious, too, is that patrons tip even in the face of further erosion of these incentives—if their service was less than desirable or if they don't plan to return to the same restaurant. Studies have shown the link between tip and service quality to be slight. People seem to tip because it's seen as the right thing to do, or because they don't want it known that they've not done the right thing. There's no law that says that patrons have to tip; they simply follow the norm.

In traffic, norms represent some kind of subtle dance with the law. Either the norms and laws move in time or one partner is out of step. In Florence, observes the writer Beppe Severigni, the locals have a phrase, *rosso pieno*, or "full red," for a traffic signal. This implies that there are other reds that are less "full." These distinctions are not noted by law, but they help explain actual behavior. Yet where do these norms come from? How do they adhere to or depart from the law? It seems that the most significant norm of all, as the legal scholar Amir Licht has noted, is the "deeper, more general norm of obeying the law." When you step off a curb because you have the "Walk" light or drive through a green light expecting not to be hit by another driver, it is not the law per se that protects you but other drivers' willingness to follow the law. Laws explain what we ought to do; norms explain what we actually do. In that gap dwells a key to understanding why traffic behaves the way it does in different places.

## Danger: Corruption Ahead—the Secret Indicator of Crazy Traffic

In 1951, some 852 people were killed on the roads in China. In the United States in that year, 35,309 people were killed in traffic. In 1999, traffic fatalities in China had risen to nearly 84,000. The U.S. figure, meanwhile, was 41,508. The population of both countries had almost doubled in that time. Why did fatalities rise so much higher in China than in the United States?

The answer lies in the number of vehicles in each country. In 1951, there were about 60,000 motor vehicles in China, while in the United States, there were roughly 49 million. By 1999, when China had 50 million vehicles, the United States had over 200 million—four times as many. And yet *twice* as many people were killed on Chinese roads than American ones. How could the country with so many fewer vehicles have so many more deaths?

This strange equation has become known as Smeed's law, after a 1949 paper, humbly titled "Some Statistical Aspects of Road Safety Research," by the British statistician and road-safety expert R. J. Smeed. What Smeed's law showed was that, across a number of countries, ranging from the United States to New Zealand, the number of people killed on the roads tended to rise as the number of cars on the road began to rise—*up to a point*—and then, gradually if not totally uniformly, the fatality rates began to drop, as, generally, did the absolute numbers of fatalities.

Smeed suspected that two things were going on: One, as the number of deaths grows higher, so too do people begin to clamor for something to be done about it (as began to happen in the United States in the 1960s, when fatalities were topping 50,000 people a year). Second, Smeed proposed that a sort of national learning curve was at work. The more cars on the road, the more people are "growing up" and learning how to sort out the problems of traffic—with better highway engineering, stronger laws, safer vehicles, and a more developed traffic culture itself (and perhaps more congestion, which tends to lower traffic fatalities).

In China, one sees things that make the hair stand on end—like bicycles traveling on restricted highways, scooter drivers carrying several children without helmets, and drivers stopping on the highway to urinate—but presumably, a number of years down the road, these things will largely be only memories. The dynamics of Smeed's law may help explain a curious phenomenon noted by Rong Jiang, a Beijing Institute of Technology transportation engineer. Studies had suggested that the crash rate was actually higher on the high-speed, divided "luxury roads" of the new China, he said, than on the two-lane rural highways. This is exactly the opposite of what happens almost everywhere else. He suspected that drivers were not adequately trained for the new high-speed roads. "The drivers were used to low speed on the open road," he explained. "But if they travel along the freeway, they keep the same habits. If their vehicle has a malfunction they will just park on the shoulder, without any alerting equipment. There are many such collisions."

Smeed's law, if history serves as a guide, is why one cannot simply look at the current horrific numbers of road deaths in countries like China and India and the relatively low levels of car ownership and assume that fatalities will continue to rise proportionally as more people get more cars. It may seem hard to imagine, but there is already progress of sorts even in China's massive death toll: While more people are dying on China's roads than ever before, the Chinese fatality *rate*, as measured in number of deaths per thousand registered vehicles, has actually been dropping.

Smeed's law is complicated, however, by a few factors that make China and India different from the countries Smeed considered. The first is that most people dying in traffic in the developing world are dying not in cars but outside cars. More than half of the people killed on the road in the United States are drivers or passengers, whereas in a country such as Kenya the figure can be as low as 10 percent. In Delhi, the occupants of cars represent only 5 percent of fatalities, while pedestrians, cyclists, and motorcyclists make up a staggering 80 percent. In places like the United States and England, motorization was an evolutionary process. However novel they may have been, the first automobiles, the "horseless carriages," could still be understood in terms of what had come before. The speeds were slow, the number of cars few.

China and India, by contrast, are seeing a vast flood of modern cars surging onto what are, in some cases, premodern roads. The Lexus and the rickshaw are thrust onto the same thoroughfare. Another consequence of this dizzyingly fast motorization is that people of all ages who have never before driven in their lives are being put on the road at once. In 2004 it was estimated that nearly one out of every seven drivers on the road in Beijing was a novice. The rapidly evolving Chinese insurance industry was dealing with customers who were reporting as many as *thirty claims* in a multiyear period. Some insurers reported accident risk for certain classes of individuals at nearly 100 percent—virtually moving them from the category of "accident risk" to the paradoxical "accident certainty."

In the harsh language of economics, the massive traffic fatalities and unsafe road systems in developing countries might be seen as temporarily necessary "negative externalities." In other words, like pollution or poor working conditions, they are just another price those countries have to pay in order to "catch up." Indeed, one might read the frenetic traffic behavior as somehow expressing the soul of noisy, dirty, clamoring entrepreneurial and industrial cities. Calm and safe traffic, the argument might go, is fine for those who can afford it (e.g., Switzerland). Let us get the cars and motorcycles on the road first, let us get people commuting to jobs, and then we can worry about safety. This is why, even as the rates for things like diseases begin to drop as countries get wealthier, traffic fatalities—a "disease of development"—rise until that point, as formulated by Smeed's law, where they begin to drop. When East Germany was reunited with West Germany in 1990, the traffic fatality rate in the former Communist country quadrupled: More people bought cars, drove them more often, and at higher speeds (the East German speed limit of 100 kilometers per hour on autobahns was raised to West Germany's 130). While the fatality rate is still higher in the eastern half of the country, it began to drop again after 1991.

It is eerily striking how closely fatalities can be tracked in economic terms. A country's motorization rate is linked in a somewhat linear fashion to its gross domestic product: the more money, the more cars. Researchers use the rough benchmark of a $5,000 per capita GDP as the point at which car ownership rates begin to accelerate. As work by the World Bank economists Elizabeth Kopits and Maureen Cropper shows, countries with very low GDPs have low numbers of fatalities per population (there are simply not that many cars, even if the rates per *vehicle* might be high). As the GDP grows, there is a sharp upward curve in fatalities. The rate per vehicle begins to drop with minor increments in the GDP—for example, when per capita GDP climbs from $1,200 to $4,400, the fatality risk per vehicle drops by a factor of three. After studying the data from eighty-eight countries from 1963 to 1999, Kopits and Cropper concluded that the fatalities per person begin to drop only when a country's GDP hits $8,600 (in 1985 dollars); they eventually hit levels lower than those of countries with much smaller per capita GDPs. Projecting these numbers outward, Kopits and Cropper concluded that India, for example, where the GDP (using that same 1985 standard) was $2,900 in 2000, will not see its road death rate decline until 2042.

Must this be so? Must history be a guide, must it preordain the future? Must that many people die on the roads? When one compares the rankings of per capita GDP and the traffic fatality rate, they gloomily do seem to correspond. Norway, for example, ranked as having the world's third-highest GDP in 2005 by the International Monetary Fund, was among the world's top three countries that year in terms of traffic safety. Uganda, on the other hand, ranked 154th in the world in terms of GDP, has one of the world's highest traffic fatality rates, some 160 deaths per 10,000 vehicles (a rate that will presumably rise, up to a point, as its GDP rises). The reasons are not hard to understand: lower-quality roads and infrastructure, fewer hospitals and doctors, less-safe vehicles. In Nigeria, where the buses are nicknamed "moving morgues" and "flying coffins," the situation was summed up by one commuter: "Many of us know most of the buses are death traps but since we can't afford the expensive taxi fares, we have no choice but to use the buses."

Sometimes, however, countries that have very similar levels of GDP can have varying levels of traffic risk. One of the most striking cases of this involves Belgium and the Netherlands. They are virtually identical in per capita GDP, but in Belgium, the traffic fatality rate is more than twice as high as the Netherlands (even though life expectancy itself is slightly higher in Belgium). These two countries share a border, even a language—why should Belgium be so much more dangerous? Perhaps it has to do with population density. Studies have shown that the less densely populated a place, the higher the risk of traffic fatalities. And, it turns out, the Netherlands crams more people into less space, while the Belgians have more room to roam. The *nonfatal* crash rate, on the other hand, is usually higher in more densely populated places: There are more people to run into. In Belgium that rate, too, was nearly twice as high as in the Netherlands. What about motorization levels? The higher the level of motorization, the more the fatality risk tends to drop—but where the Netherlands had only 422 vehicles per thousand people in 1999, Belgium had 522. Traffic laws might seem a good explanation, except that both Belgium and the Netherlands have similar speed limits and blood alcohol concentration restrictions.

So why is Belgium a more dangerous place to drive? An answer of sorts may be found in another kind of index, one that more or less aligns with the GDP but often diverges in interesting ways: corruption. According to indices compiled by the anticorruption watchdog Transparency International, the Netherlands was ranked number nine in 2006, while Belgium appeared much farther down the list, at number twenty.

What does this have to do with traffic? Most people tend to think of corruption by the standard definition of the use of public office for private gain. Working out from that, however, we might consider corruption as being indicative of a larger lack of faith in the law. In his book *Why People Obey the Law*, the legal scholar Tom Tyler posits that people generally comply with laws less because they are deterred by the penalties of not doing so, or because they have calculated it's in their best self-interest, and more because they think it's the right thing to do. Yet they are more likely to think it is the right thing, argues Tyler, if they perceive that the legal authorities are legitimate. People who go to traffic court, Tyler found, are less concerned with the outcome—even when it is a costly ticket or fine—than with the fairness of the process. When there is less respect for the law, there is a lesser cost (or greater gain) for not following it. Less effective governance means that laws are less effective, which means that people are less likely to follow them.

In Belgium, it may be no coincidence that the country both ranks comparatively poorer on the corruption index and has a public that seems less interested in following traffic laws. Lode Vereeck, a Belgian economist at Hasselt University, has noted that in survey after survey of people's attitudes toward traffic regulations, Belgians seem resistant; they're more hostile than their neighbors to things like seat-belt laws, lower speed limits, and drunk-driving laws (and also more likely to drink before driving, if surveys can be believed). At the same time, according to Vereeck, the number of violations recorded by Belgian police dropped from 1993 to 1999—even though Belgium's roads clearly did not get safer. A driver was also less likely to get a traffic fine in Belgium than in its neighbor to the north: The Netherlands, with roughly 50 percent more people (and a lower motorization rate), issued nearly *eight* times the number of tickets in 2000.

While the laws in Belgium and the Netherlands may be similar, it seems there is a different attitude to following and enforcing those laws. Some researchers have argued that Belgium's system of yielding at unmarked intersections, known as *priorité de droite*, or "yield to the right," is the real reason for Belgium's extraordinary fatality rates. But most of these intersections are in urban areas, which typically see nonfatal crashes. In any case, the *priorité de droite* itself simply speaks to the larger issue: a resistance against regulation (in the form of stop signs or traffic lights) and a lack of interest in following the existing rules. As the examples in Chapter 7 pointed out, traffic can be made to move well and safely with no signs at all, *if* strong enough social norms are in place.

The nations that rank as the least corrupt—such countries as Finland, Norway, New Zealand, Sweden, and Singapore—are also the safest places in the world to drive. Sweden, of course, practically oozes safety, from its flagship Volvos to its "Vision Zero" policy, which seeks the eventual elimination of all traffic fatalities (it passed this even after it already had the world's lowest traffic fatality rate). The British traffic psychologist Ian Walker tells the story about how a group of researchers equipped a car with cameras and got a group of Swedish military conscripts to drive around for a while. The purpose was to see how having passengers would affect a recruit's driving. "They thought, Put four young guys in a car and give them free rein—they'll go nuts," Walker says. "Actually, the guys were saying, 'Careful, slow down.' "

In Finland, which has one of the lowest crash rates in the world, drivers are given fines based on a complicated calculus primarily involving their after-tax income. The law, intended to counter the regressive nature of speeding tickets (they take up a larger part of a poor person's income than a rich person's), has led to some very high-profile speeding tickets, such as Internet entrepreneur Jaakko Rytsölä's $71,400 tab for going 43 miles per hour in a 25-mile-per-hour zone. There has been some grumbling, especially among the wealthy, but the law remains popular; in 2001, the legislature overwhelmingly rejected a cap on fines. Women seem to find the fine more fair than men (this is interesting for several reasons, which I will return to shortly). But what's remarkable about sliding-scale speeding tickets is not necessarily whether they get people to slow down. It's that in Finland legislators have the confidence to pass laws that unilaterally impose high costs on breaking the law, that traffic police will actually issue the fines rather than accept what in theory could be a huge bribe, and that the public, by and large, feels all this is fair.

It's true that Norway and Sweden are among the wealthiest countries in the world and, having taken care of the basic needs of their societies (e.g., getting everyone food and running water, establishing political stability), they can move on to things like safer roads. But as the case of Belgium shows, GDP itself is not necessarily a predictor for the safeness of the roads. France, traditionally one of the more dangerous countries in Europe to drive in, lowered the number of people killed on its roads from 7,721 in 2001 to just under 5,000 in 2005. It is not as if the French GDP soared during this period; in fact, it was rather stagnant.

What France did was buy Breathalyzers and automated speed cameras by the thousands and overhaul its points system for violations. It brought accountability to a system that had been plagued by chronic traffic ticket "fixing." (One study found that a third of the male employees at a national utility company had had tickets fixed and that those who had were also more likely to have been in a crash). Ticket fixing is so endemic in France that starting in 1958, incoming presidents declared amnesty on a range of traffic violations, from minor to fairly serious—a rather self-defeating measure that itself has been blamed for hundreds of traffic fatalities. The traffic-ticket holiday was curtailed by Jacques Chirac and seems to be on its way out altogether. France, in at least one way, is becoming less corrupt (indeed, it did drop a few places on the index during those same years).

The lesson is that wealth seems to affect traffic fatalities but corruption may affect them even more. It could just be that lifting GDP lowers corruption *and* traffic fatalities. But a study by a group of U.S. economists concluded that the statistical relationship between corruption (as measured by the International Country Risk Guide) and traffic fatalities was actually stronger than the link between income and traffic fatalities. What they were saying, essentially, is that money is not enough. Even when countries become wealthy enough to start shifting attention to things like traffic safety, one still needs credible laws and credible people to enforce the laws. New Zealand, which is one of the five least corrupt countries in the world, is below countries like Austria and Spain in GDP but has safer roads, as measured by fatalities per 10,000 vehicles. Russia, on the other hand, is ranked as more corrupt than other countries at similar development levels, and its roads reflect that fact: Moscow is filled with notoriously corrupt traffic cops and cars blazing through traffic jams with ersatz blue sirens. Russia itself reportedly accounts for two-thirds of Europe's road fatalities.

The complex question of why poorer countries seem to suffer from more corruption and whether that corruption is a bad thing in itself has long been debated among economists and social scientists. Some argue that "efficient corruption" is a useful and necessary cost of rapid economic development, that bribes and rule skirting can be used to outwit creaky centralized bureaucracies. Others counter that corrupt politicians are not necessarily faster politicians, in terms of hustling development projects through, and may actually slow things down to get even more money. Corruption is a brake on development, they say. Countries like China, which are booming *and* have relatively widespread corruption, could be developing even faster if corruption were tamed, they contend. The first group argues that a system in which firms have to pay kickbacks to corrupt government officials means that the firm with the most "efficient" bid will also be able to afford the highest bribe, while the second group maintains that this system rewards inefficient firms. Daniel Kaufmann, an economist with the World Bank and a leading critic of corruption, uses the example of a firm that was disqualified because its bid was beneath the acceptable "minimum."

With traffic, it's arguably corruption that gets in the way of economic growth, not the other way around. While no economist would view a traffic jam as an efficient use of resources, traffic congestion can symbolize the economic vitality of a country (simply because miles driven usually increase in stronger economic times). "Bad" traffic can be seen as just an outcome of that success. But corruption itself can cause traffic problems, the sort that represent a drain on economic growth, not an outcome. Take, for example, the myriad roadblocks that are a daily fact of life in many developing countries. The process typically has little to do with vehicle inspection or safety and a lot to do with police or soldiers trying to extract something "for the boys." Corruption does not speed a driver's way through some bureaucratic tangle; rather, the tangle is formed *because* of corruption.

In some places, these systems are so entrenched that they can take on the logic of an economic system, a kind of "corruption pricing" instead of "congestion pricing." A study of the bribes that Indonesian truckers had to pay at military checkpoints showed that the closer the truckers got to their destination, the higher the bribe. (The officials also charged more for newer trucks and trucks carrying valuable cargo.) When the number of checkpoints dropped after the military scaled back its forces, the average bribe per checkpoint increased, leaving the researchers to conclude that fewer traffic officials may be better (although their absence may invite criminals to take their place).

As the economist Tim Harford observed after a visit to Cameroon (one of the world's poorest and most corruption-plagued countries), corruption in traffic is tremendously unfair and inefficient. Protracted "inspections" and bartering over small amounts slows the flow of goods and people. The money goes into the pockets of underpaid officials, not to fixing roads or making them safer. Trip times and costs become wildly unpredictable. Robert Guest, Africa correspondent for the *Economist*, wrote of once accompanying the driver of a Guinness beer truck on a three-hundred-mile journey in Cameroon. The trip, which might have taken twenty hours elsewhere, took four days. The reason was in part the crumbling roads, but also the forty-seven checkpoints at which they were forced to stop for dubious safety inspections and petty bribes. Drivers suffer not only the hardship of bad roads but the privilege of paying to use them. The bribes paid and the ensuing delays get passed on to beer consumers in the form of higher prices. Guest's suggestion: "Lift those roadblocks and put the police to work repairing potholes."

Corruption begins at street level. The traffic cop is its foot soldier, the agent of bad traffic. He pulls over motorists for phantom violations, reducing not only traffic flow but the incentive for any driver to follow the law. Some argue that corrupt cops *increase* the incentive to follow the law because these cops are that much more on the lookout for excuses to issue a fine, but this presumes they are actually pulling people over for legitimate reasons. As one of the average person's primary interfaces with the legal system, the traffic cop becomes a symbol of the legitimacy of the regime. And what about the traffic he's directing? Corruption casts its shadow there as well.

To return to the frenetic streets of Delhi: My impression was that many drivers did not seem to be particularly qualified for a license. There's a good reason for this. A study conducted by a team of researchers for the U.S. National Bureau of Economic Research looked at the process of getting a driver's license in Delhi. The group tracked 822 individuals in three groups: a "bonus" group, whose members would get a financial reward if they could obtain a license in the fastest time legally possible; a "lesson" group, whose members were given free driving lessons before they attempted to get the license; and a "comparison" group, which was given no special instructions.

The researchers found that those who wanted the license soonest—that is, the members of the bonus group—got it more often, and faster, than people in the other groups. The reason, it turned out, was that like many drivers in Delhi, they used an "agent" to speed the process. But when the researchers later gave all the survey participants a driving test, 69 percent of the bonus group failed, compared to just 11 percent of the drivers who had taken lessons. But learning to drive properly clearly did not pay off: The people who had the best driving skills were 29 percent less likely to get a license than the people with the worst driving skills. Corruption did indeed grease the wheels, but at the expense of the quality of those behind the wheel. "Corruption," the authors wrote, "appears to substitute for actual driving skill."

This study provides a hint about how the norms discussed in the previous section evolve and flourish. The scores of new drivers who land on Delhi streets each month learn the norms of a system made up of the collective experience of all the previous drivers who bribed their way through the Regional Transport Office. No small wonder this traffic system isn't marked by scrupulous attention to formal rules. In the writer Pavan Varma's description of what motivates corruption in India, it is not hard to see a metaphor for the country's traffic behavior: "In a cut-throat world, the immediate task is to get on with the job, to reach a desired goal, to finesse an obstacle. The premium is on pragmatism and agility, the capacity to seize an opportunity when it comes, and to profit when possible. What matters is not fixity of principle but clarity of purpose."

What *is* surprising is how strong these corruption norms can be, even in a different context. In one study, the economists Ray Fisman and Edward Miguel looked at the number of parking tickets issued to diplomats in New York City between 1997 and 2002. During this time, diplomats could be given parking tickets, but there was no enforceable punishment for not paying them. Thus empowered, diplomats racked up some 150,000 tickets.

The tickets were not acquired randomly. The diplomats who got the most tickets tended to be from the countries deemed to be more corrupt by the Transparency International index (those countries also got more "egregious" tickets, such as for blocking fire hydrants). The countries whose diplomats received *no* tickets included Sweden, Norway, Japan, and Denmark—judged among the least corrupt countries. These countries were scrupulous in following the law, even when it was clearly not necessary. India, in case you were wondering, was roughly halfway down the list, just as it is on the corruption index. Lest you think I am singling out India, I might add that the United States embassy in London, as of 2007, owed the highest amount (ahead of even corruption-plagued Nigeria) of unpaid traffic congestion-pricing fees to the city of London. The United States, which claims that its diplomats are exempt from the congestion-pricing "tax," is not one of the ten least corrupt countries (it was ranked twentieth in 2007). (The least-corrupt country, Finland, whose diplomats are also exempt from taxes, pays the charge.)

In traffic, laws are only as good as the norms regarding them. This may be why, as I discussed in Chapter 7, the engineer Hans Monderman could strip the signs from a roundabout and Dutch drivers would still act in a responsible, safe manner; and why, in other countries, a roundabout can be filled with signs and drivers will still act in an irresponsible, dangerous manner. Which brings us back to two questions: Are developing countries fated to have a disproportionate share of traffic fatalities? And how many of these fatalities come from lack of money, how many from laws or norms weakened by corruption? The passengers crowded into unsafely overloaded buses may be there because it's the only transportation they can afford or because there is no one to stop the bus from being overloaded—perhaps because the government thinks it can't afford to *not* let people ride the overcrowded bus.

The vexing, intertwined nature of this dilemma is reflected in a piece of Hindi slang I learned while in Delhi: *jugad.* The word has a shifting palette of meanings, mostly arrayed around the central idea of "creative improvisation." It can refer, on the one hand, to the jury-rigged vehicles one finds in India, especially in rural areas. Lacking money for a car, say, a farmer will craft a functioning vehicle out of an old motorcycle, a car axle, and a diesel engine. That this *jugad* vehicle might not be safe, at least when it's sharing the road with newer cars, is one of the clear kinds of traffic risks that come with lack of money.

But *jugad* is also used as a kind of surrogate for "bribe"; here it refers to doing whatever needs to be done to get something accomplished. The case of the Delhi drivers who acquired licenses quickly is a form of *jugad* in practice. Would-be drivers know that corrupt bureaucrats respond more to money than driving skills. Is this kind of corruption, which has a ripple effect that translates into the myriad traffic violations that occur in Delhi every day—and studies suggest that the more traffic laws are violated, the more casualties there will be—purely an effect of lack of resources? Or is it, as many would argue, precisely the sort of thing that holds up the development of a country? If GDP and traffic fatalities are somewhat related, and GDP and corruption are somewhat related, and traffic fatalities and corruption seem to be the most clearly related, then fighting corruption may be the best way to lower traffic fatalities and raise GDP.

There are, after all, creative ways of combating corruption that do not require huge amounts of money. In Mexico City, Alfredo Hernández García, the city's traffic czar, described a novel plan to fight corruption and improve traffic safety. In 2007, he noted, the last of the city's male traffic officers had been phased out, replaced entirely by women (known as *cisnes*, or "swans"). Why? "Because women are less likely to be corrupted," he explained in his office in the Secretaría de Seguridad Pública. Previously, Mexico City traffic cops were famous for soliciting *refrescos* or "soft drinks"—that is, bribes in lieu of a ticket. According to Hernández García, the *cisnes* have increased the number of tickets written on the order of 300 percent. They have been given handheld units to issue tickets and ensure payment—drivers can use credit cards—and take photographs. "People do not accept they are breaking the law," he said. "We have to provide evidence."

The theory of women as less corruptible may be based on more than the hunches of a few higher-ups in the police department. A study by a group of U.S. economists found that women were less likely to engage in hypothetical corruption, that female managers in one country they studied were less likely to engage in actual corruption, and that the countries that rank as least corrupt on the global indices tend to have more women in government. Indeed, they may be onto something: Finland, ranked as the least-corrupt country in the world, set the record in 2007 for having the government with the most women in cabinet-level positions. As you will recall, they do not mess about with their traffic tickets.

## Chapter Nine

# Why You Shouldn't Drive with a Beer-Drinking Divorced Doctor Named Fred on Super Bowl Sunday in a Pickup Truck in Rural Montana: What's Risky on the Road and Why

### Semiconscious Fear: How We Misunderstand the Risks of the Road

In a basement laboratory in the looming red-brick Henry Ford Hospital in Detroit, Michigan, a team of researchers has, for the past few years, been looking at what happens to our brains as we drive. The device that measures the faint magnetic fields the brain emits is too massive to fit inside of a car, so research subjects are instead studied in the hospital's Neuromagnetism Laboratory, where they watch film clips of a car navigating through traffic. As I lay back on the cozy bed inside the magnetically shielded lab to get a feel for the procedure, Richard Young, a scientist with General Motors who leads the research team, told me, "Our biggest problem is people falling asleep in the bed."

To keep people awake as they play passenger to the filmed driving, subjects are given a simple "event-detection task." When a red light near the screen goes on, the subject, attached to a neuromagnetometer, presses a simulated brake pedal. This simple habit of braking in response to a red light (i.e, brake lights), something drivers do an estimated fifty thousand times a year, triggers a burst of activity in the brain. The visual cortex lights up about 80 to 110 milliseconds after the red signal comes on. This indicates that you have seen the signal. The left prefrontal lobe, an area of the brain linked to decision making, begins to buzz with activity. This is the microinstant during which you're deciding what to do with the information you have acquired—here, the rather simple response of simply pressing the brake. It comes about 300 milliseconds before you actually do it. About 180 milliseconds before braking begins, the motor cortex sees action—your foot is about to be told to move. About 80 milliseconds after you have pressed the brake, the visual cortex is again activated. You're registering that the red signal has been turned off.

The scientists are probing the neural pathways involved in what they call the "mind on the drive," in part to learn what cell phone conversations and other activities do to our brains as we drive. But sometimes, as they watch these real-time movies of people's brains in traffic, there are strange and unanticipated plot twists.

Once, while watching the real-time fMRI (functional magnetic resonance imaging) readings of a subject, Young noticed a burst of brain activity, not during the braking event but during "normal" driving. "There was a spike. There were brain areas lighting up in the emotional cortex, the amygdala, the limbic cortex, the lower brain," Young recalled. This hinted at more complex responses than what usually showed up in the fairly well-conditioned responses to braking or keeping the vehicle on the road at a certain speed. What was going on? Young compared the activity to the actual video of the drive. At the moment his brain went on the boil, the driver was passing a semitrailer. After the trial, Young asked the subject if he had noticed "anything unusual during the last run." He had. According to Young, "The person said, 'Oh yes, I was passing that eighteen-wheeler and every time I pass one of those things I get real nervous.' "

That small peek into the brain of the driver revealed a simple, if underappreciated truth about driving: When we are in traffic, we all become on-the-fly risk analysts. We are endlessly having to make snap decisions in fragments of moments, about whether it is safe to turn in front of an oncoming car, about the right speed to travel on a curve, about how soon we should apply the brakes when we see a cluster of brake lights in the distance. We make these decisions not with some kind of mathematical probability in the back of our heads—I have a 97.5 percent chance of passing this car successfully—but with a complicated set of human tools. These could be cobbled from the most primeval instincts lurking in the ancient brain, the experience from a lifetime of driving, or something we heard yesterday on the television news.

On the one hand, it was perfectly natural, normal, and wise for the driver in Detroit to show fear in the face of an eighteen-wheeler. Large trucks, from the point of view of a car, are dangerous. Because of the staggering differences in mass—trucks weigh twenty to thirty times more than a car—the simple physics of a collision are horrifically skewed against the car. When trucks and cars collide, nearly nine of ten times it's the truck driver who walks away alive.

As the driver's brain activity would seem to indicate, we know this on some instinctual level, as if our discomfort in driving next to a looming truck on a highway is some modern version of the moment our prehistoric ancestor felt the hairs on the back of his neck raise when confronted with a large predator. Indeed, the amygdala, one of the areas that lit up in the Detroit driver, is thought to be linked with fear. It can be activated even before the cognitive regions kick in—neuroscientists have described the amygdala as a kind of alarm that triggers our attention to things we should probably fear. And we all likely have proof of the dangerous nature of trucks. We have seen cars crumpled on the roadside. We've heard news stories of truck drivers, wired on stimulants, forced to drive the deregulated trucking industry's increasingly long shifts. We can easily recall being tailgated or cut off by some crazy trucker.

Just one thing complicates this image of trucks as the biggest hazard on the road today: In most cases, when cars and trucks collide, the car bears the greater share of what are called "contributory factors." This was the surprising conclusion that Daniel Blower, a researcher at the University of Michigan Transport Research Institute, came to after sifting through two years' worth of federal crash data.

It was a controversial finding. Blower, to begin with, had to determine that it did not simply stem from "survivor bias": "The truck driver is the only one that survives these eighty-five percent of the time," he explained. "He's the one who gets to tell the story. That's what's reflected in the police report." So he dug deeper into the records, analyzing the relative position and motion of the vehicles before a crash. Instead of relying on drivers' accounts, he looked at "unmistakable" physical evidence. "In certain crash types like head-ons, the vehicle that crosses the center much more likely contributed to the crash than the vehicle that didn't cross the center line," he said. "Similarly, in rear-end crashes, the striking vehicle in the crash is much more likely to have contributed to the crash in a major way than the vehicle that was struck." After examining more than five thousand fatal truck-car crashes, Blower found that in 70 percent of cases, the driver of the car had the *sole* contributing responsibility in the crash.

This hardly means trucks are not dangerous. But the reason trucks are dangerous seems to have more to do with the actions of car drivers combined with the physical characteristics of trucks (in head-on collisions, for example, they are obviously less able to get out of the way) and less to do with the actions of truck drivers. "The caricature that we have that the highways are thronged with fatigued, drug-addled truck drivers is, I think, just wrong," Blower said. Certainly there are aggressive truck drivers and truckers jacked up on methamphetamine, but the more pressing problem, the evidence tells us, seems to be that car drivers do not fully understand the risk of heavy trucks as they drive in their presence. This is not something we are necessarily taught when we learn to drive. "In a light vehicle you are correct to be afraid of them, but it's not because the drivers are disproportionately aggressive or bad drivers," Blower said. "It's because of physics, truck design, the different performance characteristics. You can make a mistake around a Geo Metro and live to tell about it. You make that same mistake around a truck and you could easily be dead."

What all this seems to suggest is that car drivers have less to fear from trucks than from what they themselves do around trucks. I had a glimpse of this a few years back when I rode in an eighteen-wheel tractor-trailer for the first time, watching in horror as cars darted in front of the truck with dangerous proximity, sometimes disappearing from sight beneath the truck's long, high hood. So why does it seem that virtually everyone, like my Latin-teacher friend in the Prologue, has some horror story about crazy truckers?

One possible answer goes back to the spike in brain activity of the

Detroit driver. He was afraid, probably before he even knew why. The size of trucks makes most of us nervous—and rightfully so. When we have a close brush with a truck or we see the horrific results of a crash between a car and a truck, it undoubtedly leaves a greater impression on our consciousness, which can skew our view of the world. "Being tailgated by a big truck is worth getting tailgated by fifty Geo Metros," as Blower put it. "It stays with you, and you generalize with that." (Studies have suggested that people think there are more trucks on the road than is actually the case.)

Here's the conundrum: If, on both an instinctual level and a more intellectual level, the drivers of cars fear trucks, why do car drivers, in so many cases, act so dangerously around them? The answer, as we are about to see, is that on the road we make imperfect guesses as to exactly what is risky and why, and we act on those biases in ways we may not even be aware of.

## Should I Stay or Should I Go?
## Why Risk on the Road Is So Complicated

Psychologists have suggested that we generally think about risk in two different ways. One way, called "risk as analysis," involves reason, logic, and careful consideration about the consequences of choices. This is what we do when we tell ourselves, on the way to the airport with a nervous stomach, "Statistically, flying is much safer than driving."

The second way has been called "risk as feelings." This is why you have the nervous stomach in the first place. Perhaps it's the act of leaving the ground: Flying just *seems* more dangerous than driving, even though you keep telling yourself it isn't. Studies have suggested that we tend to lean more on "risk as feelings" when we have less time to make a decision, which seems like a survival instinct. It was smart of the Detroit driver to feel risk from the truck next to him, but the instinctual fear response doesn't always help us. In collisions between cars and deer, for example, the greatest risk to the driver comes in trying to avoid hitting the animal. No one with a conscience wants to hit a deer, but we may also be fooled into thinking that the deer itself presents the greatest hazard. Hence the traffic signs that say DON'T VEER WHEN YOU SEE A DEER.

One good reason why we rely on our feelings in thinking about risk is that "risk as analysis" is an incredibly complex and daunting process, more familiar to mathematicians and actuaries than the average driver. Even when we're given actual probabilities of risk on the road, often the picture just gets muddier. Take the simple question of whether driving is safe or dangerous. Consider two sets of statistics: For every 100 million miles that are driven in vehicles in the United States, there are 1.3 deaths. One hundred million miles is a massive distance, the rough equivalent of crisscrossing the country more than thirty thousand times. Now consider another number: If you drive an average of 15,500 miles per year, as many Americans do, there is a roughly 1 in 100 chance you'll die in a fatal car crash over a lifetime of 50 years of driving.

To most people, the first statistic sounds a whole lot better than the second. Each trip taken is incredibly safe. On an average drive to work or the mall, you'd have a 1 in 100 million chance of dying in a car crash. Over a lifetime of trips, however, it doesn't sound as good: 1 in 100. How do you know if this one trip is going to be *the* trip? Psychologists, as you may suspect, have found that we are more sensitive to the latter sorts of statistics. When subjects in one study were given odds, similar to the aforementioned ones, of dying in a car crash on a "per trip" versus a "per lifetime" basis, more people said they were in favor of seat-belt laws when given the lifetime probability.

This is why, it has been argued, it has long been difficult to convince people to drive in a safer manner. Each safe trip we take reinforces the image of a safe trip. It sometimes hardly seems worth the bother to wear a seat belt for a short trip to a local store, given that the odds are so low. But events that the odds say will almost certainly never happen have a strange way of happening sometimes (risk scholars call these moments "black swans"). Or, perhaps more accurately, when they do happen we are utterly unprepared for them—suddenly, there's a train at the always empty railroad crossing.

The risk of driving can be framed in several ways. One way is that most people get through a lifetime without a fatal car crash. Another way, as described by one study, is that "traffic fatalities are by far the most important contributor to the danger of leaving home." If you considered only the first line of thinking, you might drive without much of a sense of risk. If you listened to only the second, you might never again get in a car. There is a built-in dilemma to how societies think about the risk of driving; driving *is* relatively safe, considering how much it is done, but it could be much safer. How much safer? If the number of deaths on the road were held to the acceptable-risk standards that the U.S. Occupational Safety and Health Administration maintains for service-industry fatalities, it has been estimated, there would be just under four thousand deaths a year; instead, the number is eleven times that. Does telling people it is dangerous make it safer?

One often hears, on television or the radio, such slogans as "Every fifteen minutes, a driver is killed in an alcohol-related crash" or "Every thirteen minutes, someone dies in a fatal car crash." This is meant, presumably, to suggest not just the magnitude of the problem but the idea that a fatal crash can happen to anyone, anywhere. And it can. Yet even when these slogans leave out the words "on average," as they often do, we still do not take it to mean that someone is actually dying, like clockwork, every fifteen minutes.

These kinds of averages obscure the startling extent to which risk on the road is not average. Take the late-night hours on weekends. How dangerous are they? In an average year, more people were killed in the United States on Saturday and Sunday from midnight to three a.m. than all those who were killed from midnight to three a.m. the rest of the week. In other words, just two nights accounted for a majority of the week's deaths in that time period. On Sunday mornings from twelve a.m. to three a.m., there was not one driver dying every thirteen minutes but one driver dying every seven minutes. By contrast, on Wednesday mornings from three a.m. to six a.m., a driver was killed every thirty-two minutes.

Time of day has a huge influence on what kinds of crashes occur. The average driver faces the highest risk of a crash during the morning and evening rush hours, simply because the volume of traffic is highest. But fatal crashes occur much less often during rush hours; one study found that 8 of every 1,000 crashes that happened outside the peak hours were fatal, while during the rush hour the number dropped to 3 out of every 1,000. During the weekdays, one theory goes, a kind of "commuters' code" is in effect. The roads are filled with people going to work, driving in heavy congestion (one of the best road-safety measures, with respect to fatalities), by and large sober. The morning rush hour in the United States is twice as safe as the evening rush hour, in terms of fatal and nonfatal crashes. In the afternoon, the roads get more crowded with drivers out shopping, picking up the kids or the dry cleaning. Drivers are also more likely to have had a drink or two. The "afternoon dip," or the circadian fatigue that typically sets in around two p.m., also raises the crash risk.

What's so striking about the massive numbers of fatalities on weekend mornings is the fact that so few people are on the roads, and so many—estimates are as high as 25 percent—have been drinking. Or think of the Fourth of July, one of the busiest travel days in the country and also, statistically, the most dangerous day to be on the road. It isn't simply that more people are out driving, in which case more fatalities would be expected—and thus the day would not necessarily be more dangerous in terms of crash rate. It has more to do with what people are doing on the Fourth: Studies have shown there are more alcohol-related crashes on the Fourth of July than on the same days the week before or after—and, as it happens, many more than during any other holiday.

What's the actual risk imposed by a drunk driver, and what should the penalty be to offset that risk? The economists Steven D. Levitt and Jack Porter have argued that legally drunk drivers between the hours of eight p.m. and five a.m. are thirteen times more likely than sober drivers to cause a fatal crash, and those with legally acceptable amounts of alcohol are seven times more likely. Of the 11,000 drunk-driving fatalities in the period they studied, the majority—8,000—were the drivers and the passengers, while 3,000 were other drivers (the vast majority of whom were sober). Levitt and Porter argue that the appropriate fine for drunk driving in the United States, tallying up the externalities that it causes, should be about $8,000.

Risk is not distributed randomly on the road. In traffic, the roulette wheel is loaded. Who you are, where you are, how old you are, how you are driving, when you are driving, and what you are driving all exert their forces on the spinning wheel. Some of these are as you might expect; some may surprise you.

Imagine, if you will, Fred, the pickup-driving divorced Montana doctor out for a spin after the Super Bowl who is mentioned in this chapter's title. Obviously, Fred is a fictional creation, and even if he did exist there'd be no way to judge the actual risk of driving with him. But each of the little things about Fred, and the way those things interact, play their own part in building a profile of Fred's risk on the road.

The most important risk factor, one that is subtly implicated in all the others, is speed. In a crash, the risk of dying rises with speed. This is common sense, and has been demonstrated in any number of studies. In a crash at 50 miles per hour, you're fifteen times more likely to die than in a crash at 25 miles per hour—not twice as likely, as you might innocently expect from the doubling of the speed. The relationships are not proportional but exponential: Risk begins to accelerate much faster than speed. A crash when you're driving 35 miles per hour causes a *third* more frontal damage than one where you're doing 30 miles per hour.

Somewhat more controversial is the relationship between speed and the potential for a crash. It is known that drivers who have more speeding violations tend to get into more crashes. But studies have also looked at the speeds of vehicles that crashed on a given road, compared them to the speeds of vehicles that did not crash, and tried to figure out how speed affects the likelihood that one will crash. (One problem is that it's extremely hard to tell how fast cars in crashes were actually going.) Some rough guidelines have been offered. An Australian study found that for a mean speed—not a speed limit—of 60 kilometers per hour (about 37 miles per hour), the risk of a crash doubled for every additional 5 kilometers per hour.

In 1964, one of the first and most famous studies of crash risk based on speed was published, giving rise to the so-called Solomon curve, after its author, David Solomon, a researcher with the U.S. Federal Highway Administration. Crash rates, Solomon found after examining crash records on various sections of rural highway, seemed to follow a U-shaped curve: They were lowest for drivers traveling at the median speed and sloped upward for those going more or less than the median speed. Most strikingly, Solomon reported that "low speed drivers are more likely to be involved in accidents than relatively high speed drivers."

Solomon's finding, despite being almost a half century old, has become a sort of mythic (and misunderstood) touchstone in the speed-limit debate, a hoary banner waved by those arguing in favor of higher speed limits. It's not the actual speed itself that's the safety problem, they insist, it's *speed variance*. If those slower drivers would just get up to speed, the roads would flow in smooth harmony. It's not speed that kills, it's variance. (This belief, studies have indicated, is most strongly held by young males—who are, after all, experts, given that they get in the most crashes.) And what causes the most variance? Speed limits that are too low!

Dear reader, much as I—as guilty as anyone of an occasional craving for speed—would like to believe this, the arguments against it are too compelling. For one, it assumes that the drivers who are going slow want to be driving slowly, and are not simply slowing for congested traffic, or entering a road from a turn, when they are suddenly hit by one of those drivers traveling the mean speed or higher. Solomon himself acknowledged (but downplayed) that these kinds of events might account for nearly half of the rear-end crashes at low speeds. Studies have found that a majority of rear-end crashes involved a stopped vehicle, which presumably had stopped for a good reason—and not to get in the way of the would-be speed maven behind him. Further, Gary Davis, an engineering professor at the University of Minnesota, proving yet again that statistics are one of the most dangerous things about traffic, has suggested there is a disconnect—what statisticians call an "ecological fallacy"—at work in speed-variance studies. Individual risk is conflated with the "aggregate" risk, even if in reality, he suggests, what holds for the whole group might not hold for individuals.

In pure traffic-engineering theory, a world that really exists only on computer screens and in the dreams of traffic engineers and bears little resemblance to how drivers actually behave, a highway of cars all flowing at the same speed is a good thing. The fewer cars you overtake, the lower your chance of hitting someone or being hit. But this requires a world without cars slowing to change lanes to enter the highway, because they are momentarily lost, or because they're hitting the tail end of a traffic jam. In any case, if faster cars being put at risk by slower cars were the mythical problem some have made it out to be, highway carnage would be dominated by cars trying to pass—but in fact, one study found that in 1996, a mere 5 percent of fatal crashes involved two vehicles traveling in the same direction. A much more common fatal crash is a driver moving at high speed leaving the road and hitting an object that isn't moving at all. That is a case where speed variance really does kill.

Let us move on to perhaps the oddest risk factor: Super Bowl Sunday.

In one study, researchers compared crash data with the start and end times of all prior Super Bowl broadcasts. They divided all the Super Bowl Sundays into three intervals (before, during, and after). They then compared Super Bowl Sundays to non–Super Bowl Sundays. They found that in the before-the-game period, there was no discernible change in fatalities. During the game, when presumably more people would be off the roads, the fatal crash rate was 11 percent less than on a normal Sunday. After the game, they reported a relative increase in fatalities of 41 percent. The relative risks were higher in the places whose teams had lost.

The primary reason for the increased postgame risk is one that I have already discussed: drinking. Nearly twenty times more beer is drunk in total on Super Bowl Sunday than on an average day. Fred's risk would obviously be influenced by how many beers he had downed (beer, at least in the United States, is what most drivers pulled over for DUIs have been drinking) and the other factors that determine blood alcohol concentration (BAC). Increases in crash risk, as a number of studies have shown, begin to kick in with as little as .02 percent BAC level, start to crest significantly at .05 percent, and spike sharply at .08 to .1 percent.

Determining crash risk based on a person's BAC depends, of course, on the person. A famous study in Grand Rapids, Michigan, in the 1960s (one that would help establish the legal BAC limits in many countries), which pulled over drivers at random, found that drivers who had a .01 to .04 percent BAC level actually had *fewer* crashes than drivers with a BAC of zero. This so-called Grand Rapids dip led to the controversial speculation that drivers who had had "just a few" were more aware of the risks of driving, or of getting pulled over, and so drove more safely; others argued that regular drinkers were more capable of "handling" a small intake.

The Grand Rapids dip has shown up in other studies, but it has been downplayed as another statistical fallacy—the "zero BAC" group in Michigan, for example, had more younger and older drivers, who are statistically less safe. Even critics of the study, however, noted that people who reported drinking with greater frequency had *safer* driving records than their teetotaler counterparts at every level of BAC, including zero. This does not mean that drinkers are better drivers per se, or that having a beer makes you a better driver. But the question of what makes a person a safe driver is more complicated than the mere absence of alcohol. As

Leonard Evans notes, the effects of alcohol on driver *performance* are well known, but the effects of alcohol on driver *behavior* are not empirically predictable. Here is where the tangled paths of the cautious driver who has had a few, carefully obeying the speed limit, and the distracted sober driver, blazing over the limit and talking on the phone, intersect. Neither may be driving as well as they think they are, and one's poorer reflexes may be mirrored by the other's slower time to notice a hazard. Only one is demonized, but they're both dangerous.

The second key risk is Fred himself. Not because he is Fred, for there is no evidence that people named Fred get in more crashes than people named Max or Jerry. It is the fact that Fred is male. Across every age group in the United States, men are more likely than women to be involved in fatal crashes—in fact, in the average year, more than twice as many men as women are likely to be killed in a car, even though there are more women than men in the country. The global ratio is even higher. Men do drive more, but after that difference is taken into account, their fatal crash rates are still higher.

According to estimates by researchers at Carnegie Mellon University, men die at the rate of 1.3 deaths per 100 million miles; for women the rate is .73. Men die at the rate of 14.51 deaths per 100 million trips, while for women it is 6.55. And crucially, men face .70 deaths per 100 million minutes, while for women the rate is .36. It may be true that men drive more, and drive for longer periods when they do drive, but this does not change the fact that for each minute they're on the road, each mile they drive, and each trip they take, they are more likely to be killed—and to kill others—than women.

It is tempting to use this information to make some point about whether men or women are "better drivers," but that's complicated by the fact that in the United States, women get into nonfatal crashes at a higher rate than men. This might be at least partially the result of men driving more on roads that are more prone to fatal crashes (e.g., rural high-speed two-lane roads). What *can* be argued is that men drive more aggressively than women. Men may or may not be better drivers than women, but they seem to die more often trying to prove that they are.

As a gender, men seem particularly troubled by two potent compounds: alcohol and testosterone. Men are twice as likely as women to be involved in an alcohol-related fatal crash. They're more likely to drink, to drink more, and to drive more after they drink. On the testosterone side, men are less likely to wear seat belts; and by just about every measure, they drive more aggressively. Men do things like ride motorcycles more often than women, an activity that is twenty-two times more likely to result in death than driving a car. Male motorcyclists, from Vietnam to Greece to the United States, are less likely than women to wear a helmet. As we all know, alcohol and testosterone mix in unpleasant ways, so motorcyclists who have been drinking are less likely to wear helmets than those who have not, just as male drivers who have been drinking are less likely to wear seat belts than those who are sober.

The fact that Fred is divorced puts him in a riskier pool. A French study that looked into the experiences of some thirteen thousand company employees over eight years found that a recent divorce or separation was linked to a fourfold increase in the risk of a crash that could be at least partially attributed to the driver. One could hypothesize many reasons: There's the emotional stress (as John Hiatt once sang in a breakup song, "Don't think about her while you're trying to drive"), and perhaps more drinking. Or there may be lifestyle changes, like driving more to visit the kids on weekends. Perhaps people who get divorced are simply the type of people who take more risks. Fred might take some comfort, however, from a New Zealand study that found that people who have never been married have even a higher crash risk than those who are divorced. (The study took into account age and gender differences.)

Fred may not have a life partner, but he should be glad if you chose to join him in his truck: Passengers seem to be a life-saving device. Studies from Spain to California have come to the conclusion that a driver has a lower chance of being in a fatal crash if there's a passenger. This holds particularly true for middle-aged drivers—especially when the passenger is a woman and the driver is a man. (Whether this stems from men looking out for women or women telling men to drive more safely is open to debate.)

The exception here is teenage drivers. Teens are less likely to be wearing seat belts and more likely to be drinking when driving when there *are* passengers in the car. Many studies have found that teen drivers are more likely to crash with passengers onboard, which is why, in many places, teens are restricted from carrying passengers of their own age during their first few years of driving.

Researchers are beginning to uncover fascinating things about how that risk plays out. A study that looked at the drivers exiting the parking lot at ten different high schools found that teenage drivers seemed to drive faster and follow cars at closer distances than other drivers did. Males drove more riskily than females. This is common knowledge, verified by insurance rates. But their risk-taking varied: Male drivers drove faster and followed closer when they had a male riding shotgun. When they had a female in the front seat, they actually behaved *less* riskily, and they were safer still when they drove by themselves (a pattern that also held for female drivers).

What seems to be a need to impress in the presence of males turns into a protective impulse when a female passenger (possibly a girlfriend) is in the car—or it could be that the female passenger serves as the voice of reason. This "girlfriend effect" seems to take root early and persist through later life. It need not be a romantic partner: The Israel Defense Forces, in an effort to reduce road deaths for soldiers on leave, trains female soldiers (dubbed "angels") to act as a "calming" influence on their male comrades.

Now consider *where* Fred is driving. What's the matter with Montana? In 2005, 205 people were killed on Montana's roads, roughly one-third the number that were killed in New Jersey. But Montana has just under one-tenth the population of New Jersey. People clearly drive more in Montana, but even adjusting for what is known as VMT (or "vehicle miles traveled"), Montana drivers are still twice as likely as New Jersey drivers to die on the roads. The big culprit is alcohol: Montana drivers were nearly three times as likely as New Jersey drivers to be involved in an alcohol-related fatal crash. Montana also has higher speed limits than New Jersey, and fewer chances to get caught violating traffic laws. And, most importantly, most Montana roads are rural.

There is, in theory, nothing nicer than a drive in the country, away from the "crazy traffic" of the city. But there is also nothing more dangerous. We would all do well to heed what the sign says: IT'S GOD COUNTRY, DON'T DRIVE LIKE HELL THROUGH IT. Rural, noninterstate roads have a death rate more than two and half times higher than all other roads—even after adjusting for the fewer vehicles found on rural roads. Taking a curve on a rural, noninterstate road is more than six times as dangerous as doing so on any other road. Most crashes involve single cars leaving the roadway, which suggests poorly marked roads, high speeds, fatigue or falling asleep, or alcohol—or some combination of any or all of these. When crashes do happen, medical help is often far away.

In Fred's case, he *is* the medical assistance. But what of the fact that he is a doctor? Why should that be a risk? Doctors are usually well-educated, affluent, upstanding members of the community; they drive expensive cars in good condition. But a study by Quality Planning Corporation, a San Francisco–based insurance research firm, found doctors to have the second-highest crash risk in an eight-month sample of a million drivers, just after students (whose risk is largely influenced by their young age). Why is that? Are doctors overconfident, type A drivers racing from open-heart surgery to the golf course?

One simple contributing factor may be that, in the United States at least, many doctors are male (nearly 75 percent in 2005). But firefighters and pilots are usually male as well, and those two professions were at the bottom of the risk list. Firefighters spend a lot of time in fire stations, not on the road, and pilots spend much of their time in the air. Exposure matters, which is seemingly why real estate agents, always driving from house to house, showed up high on the list. (Architects ranked high as well, prompting QPC's vice president to speculate that they're often distracted by looking at buildings!) Doctors drive a lot, often in urban settings, often with a certain urgency, perhaps dispensing advice via cell phone. Most important, they may also be tired. A report in the *New England Journal of Medicine* suggested that every time in a given month interns at Harvard Medical School pulled an extended shift, their crash risk rose by 9.1 percent. The more shifts they worked, the greater the risk that they would fall asleep while stopped in traffic, or even while driving.

Now let's talk about Dr. Fred's vehicle of choice, the pickup truck. It's an increasingly popular vehicle in the United States. The number of households owning pickups rose by nearly 50 percent from 1977 to 1990, and pickup registrations continue to rise every year. It is also the most dangerous vehicle on the road: More people in the United States die in pickups per 100 million vehicles registered than in any other kind of vehicle.

Pickups also impose the most risk on drivers of other vehicles. One study showed that the Ford F-350 presents nearly seven times the risk to other cars as the Dodge Caravan, a minivan. From a vehicular point of view, pickups are high, heavy, and have very stiff front ends—meaning other vehicles have to absorb more energy in a crash. When drivers of pickups crash into other cars, they die at a lower rate than the drivers of smaller cars. Because of simple physics, larger vehicles, with larger crush zones and, often, higher-quality materials, are better able to sustain a collision.

Though not always. As some crash tests have shown, weight is often no help at all when a vehicle hits a fixed object like a wall or a large tree. Marc Ross, a physicist at the University of Michigan, told me that "mass sort of drops out of the calculation for a fixed barrier." The car's design—its ability to absorb its own kinetic energy—is as important as its size. In crash testing done several years ago by the Insurance Institute for Highway Safety, vehicles with crash-test dummies were sent into a barrier at 40 miles per hour. Consider two vehicles: the big and brawny Ford F-150 pickup truck, weighing in at nearly 5,000 pounds, and the tiny Mini Cooper, at just under 2,500 pounds. Which would you have rather been in? The test photos make the answer clear: the Mini Cooper. The Ford, despite having more space between the obstacle and the driver, saw a "major collapse of the occupant compartment" that "left little survival space for the driver." In the Mini, meanwhile, "the dummy's position in relation to the steering wheel and instrument panel after the crash test indicates that the driver's survival space was maintained very well."

As Malcolm Gladwell argued in the *New Yorker*, larger, heavier vehicles, which are more difficult to maneuver and slower to stop, may also make it harder for a driver to avoid a crash in the first place. What complicates this is the finding that, in the United States, small cars are involved in more *single-car* fatal crashes than large cars—and it's single-car crashes that the greater maneuverability of smaller, lighter cars should help prevent. Smaller cars may be more maneuverable, but they also tend to be driven by riskier younger drivers, while sports cars that handle well may be "self-selected" by more adventurous drivers. Researchers with the National Highway Traffic Safety Administration raised another question: Would the higher maneuverability of smaller cars lead drivers to take more risks? "The quicker response of light vehicles," they argued, "may give the average driver yet more opportunity to blunder."

Risk can be deceiving. The answer to "What are the riskiest vehicles on the road?" is more complicated than it seems. Assigning risk based purely on "vehicle factors" is limiting, because it neglects the idea of who is driving the vehicle and how it is being driven. Leonard Evans, the former GM researcher, notes that crash rates are higher for two-door cars than four-door cars (up to a certain weight, where the rates become equal). "The believers in vehicle factors would say, 'We've got it, you've just got to weld another couple doors on the vehicle and you've got a safe car.' "

Those two doors are often not an engineering distinction, but a lifestyle distinction: the difference, say, between a two-door Acura RSX and a four-door Toyota Corolla. In the United States from 2002 through 2005, the death rate in the "fast and furious" Acura was more than twice as high as that in the sleepy Corolla. In terms of weight, the two vehicles are virtually identical. The different crash rate owes more to the drivers of four-doors and two-doors than to the cars themselves.

The idea that who is driving (and how) affects the risk of what is being driven is well depicted in the case of the Ford Crown Victoria and the Mercury Marquis, as Marc Ross and Tom Wenzel have pointed out. The Crown Vic and its corporate cousin the Marquis, large, staid V-8 sedans both, are basically the same car—one repair manual covers both models. They both pose the same relative risk to their drivers, which is no surprise given their similarities. The Crown Vic, however, statistically poses more risk to others. Why is that? The Crown Victoria is a popular police car, meaning that it's involved in a lot more dangerous high-speed pursuits than the Marquis. (Crown Vics, it must be said, are also the taxi of choice in New York City.)

There are "safer" cars in the hands of dangerous drivers, and "more dangerous" cars in the hands of safe drivers. Small cars such as subcompacts do pose a greater risk for their occupants if involved in a crash—although more-expensive subcompacts are less risky than cheaper subcompacts—but subcompacts also tend to be driven by people (e.g., younger drivers) with higher risks of getting into a crash, because of "behavioral factors." Still, age is just one behavioral factor, and it interacts with the type of car being driven. As I will discuss in the next section, the drivers of small cars may actually act in safer ways because of the size of the car. Are large passenger cars statistically the safest because they pose less of a rollover risk than SUVs or because they weigh more than small cars? Or is it because they tend to be driven by the statistically safest demographic?

Returning to Fred and his pickup truck: It's hard to tell where the risks of one end and those of the other begin. Men tend to drive pickup trucks more than women, men tend to wear seat belts less often, men who live in rural areas are more likely to drive pickup trucks without seat belts, and, after motorcyclists, the drivers of pickup trucks are the most likely to have been drinking when involved in a fatal crash. These would be only a handful of the potential risk factors—an Australian study, for example, found that black cars were more likely to crash than white cars. Is it visibility, or the types of people who drive black cars versus white ones? We all know no one washes a rental car, but are rental cars driven more recklessly? (There is some evidence to suggest so.) A study in Israel found that fewer drivers died on the roads in the first and second days after a suicide-bombing attack but then tracked an *increase* in danger on the third day. Are people simply staying off the roads in the aftermath, then rejoining them en masse? (Or does the aftereffect of terror cause people to act with less regard for life?)

As the risk expert John Adams likes to say, understanding risk is not rocket science—it's more complicated. Looking at statistics from the United Kingdom, he notes that a young man is 100 times more likely than a middle-aged woman to be killed in traffic. Someone driving on Sunday morning at three a.m. has a risk *134 times* greater than someone driving at ten a.m. on Sunday. Someone with a personality disorder is 10 times more likely to have a serious crash, while someone 2.9 times over the BAC limit would be 20 times more likely than a sober driver to crash.

"So if these were independent variables," he told me, "you could multiply them and come to the conclusion that a disturbed, drunken young man on the road on a Sunday morning was about 2.5 million times more likely to have a serious accident than a normal, sober middle-aged woman driving to church seven hours later." They are, however, *not* independent. "There are proportionally more disturbed, drunken young men on the road at three o'clock on a Sunday morning," Adams noted. Now add other factors. Were the car's tires in good shape? Was it foggy? Was the driver tired or awake? "Once you start trying to imagine all the factors," Adams said, "that might not be an exaggeration of the disparity between one person's risk and another person's risk." He used this example to "have a go" at what he calls the "Richter scales" of risk, which show, for instance, that a person has a 1 in 8,000 chance of dying or being seriously injured in a car crash, and a 1 in 25,000 chance of the same thing happening while playing soccer. "The purveyors of these tables say they produce them to guide the lay public in making risks. The lay public is hopeless at making use of numbers like this."

There is one solid bit of advice that could be dispensed regarding whether you should take a trip with the fictional Fred: Ride in the backseat (if he had one, that is). The fatality risk in the backseat is 26 percent lower than in the front. The backseat is safer than air bags. But you run the risk of offending Fred.

## The Risks of Safety

Be wary then; best safety lies in fear.
—William Shakespeare, *Hamlet*

In the 1950s, when car fatalities in the United States were approaching their zenith, an article in the *Journal of the American Medical Association* argued that the "elimination of the mechanically hazardous features of the interior construction"—for example, metal dashboards and rigid steering columns—would prevent nearly 75 percent of the annual road fatalities, saving some 28,500 lives.

Car companies were once rightly castigated for trying to shift the blame for traffic fatalities to the "nut behind the wheel." And in the decades since, in response to public outcry and the ensuing regulations, the insides of cars have been made radically safer. In the United States (and most other places), fewer people in cars die or are injured now than in the 1960s, even though more people drive more miles. But in an oft-repeated pattern with safety devices from seat belts to air bags, the actual drop in fatalities did not live up to the early hopes. Consider the so-called chimsil. The term is slang for "center high-mounted stop lamp" (CHMSL), meaning the third rear brake light that became mandatory on cars in the 1980s, after decades of study.

On paper at least, the chimsil sounded like a great idea. It would give drivers more information that the car ahead was braking. Unlike brake lights, which go from one shade of red to a brighter shade of red (some engineers have argued that an outright change in colors would make more sense), the chimsil would illuminate only during braking. Drivers scanning through the windshield of the car ahead of them to gauge traffic would have more information. Tests had shown that high-mounted lamps improved reaction times. Experts predicted that the lamps would help reduce certain types of crashes, particularly rear-end collisions. Early studies, based on a trial that equipped some cars in taxi fleets with the lights, indicated that these incidents could be cut by 50 percent. Later estimates, however, dropped the benefit to around 15 percent. Studies now estimate that the chimsil has "reached a plateau" of reducing rear-end crashes by 4.3 percent. This arguably justifies the effort and cost of having them installed, but the chimsil clearly has not had the effect for which its inventors had hoped.

Similar hopes greeted the arrival of the antilock braking system, or ABS, which helps avoid "locked brakes" and allows for greater steering control during braking, particularly in wet conditions. But problems arose. A famous, well-controlled study of taxi drivers in Munich, Germany, found that cars equipped with ABS drove faster, and closer to other vehicles, than those without. They also got into more crashes than cars without ABS. Other studies suggested that drivers with ABS were less likely to rear-end someone but more likely to be rear-ended *by* someone else.

Were drivers trading a feeling of greater safety for more risk? Perhaps they were simply swapping collisions with other vehicles for potentially more dangerous "single-vehicle road-departure" crashes—studies on test tracks have shown that drivers in ABS-equipped cars more often veered off the road when trying to avoid a crash than non-ABS drivers did. Other studies revealed that many drivers didn't know how to use ABS brakes correctly. Rather than exploiting ABS to drive more aggressively, they may have been braking the wrong way. Finally, drivers with ABS may simply have been racking up more miles. Whatever the case, a 1994 report by the National Highway Traffic Safety Administration concluded that the "overall, net effect of ABS" on crashes—fatal and otherwise—was "close to zero." (The reason why is still rather a mystery, as the Insurance Institute for Highway Safety concluded in 2000: "The poor early experience of cars with antilocks has never been explained.")

There always seems to be something else to protect us on the horizon.

The latest supposed silver bullet for traffic safety is electronic stability control, the rollover-busting technology that, it is said, can help save nearly ten thousand lives per year. It would be a good thing if it did, but if history is a guide, it will not.

Why do these changes in safety never seem to have the predicted impact? Is it just overambitious forecasting? The most troublesome possible answer, one that has been haunting traffic safety for decades, suggests that, as with the roads in Chapter 7, the safer cars get, the more risks drivers choose to take.

While this idea has been around in one form or another since the early days of the automobile—indeed, it was used to argue against railroad safety improvements—it was most famously, and controversially, raised in a 1976 article by Sam Peltzman, an economist at the University of Chicago. Describing what has since become known as the "Peltzman effect," he argued that despite the fact that a host of new safety technologies—most notably, the seat belt—had become legally required in new cars, the roads were no safer. "Auto safety regulation," he concluded, "has not affected the highway death rate." Drivers, he contended, were trading a decrease in accident risk with an increase in "driving intensity." Even if the occupants of cars themselves were safer, he maintained, the increase in car safety had been "offset" by an increase in the fatality rate of people who did not benefit from the safety features—pedestrians, bicyclists, and motorcyclists. As drivers felt safer, everyone else had reason to feel less safe.

Because of the twisting, entwined nature of car crashes and their contributing factors, it is exceedingly difficult to come to any certain conclusions about how crashes may have been affected by changes to any one variable of driving. The median age of the driving population, the state of the economy, changes in law enforcement, insurance factors, weather conditions, vehicle and modal mix, alterations in commuting patterns, hazy crash investigations—all of these things, and others, play their subtle part. In many cases, the figures are simply estimates.

This gap between expected and achieved safety results might be explained by another theory, one that turns the risk hypothesis rather on its head. This theory, known as "selective recruitment," says that when a seat-belt law is passed, the pattern of drivers who switch from not wearing seat belts to wearing seat belts is decidedly not random. The people who will be first in line are likely to be those who are already the safest drivers. The drivers who do not choose to wear seat belts, who have been shown in studies to be riskier drivers, will be "captured" at a smaller rate—and even when they are, they will still be riskier.

Looking at the crash statistics, one finds that in the United States in 2004, more people not wearing their seat belts were killed in passenger-car accidents than those who were wearing belts—even though, if federal figures can be believed, more than 80 percent of drivers wear seat belts. It is not simply that drivers are less likely to survive a severe crash when not wearing their belts; as Leonard Evans has noted, the most severe crashes *happen* to those not wearing their belts. So while one can make a prediction about the estimated reduction in risk due to wearing a seat belt, this cannot simply be applied to the total number of drivers for an "expected" reduction in fatalities.

Economists have a clichéd joke: The most effective car-safety instrument would be a dagger mounted on the steering wheel and aimed at the driver. The incentive to drive safely would be quite high. Given that you are twice as likely to die in a severe crash if you're not wearing a seat belt, it seems that *not* wearing a seat belt is essentially the same as installing a dangerous dagger in your car.

And yet what if, as the economists Russell Sobel and Todd Nesbit ask, you had a car so safe you could usually walk away unharmed after hitting a concrete wall at high speed? Why, you would "race it at 200 miles per hour around tiny oval racetracks only inches away from other automobiles and frequently get into accidents." This was what they concluded after tracking five NASCAR drivers over more than a decade's worth of races, as cars gradually became safer. The number of crashes went up, they found, while injuries went down.

Naturally, this does not mean that the average driver, less risk-seeking than a race-car driver, is going to do the same. For one, average drivers do not get prize money; for another, race-car drivers wear flame-retardant suits and helmets. This raises the interesting, if seemingly outlandish, question of why car drivers, virtually alone among users of wheeled transport, do not wear helmets. Yes, cars do provide a nice metal cocoon with inflatable cushions. But in Australia, for example, head injuries among car occupants, according to research by the Federal Office of Road Safety, make up *half* the country's traffic-injury costs. Helmets, cheaper and more reliable than side-impact air bags, would reduce injuries and cut fatalities by some 25 percent. A crazy idea, perhaps, but so were air bags once.

Seat belts and their effects are more complicated than allowed for by the economist's language of incentives, which sees us all as rational actors making predictable decisions. I have always considered the act of wearing my seat belt not so much an incentive to drive more riskily as a grim reminder of my own mortality (some in the car industry fought seat belts early on for this reason). This doesn't mean I'm immune from behavioral adaptation. Even if I cannot imagine how the seat belt makes me act more riskily, I *can* easily imagine how my behavior would change if, for some reason, I was driving a car *without* seat belts. Perhaps my ensuing alertness would cancel out the added risk.

Moving past the question of how many lives have been saved by seat belts and the like, it seems beyond doubt that increased feelings of safety can push us to take more risks, while feeling less safe makes us more cautious. This behavior may not always occur, we may do it for different reasons, we may do it with different intensities, and we may not be aware that we are doing it (or by how much); but the fact that we do it is why these arguments are still taking place. This may also explain why, as Peltzman has pointed out, car fatalities per mile still decline at roughly the same rate every year now as they did in the first half of the twentieth century, well before cars had things like seat belts and air bags.

In the first decade of the twentieth century, forty-seven men tried to climb Alaska's Mount McKinley, North America's tallest peak. They had relatively crude equipment and little chance of being rescued if something went wrong. All survived. By the end of the century, when climbers carried high-tech equipment and helicopter-assisted rescues were quite frequent, each decade saw the death of dozens of people on the mountain's slopes. Some kind of adaptation seemed to be occurring: The knowledge that one could be rescued was either driving climbers to make riskier climbs (something the British climber Joe Simpson has suggested); or it was bringing less-skilled climbers to the mountain. The National Park Service's policy of increased safety was not only costing more money, it perversely seemed to be costing more lives—which had the ironic effect of producing calls for more "safety."

In the world of skydiving, the greatest mortality risk was once the so-called low-pull or no-pull fatality. Typically, the main chute would fail to open, but the skydiver would forget to trigger the reserve chute (or would trigger it too late). In the 1990s, U.S. skydivers began using a German-designed device that automatically deploys, if necessary, the reserve chute. The number of low- or no-pull fatalities dropped dramatically, from 14 in 1991 to 0 in 1998. Meanwhile, the number of once-rare open-canopy fatalities, in which the chute deploys but the skydiver is killed upon landing, surged to become the leading cause of death. Skydivers, rather than simply aiming for a safe landing, were attempting hook turns and swoops, daring maneuvers done with the canopy open. As skydiving became safer, many skydivers, particularly younger skydivers, found new ways to make it riskier.

The psychologist Gerald Wilde would call what was happening "risk homeostasis." This theory implies that people have a "target level" of risk: Like a home thermostat set to a certain temperature, it may fluctuate a bit from time to time but generally keeps the same average setting. "With that reliable rip cord," Wilde told me at his home in Kingston, Ontario, "people would want to extend their trip in the sky as often as possible. Because a skydiver wants to be up there, not down here."

In traffic, we routinely adjust the risks we're willing to take as the expected benefit grows. Studies, as I mentioned earlier in the book, have shown that cars waiting to make left turns against oncoming traffic will accept smaller gaps in which to cross (i.e., more risk) the longer they have been waiting (i.e., as the desire for completing the turn increases). Thirty seconds seems to be the limit of human patience for left turns before we start to ramp up our willingness for risk.

We may also act more safely as things get more dangerous. Consider snowstorms. We've all seen footage of vehicles slowly spinning and sliding their way down freeways. The news talks dramatically of the numbers of traffic deaths "blamed on the snowstorm." But something interesting is revealed in the crash statistics: During snowstorms, the number of collisions, relative to those on clear days, goes up, but the number of fatal crashes goes *down*. The snow danger seems to cut both ways: It's dangerous enough that it causes more drivers to get into collisions, and dangerous enough that it forces them to drive at speeds that are less likely to produce a fatal crash. It may also, of course, force them not to drive in the first place, which itself is a form of risk adjustment.

In moments like turning left across traffic, the risk and the payoff seem quite clear and simple. But do we behave consistently, and do we really have a sense of the actual risk or safety we're looking to achieve? Are we always pushing it "to the max," and do we even know what that "max" is? Critics of risk homeostasis have said that given how little humans actually know about assessing risk and probability, and given how many misperceptions and biases we're susceptible to while driving, it's simply expecting too much of us to think we're able to hold to some perfect risk "temperature." A cyclist, for example, may feel safer riding on the sidewalk instead of the street. But several studies have found that cyclists are more likely to be involved in a crash when riding on the sidewalk. Why? Sidewalks, though separated from the road, cross not only driveways but intersections—where most car-bicycle collisions happen. The driver, having already begun her turn, is less likely to expect—and thus to see—a bicyclist emerging from the sidewalk. The cyclist, feeling safer, may also be less on the lookout for cars.

The average person, the criticism goes, is hardly aware of what their chances actually would be of surviving a severe crash while wearing a seat belt or protected by the unseen air bag lurking inside the steering wheel. Then again, as any trip to Las Vegas will demonstrate, we seem quite capable of making confident choices based on imperfect information of risk and odds. The loud, and occasionally vicious, debate over "risk compensation" and its various offshoots seems less about whether it can happen and more about whether it always happens, or exactly why.

Most researchers agree that behavioral adaptation seems more robust in response to direct feedback. When you can actually feel something, it's easier to change your behavior in response to it. We cannot feel air bags and seat belts at work, and we do not regularly test their capabilities—if they make us feel safer, that sense comes from something besides the devices themselves. Driving in snow, on the other hand, we don't have to rely on internalized risk calculations: One can feel how dangerous or safe it is through the act of driving. (Some studies have shown that drivers with studded winter tires drive faster than those without them.)

A classic way we sense feedback as drivers is through the size of the vehicle we are driving. The feedback is felt in various ways, from our closeness to the ground to the amount of road noise. Studies have suggested that drivers of small cars take fewer risks (as judged by speed, distance to the vehicle ahead of them, and seat-belt wearing) than drivers of larger cars. Many drivers, particularly in the United States, drive sport-utility vehicles for their perceived safety benefits from increased weight and visibility. There is evidence, however, that SUV drivers trade these advantages for more aggressive driving behavior. The result, studies have argued, is that SUVs are, overall, no safer than medium or large passenger cars, and less safe than minivans.

Studies have also shown that SUV drivers drive faster, which may be a result of feeling safer. They seem to behave differently in other ways as well. A study in New Zealand observed the position of passing drivers' hands on their steering wheels. This positioning has been suggested as a measure of perceived risk—research has found, for instance, that more people are likely to have their hands on the top half of the steering wheel when they're driving on roads with higher speeds and more lanes. The study found that SUV drivers, more than car drivers, tended to drive either with only one hand or with both hands on the bottom half of the steering wheel, positions that seemed to indicate lower feelings of risk. Another study looked at several locations in London. After observing more than forty thousand vehicles, researchers found that SUV drivers were more likely to be talking on a cell phone than car drivers, more likely not to be wearing a seat belt, and—no surprise—more likely not to be wearing a seat belt *while* talking on a cell phone.

It could just be that the types of people who talk on cell phones and disdain seat belts while driving also like to drive SUVs. But do they like to drive an SUV because they think it's a safer vehicle or because it gives them license to act more adventurously on the road? To return to the mythical Fred, pickup drivers are less likely than other drivers to wear their seat belts. Under risk-compensation theory, he is doing this because he feels safer in the large pickup truck. But could he not drive in an even more risky fashion yet lower the "cost" of that risky driving by buckling up? It all leads to questions of where we get our information about what is risky and safe, and how we act upon it. Since relatively few of us have firsthand experience with severe crashes in which the air bags deployed, can we really have an accurate sense of how safe we are in a car with air bags versus one without—enough to get us to change our behavior?

Risk is never as simple as it seems. One might think the safest course of action on the road would be to drive the newest car possible, one filled with the latest safety improvements and stuffed full of technological wonders. This car must be safer than your previous model. But, as a study in Norway found, *new cars crash most.* It's not simply that there are more new cars on the road—the *rate* is higher. After studying the records of more than two hundred thousand cars, the researchers concluded: "If you drive a newer car, the probability of both damage and injury is higher than if you drive an older car."

Given that a newer car would seem to offer more protection in a crash, the researchers suggested that the most likely explanation was drivers changing the way they drive in response to the new car. "When using an older car which may not feel very safe," they argued, "a driver probably drives more slowly and is more concentrated and cautious, possibly keeping a greater distance to the car in front." The finding that new cars crash most has shown up elsewhere, including in the United States, although another explanation has been offered: When people buy new cars, they drive them more than old cars. This in itself, however, may be a subtle form of risk compensation: I feel safer in my new car, thus I am going to drive it more often.

Studying risk is not rocket science; it's more complicated. Cars keep getting objectively safer, but the challenge is to design a car that can overcome the inherent risks of human nature.

In most places in the world, there are more suicides than homicides. Globally, more people take their own lives in an average year—roughly a million—than the total murdered *and* killed in war. We always find these sorts of statistics surprising, even if we are simultaneously aware of one of the major reasons for our misconception: Homicides and war receive much more media coverage than suicides, so they seem more prevalent.

A similar bias helps explain why, in countries like the United States, the annual death toll from car crashes does not elicit more attention. If the media can be taken as some version of the authentic voice of public concern, one might assume that, over the last few years, the biggest threat to life in this country is terrorism. This is reinforced all the time. We hear constant talk about "suspicious packages" left in public buildings. We're searched at airports and we watch other people being searched. We live under coded warnings from the Department of Homeland Security. The occasional terrorist cell is broken up, even if it often seems to be a hapless group of wannabes.

Grimly tally the number of people who have been killed by terrorism in the United States since the State Department began keeping records in the 1960s, and you'll get a total of less than 5,000—roughly the same number, it has been pointed out, as those who have been struck by lightning. But each year, with some fluctuation, the number of people killed in car crashes in the United States tops 40,000. More people are killed on the roads each month than were killed in the September 11 attacks. In the wake of those attacks, polls found that many citizens thought it was acceptable to curtail civil liberties to help counter the threat of terrorism, to help preserve our "way of life." Those same citizens, meanwhile, in polls and in personal behavior, have routinely resisted traffic measures designed to reduce the annual death toll (e.g., lowering speed limits, introducing more red-light cameras, stiffer blood alcohol limits, stricter cell phone laws).

Ironically, the normal business of life that we are so dedicated to preserving is actually more dangerous to the average person than the threats against it. Road deaths in the three months after 9/11, for example, were 9 percent higher than those during the similar periods in the two years before. Given that airline passenger numbers dropped during this same period, it can be assumed some people chose to drive rather than fly. It might be precisely because of all the vigilance that no further deaths due to terrorism have occurred in the United States since 9/11—even as more than two hundred thousand people have died on the roads. This raises the question of why we do not mount a similarly concerted effort to improve the "security" of the nation's roads; instead, in the wake of 9/11, newspapers have been filled with stories of traffic police being taken off the roads and assigned to counterterrorism.

In the 1990s, the United Kingdom dropped its road fatalities by 34 percent. The United States managed a 6.5 percent reduction. Why the difference? Better air bags, safer cars? It was mostly speed, one study concluded (although U.S. drivers also rack up many more miles each year). While the United Kingdom was introducing speed cameras, the United States was resisting cameras and raising speed limits. Had the United States pulled off what the United Kingdom did, it is suggested, 10,000 fewer people would have been killed.

Why doesn't the annual road death toll elicit the proportionate amount of concern? One reason may simply be the trouble we have in making sense of large numbers, because of what has been called "psychophysical numbing." Studies have shown that people think it's more important to save the same number of lives in a small refugee camp than a large refugee camp: Saving ten lives in a fifty-person camp seems more desirable than saving ten lives in a two-hundred-person camp, even though ten lives is ten lives. We seem less sensitive to changes when the numbers are larger.

By contrast, in what is called the "identifiable victim effect," we can be quite sensitive to the suffering of one person, like the victim of a terrible disease. We are, in fact, so sensitive to the suffering of one person that, as work by the American psychologist and risk-analysis expert Paul Slovic has shown, people are more likely to give more money to charity campaigns that feature one child rather than those that show multiple children—even when the appeal features only *one* more child.

Numbers, rather than commanding more attention for a problem, just seem to push us toward paralysis. (Perhaps this goes back to that evolutionary small-group hypothesis.) Traffic deaths present a further problem: Whereas a person in jeopardy can possibly be saved, we cannot know with certainty ahead of time who will be a crash victim—even most legally drunk drivers, after all, make it home safely. In fatal crashes, victims usually die instantly, out of sight. Their deaths are dispersed in space and time, with no regular accumulated reporting of all who died. There are no vigils or pledge drives for fatal car-crash victims, just eulogies, condolences, and thoughts about how "it can happen to anyone," even if fatal car crashes are not as statistically random as we might think.

Psychologists have argued that our fears tend to be amplified by "dread" and "novelty." A bioterrorism attack is a new threat that we dread because it seems beyond our control. People have been dying in cars, on the other hand, for more than a century, often by factors presumably within their control. We also seem to think things are somehow less risky when we can feel a personal benefit they provide (like cars) than when we cannot (like nuclear power). Still, even within the realm of traffic, risks seem to be misperceived. Take so-called road rage. The number of people shot and killed on the road every year, even in gun-happy America, unofficially numbers around a dozen (far fewer than those killed by lightning). Fatigue, meanwhile, contributes to some 12 percent of crashes. We are better advised to watch out for yawning drivers than pistol-packing drivers.

Our feelings about which risks we should fear, as the English risk expert John Adams argues, are colored by several important factors. Is something voluntary or not? Do we feel that something is in our control or beyond our control? What is the potential reward? Some risks are voluntary, in our control (we think), and there is a reward. "A pure self-imposed, self-controlled voluntary risk might be something like rock climbing," Adams said. "The risk is the reward." No one forces a rock climber to take risks, and when rock climbers die, no one else feels threatened. (The same might be said of suicide versus murder.) Other risks are voluntary but we cede control—for example, taking a cross-country bus trip. We have no sway over the situation. Imagine that you are at the bus station and see a driver drinking a beer at the bar. Then imagine you see the same driver at the wheel as you board your bus. How would you feel? Nervous, I would guess.

Now imagine yourself at a bar having a beer. Then imagine yourself getting in your car to drive home. Did you envision the same dread and panic? Probably not, because you were, at least in your own mind, in control. You're the manager of your own risk. This is why people think they have a better chance of winning the lottery if they pick the numbers (it is also, admittedly, more fun that way). We get nervous about ceding control over risk to other people. Not surprisingly, we tend to inflate risk most dramatically for things that are involuntary, out of our control, and offer no reward. "The July 7 bombings here in London killed six days' worth of death on the road," Adams said. "After this event, ten thousand people gathered in Trafalgar Square. You don't get ten thousand people in Trafalgar Square lamenting last week's road death toll."

Why *is* there no outrage? Driving is voluntary, it's in our control, and there's a reward. And so we fail to recognize the real danger cars present. Research in the United States has shown, for example, that exurban areas—the sprawling regions beyond the old inner-ring suburbs—pose greater risks to their inhabitants than central cities as a whole. This despite a cultural preconception that the opposite is true. The key culprit? Traffic fatalities. The less dense the environment, the more dangerous it is. If we wanted dramatically safer roads overnight—virtually fatality-free—it wouldn't actually be difficult. We could simply lower the speed limit to ten miles per hour (as in those Dutch *woonerven*). Does that seem absurd? In the early 1900s that *was* the speed limit. In Bermuda, very few people die in cars each year. The island-wide speed limit is 35 kilometers per hour (roughly 22 miles per hour). In the United States, to take one example, Sanibel Island, Florida, which has a 35 mph maximum, has not seen a traffic fatality this century, despite a heavy volume of cars and cyclists. But merely lowering mean speeds as little as one mile per hour, as Australian researchers have found, lowers crash risks.

As societies, we have gradually accepted faster and faster speeds as a necessary part of a life of increasing distances, what Adams calls "hypermobility." Higher speeds enable life to be lived at a scale in which time is more important than distance. Ask someone what their commute is, and they will inevitably give an answer in minutes, as if they were driving across a clock face. Our cars have been engineered to bring a certain level of safety to these speeds, but even this is rather arbitrary, for what is safe about an activity that kills tens of thousands of people a year and seriously injures many more than that? We drive with a certain air of invincibility, even though air bags and seat belts will not save us in roughly half the crashes we might get into, and despite the fact that, as Australian crash researcher Michael Paine has pointed out, half of all traffic fatalities to seat-belt-wearing drivers in frontal collisions happen at impact speeds at or below the seemingly slow level of 35 miles per hour.

We have deemed the rewards of mobility worth the risk. The fact that we're at the wheel skews our view. Not only do we think we're better than the average driver—that "optimistic bias" again—but studies show that we think we're less likely than the average driver to be involved in a crash. The feeling of control lowers our sense of risk. What's beyond our control comes to seem riskier, even though it is "human factors," not malfunctioning vehicles, faulty roads, or the weather, that are responsible for an estimated 90 percent of crashes.

On the road, we make our judgments about what's risky and what's safe using our own imperfect human calculus. We think large trucks are dangerous, but then we drive unsafely around them. We think roundabouts are more dangerous than intersections, although they're more safe. We think the sidewalk is a safer place to ride a bike, even though it's not. We worry about getting into a crash on "dangerous" holiday weekends but stop worrying during the week. We do not let children walk to school even though driving in a car presents a greater hazard. We use hands-free cell phones to avoid risky dialing and then spend more time on risky calls (among other things). We carefully stop at red lights when there are no other cars, but exceed the speed limit during the rest of the trip. We buy SUVs because we think they're safer and then drive them in more dangerous ways. We drive at a minuscule following distance to the car ahead, exceeding our ability to avoid a crash, with a blind faith that the driver ahead will never have a reason to suddenly stop. We have gotten to the point where cars are safer than ever, yet traffic fatalities cling to stubbornly high levels. We know all this, and act as if we don't.

Epilogue

# Driving Lessons

Before embarking on this book, I hadn't thought much about driving since first learning to do it and acquiring my license on the, ahem, second try. Since then, I've logged a few hundred thousand miles or so, had several minor crashes ("accidents" if you must, though both were easily my fault, because of careless behavior whose specifics shall be withheld), and dropped by to the Department of Motor Vehicles every decade or so to glance at an eye chart and get renewed by a grumpy clerk. I mostly just got behind the wheel, fussed over the radio, and hit the road with a mixture of anxiety and wonder: anxiety over the danger of it all, the crumpled cars on the roadside, the shockingly poor behavior, the nervous way people say, "Drive safely" as you leave them; and a simultaneous sense of wonder that we're all able to move about at high speeds, in such great numbers, with such fluidity.

After spending a long time sifting through the theories and science of traffic, I wondered if there was not still more to be learned about driving a car. I thought, Why not go to those people who, for sport and for a living, drive cars at the absolute limits, in conditions that make even the most frantic traffic seem sedentary? What could race-car drivers have to teach civilians about driving? And so one morning I found myself hunched into one of those small chairs with an attached desk, part of a group including gum-chewing teens and graying sixtysomethings, in a brightly lit classroom at the Bob Bondurant School of High Performance

Driving, just south of Phoenix. At the front of the class stood Les Betchner, jauntily tanned and with spiky blond hair, a sometime stock-car racer who exuded the easy patter and ridiculously innate competence that just seems the birthright of people like airline pilots and sports instructors.

Drivers, as you well know by now, tend to self-enhance. We are thin-skinned about our sense of driving competence. One is loath to admit, at age forty, that there are new things to be learned. And yet this is just what was happening. "A steering wheel doesn't do much," Betchner was saying. "You steer with the pedals." *What?* I snapped to attention. Steer with the pedals? He was PowerPointing his way through the problems of skidding around corners. Racers loathe skidding, not because it means they are out of control but because they are, as they say, "scrubbing speed." "We never want to slide," Betchner said. "That's the slow way around the track."

As you may recall from your driving lessons, there are two kinds of corner skids, an "understeer skid" and an "oversteer skid." On the race track they say an understeer skid means it's your front end that's smacking the guardrail, while in an oversteer skid your rear end hits first. Despite the word *steer*, steering is only part of knowing how to react to and correct for under- or oversteer situations. It can often hurt more than help. "Add a bunch of steering, you go right off the road," Betchner said. "Physics is now part of your life."

The real key to skid control, he explained, is "weight transfer." In an understeer skid, the car's front wheels have lost traction. Attempting to steer will only make matters worse. Braking shifts weight to the front and adds grip. In an oversteer skid, meanwhile, the rear of the car has lost traction and wants to pass the front. The slip angle, or the difference between the direction the tires are pointed and the direction they are actually moving, is greater in the rear tires than the front. The first step in taming the rear wheels is, essentially, taking the turn more widely. So instead of moving the steering wheel in the direction of the turn, increasing the slip angle, you must "steer into the skid"—move the steering wheel in the direction the rear of the car is moving. Many of us know what "steer into the skid" means without really knowing what it means. The larger problem, Betchner pointed out, is that no one is ever taught what to do next. He queried the room. There were some half-mumbled answers. No two seemed to be the same. "Pray?" someone joked.

The answer is the opposite of what you might expect: Hit the gas. "When in doubt, flat out," instructed Betchner. (Actually, he added, you want to add just a touch of throttle input.) The natural instinct, of course, is to hit the brakes. The problem is that this shifts weight to the front end of the car—exactly where you don't want it to be. As your car dips toward the front end, you're helping your rear wheels lose their already tenuous grip on the road. They need every ounce of pressure they can get. Then there is the final problem. You can't just keep steering into a skid. "That's where we find ourselves getting into trouble," said Mike McGovern, another longtime Bondurant instructor. "We do that first part well, but when the car hooks up and comes back to straight, we hold the steering. We don't unwind it. We're telling the car to turn again, and that's when you get into a secondary skid." This is another somewhat counterintuitive lesson: To fully reassert control, you need to relinquish the steering, letting the pull of the realigned tires do the work as the steering wheel spins through your hands.

Another lesson that seemed rather obvious—but proved curiously powerful once tried out on the test track—was the Bondurant mantra "Look where you want to go." This recalls the "moth effect" phenomenon and brings up a chicken-and-egg sort of problem that vision researchers still debate: Do we automatically travel in the direction we are looking, or do we first search for a target destination and then keep looking in that direction to maintain our course? Do we drive where we look or look where we drive? The former, arguably: As one study found, "there is a systematic and reliable tendency for [drivers] to follow their direction of gaze with their direction of travel, in many cases without the conscious awareness of doing so at all."

This might seem rather academic and of little concern to you, but consider what happens when a car suddenly pulls out in front of you as you're speeding down a rural road. If you "target fixate," as the Bondurant instructors call it—that is, look at the car that pulled out instead of where you need to be to evade the crash—do you have less chance of avoiding the accident? Does your "gaze eccentricity," as vision people call it, negatively affect your ability to steer away from the obstacle?

The science is still inconclusive, but on the Bondurant "skid pad" the effectiveness of the racer's maxim "Look where you want to go" was made strikingly clear. I was driving a Pontiac Grand Prix equipped with outrigger wheels attached to the back end. At the flick of a switch, the instructor could raise the car to simulate a skid at much faster speeds. As I repeatedly drove in loops and practiced getting out of oversteer skids, I found I corrected more easily by concentrating not on the giant barrier of rubber tires I was sliding toward (admittedly not an easy thing to ignore) but on that place around the corner where I wanted to be.

It would be easy to dismiss the school, with its fleets of Corvettes, its acrid tang of burned rubber and exhaust, its looping Grand Prix–style track, as a playground for the unhinged libidinal fantasies of people normally shackled by the world of everyday driving. Indeed, there was a heavy midlife-crisis vibe about the place. And yet there were myriad moments where I thought to myself, Why didn't I know this before?

"Driver's ed taught you how to get a license," Bob Bondurant told me in his office, his ever-present dog Rusty, a Queensland heeler, panting nearby. "It didn't teach you skid control or evasive emergency maneuvers." In 1967, Bondurant's promising racing career was cut short when the steering arm on his McLaren Mk II broke at 150 miles per hour, propelling him into an embankment that sent his car "as high as a telephone pole." Since then, he has been teaching people like Clint Eastwood and James Garner how to handle a car. This is not how most of us learn, of course. "The driver-ed guy might be your English teacher," Bondurant said. He or she knows as much about driving, he implied, as the average person. And mostly, this is *fine*. Despite the prediction from Karl Benz, the founder of the Mercedes-Benz company, that the global car market would be limited because only a relative few would possess the skills needed to drive, most of us, as Bondurant said, "just plunk our butt down in the seat and drive down the road."

Indeed, there is a strong argument against the idea that we should emulate the actions of people like race-car drivers in everyday life. In a well-known (but not since repeated) study conducted in the 1970s, researchers from the Insurance Institute for Highway Safety looked into the off-course driving records of a pool of stock-car drivers. These drivers were no doubt capable of handling themselves around tight turns, no doubt superior at anticipating their moves ahead of time, no doubt possessed faster reaction times than ordinary people. How had they actually performed on the road, off the track? They'd not only gotten more traffic tickets (which we would expect given their penchant for risk) but they'd also had more crashes than the average driver. Racers possess superior control of a car, to be sure, but control alone does not win races. They also need that ineffable something within that tells them to push just slightly beyond their limits, and the limits of every other driver, to win. As Mario Andretti put it, "If everything seems under control, you're just not going fast enough." They had, one might argue, put themselves into positions in which their skills were not always enough to keep them out of trouble.

In everyday traffic, "good driving" has little to do with cornering ability or navigating between tight packs of high-speed vehicles. It's more a matter of just following the rules, staying awake, and not hitting anyone. This is not to say that racing cannot teach us things about everyday driving. Racers, Betchner said, sit erect and close, alert for feedback signals that can be felt in the pedals and steering wheel. The typical driver's posture, however, is terrible. "Most of us sit back, the 'Detroit lean,' " he said. "The car communication is horrible." Some drivers, he lamented, sit so far back they cannot reliably depress the brake pedal far enough to activate the antilock system. Or consider vision, the sense that is supposed to account for 90 percent of our driving activity. The racer's dictum that you should always be looking ahead to where you want to go next, which helps them speed through turns, is just as apt for something as prosaic as navigating an intersection. One reason for the high numbers of pedestrians struck in the crosswalk by vehicles turning is that drivers are simply not looking in the right place; they may be concentrating on making the corner itself as they turn (particularly if they are on a cell phone or otherwise distracted), rather than on what the result of their turn will be. In racing, this slows you down. In real life, it means you might hit someone.

Everyday driving also presents those moments for which nothing in our previous experience can have adequately prepared us: the oncoming car crossing the line, the sudden obstacle in the headlights. At Bondurant, I went through repeated drills—for instance, driving a car as fast as I could toward a set of cones, hitting the brakes hard enough to activate the antilock system (something that actually took me several tries), and then steering off into a small lane marked by different cones. I was struck by just how much control of the vehicle I had under full braking. The ABS did not help me stop any more quickly; indeed, another exercise, one that involved steering into one of three lanes at the last moment at the command of a signal, drummed home the idea that certain crashes, inevitable if I had braked, could be rather easily avoided by simply steering. It did, however, open my eyes to the ability one has, with ABS, to stop *and* steer at the same time.

That may seem, like the other lessons at Bondurant, rather common knowledge, but the wealth of evidence derived from studies of what drivers actually do in the critical moments of emergency situations suggests otherwise. First, drivers are actually quite reluctant to steer when an obstacle suddenly looms in front of them. The majority of drivers brake first and steer last, if at all, even in tests where steering is physically the only way to avoid a crash. This may be because steering might seem to put the driver in an even more precarious position, or it may be because the driver is unaware of the way the car is capable of handling, or it may simply be a form of "operant conditioning"—pressing our brakes, like staying in our lane, has so often been the right thing to do in everyday driving, it begins to seem the only thing to do. But research has also shown that drivers rarely activate the brakes to their full power. Other studies have demonstrated that when steering is attempted, the maneuver tends to be in the same direction the obstacle is moving, which hints that drivers are not "looking where they want to go" (and moving in that direction) but are focused instead on the obstacle to be avoided.

Whether or not the "muscle memory" of my evasive actions on the test course can be sustained over years of uneventful driving is an open question. The major problem is that so many things can go wrong in traffic that it would be impossible to teach, much less remember, appropriate responses for each scenario. Add to this the problem that because these events are unexpected, our reaction times are slowed; the emotional duress of a potential crash might even further slow our reactions—sometimes, studies have shown, to the point where we do nothing.

Then there is the shifting, dynamic nature of traffic itself. It is sometimes impossible to say what a "correct" evasive maneuver would be in the moment of trying to avoid another driver, as it could be canceled out by an unexpected countermove by that other driver. In one trial, forty-nine drivers were put in a driving simulator at Daimler-Benz. As they approached an intersection, a car that had been stopped on the crossroad suddenly accelerated into the intersection, then halted in the drivers' lane. The reaction time of every driver was sufficient, in theory, to avoid a crash. But only ten of forty-nine did. Part of the problem is that they had only time enough to react to the presence of the approaching car, and not enough time to fully discern what the intruding car was going to do. It was less about a correct maneuver than a roll of the dice.

Whether advanced driver training helps drivers in the long term is one of those controversial and unresolved mysteries of the road, but my eye-opening experience at Bondurant raises the curious idea that we buy cars—for most people one of the most costly things they will ever own—with an underdeveloped sense of how to use them. This is true for many things, arguably, but not knowing what the F9 key does in Microsoft Word is less life-threatening than not knowing how to properly operate antilock brakes.

This uneasy idea is one of the many unresolved tensions and contradictions found in driving and the traffic it spawns. There is the contradiction of the car itself: With its DNA steeped in racing, today it's often just part of a loosely organized, greatly inefficient mass-transit system, a "living room on wheels." To drive safely is often to become rather bored, which may lead us to become distracted and thus less safe. On the other hand, if we drove like racers, we would have little problem becoming distracted or falling asleep, but we would inherently be driving less safely. (Even the most skilled drivers cannot overcome the fundamental physics of things like stopping distance.) We all think we're better than the average driver. We think cars are the risk when on foot; we think pedestrians act dangerously when we're behind the wheel. We want safer cars so we can drive more dangerously. Driving, with its exhilarating speed and the boundless personal mobility it grants us, is strangely life-affirming but also, for most of us, the most deadly presence in our lives. We all want to be individuals on the road, but smooth-flowing traffic requires conformity. We want all the lights to be green, unless we are on the intersecting road, in which case we want *those* lights to be green. We want little traffic on our own street but a convenient ten-lane highway blazing just nearby. We all wish the other person would not drive, so that our trip would be faster. What's best for us on the road is often not best for everyone else, and vice versa.

. . .

The reason I have avoided talking about the negative environmental consequences of the car is that I believe, as was once said, that it will be easier to remove the internal-combustion engine from the car than it will to remove the driver. With fuel economy liberated by some renewable, sustainable fuel source of the future, all the dynamics of traffic I have described will only become more amplified. As Larry Burns, vice president of R&D and strategic planning at General Motors, put it to me, "Of all the externalities of an auto that I worry about—energy, environment, equality of access, safety, and congestion—the one that I think is toughest to solve is congestion."

Even if the driver is still in the car, whether he or she will be driving in the future is another question. Virtually all of the perceptual limitations we have in driving—blind spots, overdriving our headlights, problems in detecting the rate of closure—are being addressed by scientists and car manufacturers. High-end cars already bristle with these features. An ad for BMW's xDrive system, which "uses sensors to monitor the road ahead," puts it succinctly. It says, "xDrive reaction time: 100 milliseconds. Human reaction time: unnecessary." Technologies like "gaze detection," in which the car will tell the driver that he or she is not paying attention (by tracking eye movements), are on the horizon.

The future of driving will probably look a lot less like the track at Bondurant and much more like the 200,000-square-foot parking lot at AT&T Park (ordinarily home of the San Francisco Giants) during the World Congress on Intelligent Transport Systems. The parking lot had been converted into a "Innovative Mobility Showcase" for any number of high-tech traffic devices. It looked like a kind of strange carnival of human limitations. There were "Intelligent Intersections" that could alert drivers when an approaching driver did not seem, as calculated by sensors and algorithms, intent on stopping and "Dynamic Parking" demonstrations that promised to end, through real-time sensors, the search for open parking spots.

I was riding in a Cadillac CTS with C. Christopher Kellum and Priyantha Mudalige, two researchers with General Motors. The car, via GPS technology and receivers, was communicating with the other cars, also equipped with the technology. GM calls its technology "vehicle to vehicle," and the idea is that by connecting all the cars in a kind of mobile network, this shared intelligence can help you "watch for the other guy," as Mudalige put it. A screen displayed the fact that we were connected to two other vehicles. The researchers are aware that any system released into the real world would have to contend with hundreds more at a time. "We do lots of simulations to understand what happens when there's two thousand vehicles in the same spot," said Kellum. "We need an intelligent way to parse out what information is important and what's not important. If there's an accident a mile ahead, you want that information. If it's just some guy driving a mile ahead, you don't really care."

If this sounds familiar, it's because it is: This kind of incident detection and evaluation was one of the key tasks the Stanford team had targeted in getting their robotic car Junior to drive successfully in simulated urban traffic. I was, I realized, sitting in Junior's cousin. Kellum asked me to change lanes, even though I knew, in this case, that a neighboring car had crept into my blind spot. As I put the signal on, I felt a small, Magic Fingers–style vibration in my back. This is known as a haptic warning, and it is used so that the driver will not be overwhelmed with visual or auditory information, or to underscore warnings he or she might disregard. (As you will feel when your car has drifted off the road into gravel, haptic warnings can be crudely effective.) One of the issues that haunts driver-assist technologies like "lane-departure warnings" is that these warnings can become ever more prescient, ever more sophisticated, but drivers still have to pay attention to the warning and be able to react accordingly.

Or perhaps not. Next, Kellum asked me to drive at a steady clip toward a parked car far in the distance. "Whatever feels comfortable," he said. He then asked me not to press the brakes. "We're going to go up there and our car's going to brake automatically," he said. "In real time we're constantly assessing how far away we are, the closing speeds, and when to start braking. I've done this at seventy-five miles per hour." This was essentially the same exercise as at Bondurant, but instead of being asked to lock up the ABS, I was being asked to sit back and do nothing. I was in Junior, and I was riding shotgun. The stopped car quickly loomed into view. Time seemed to slow for a moment. (In reality, as studies have suggested, it probably sped up and this was just my memory playing tricks.) A chill shot through my body; the hairs on my neck tingled. Images of blooming air bags and the buckling necks of crash-test dummies ran through my head like a fleeting nightmare. The car came to a perfect stop.

Somewhere down the road, in some distant future, humans may evolve to become perfect drivers, with highly adapted vision and reflexes for moving seamlessly at high speeds. Perhaps, like the ants, we will turn the highways into blissfully cooperative, ultraefficient streams of movement, with no merging or tailgating or finger flipping. Long before that happens, however, a sooner future seems likely: cars driving themselves, at smoothly synchronized speeds to ensure maximum traffic flow and safe following distances, equipped with merging algorithms set for highest throughput, all overseen by network routers that guide cars down the most efficient paths on these information superhighways. Maybe this will be the traffic nirvana for which we have been searching. We would do well, though, to remember the warning from the mid-twentieth-century traffic engineer Henry Barnes: "As time goes on the technical problems become more automatic, while the people problems become more surrealistic." Even if drivers are taken away from the wheel, can we ever take the mere fact of being human out of traffic?

# Acknowledgments

Despite possessing the small diploma known as a driver's license, I was, throughout the course of this endeavor, a novice in a complex field. I relied on the help of many people in many places, without whom this book would have been impossible.

In no logical order, then, and with any omissions purely unintended, allow me to unravel the roster of gratitude, beginning geographically with the American Middle West. At the University of Iowa and at its National Advanced Driving Simulator, Daniel McGehee, John Lee, Omar Ahmad, and Tara Smyser patiently explained their findings and looked the other way as I skidded out of control in Virtual Iowa on the world's most advanced driving simulator. At the University of Michigan, Michael Flannagan and Daniel Blower at the Transportation Research Institute, and Barry Kantowitz in the Department of Engineering, walked me through ergonomics, vision, and other topics. Over in Warren, Michigan, and in Detroit, Richard A. Young, Larry Burns, and Linda S. Angell of General Motors popped open the hood on the automaker's research. In Chicago, Howard Hayes and Larry Peterson of Navteq walked me through the company's traffic monitoring operations, while Jean Gornicki took me on a Navteq mapping drive of the suburbs. At the University of North Dakota, Mark Nawrot taught me Motion Parallax 101, among other things.

In Los Angeles, special thanks are due to John E. Fisher, Assistant General Manager of the Los Angeles Department of Transportation, and Frank Quon, Deputy District Director of Operations for District 7, for sharing their extensive knowledge and insight into how traffic in L.A. functions. Thanks also to Marco Ruano, Dawn Helou, Afsaneh M. Razavi, and Jeanne Bonfilio of Caltrans, and James Okazaki, Kartik Patel, and Verej Janoyan of LA DOT. Thanks to Chris Hughes, Claire Sigman, and Shane Novicki at Clear Channel's Airwatch in Orange County, as well as

Vera Jimenez at CBS2 in Los Angeles, for dishing on L.A. traffic in all its infinite varieties. Sergeant Joseph Zizi of the California Highway Patrol gave me an intimate view into patroling the highway and answered any number of statistical queries. At UCLA, a number of people across different departments shared their expertise: Donald Shoup, Jay Phelan, Brian D. Taylor, Randall Crane, and Jack Katz. At Stanford University, thanks to Sebastian Thrun and Michael Montemerlo.

In the New York region, thanks are due to Kay Sarlin, Ryan Russo, and Michael Primeggia of the New York City Department of Transportation. Sam "Gridlock Sam" Schwartz of Sam Schwartz PLLC and Michael King at Nelson/Nygaard provided invaluable insight and commentary on New York traffic. Aaron Naparstek was a constant source of traffic inspiration, and under his editorship, streetsblog.org remains the world's single best source of transportation news and opinion. At the New Jersey Department of Transportation in Trenton, Gary Toth and Yosry Bekhiet gave me a tour of the city's highway overhaul and patiently explained "Jersey jughandles" and other exotic traffic creatures of the Garden State (where this book began). In the Washington, D.C.–Beltway area, special thanks to Nancy McGuckin and Alan Pisarski; and, at the Federal Highway Administration, thanks to Tom Granda, Carl Anderson, Doug Hecox, John McCracken, Michael Trentacoste, Bill Prosser, and Ray Krammes for the tour of the Turner-Fairbank Lab, the lively roundtable discussion, and subsequent conversations. At the National Highway Safety Administration, thanks to Charles Kahane and Patricia Ellison-Potter.

In Canada, Gerry Wilde offered his theories on risk homeostasis (and top-drawer espresso). Baher Abdulhai, founder and head of the Intelligent Transportation Systems Centre and Testbed at the University of Toronto, explained the "fundamental diagrams" and other traffic intricacies to me. In Mexico City, Mario González-Román took me driving on the monumental Segundo Piso and helped in countless other ways. Thanks also to Agustín Barrios Gómez and Alan Skinner. Alfredo Hernández García, executive director of traffic control and engineering at the Secretaría de Seguridad Pública of the Gobierno del Distrito Federal, opened up the city's Traffic Management Center in the Colonia Obrera. Thanks also to Claudia Adeath at Muévete por tu Ciudad, which deserves kudos for trying to calm Mexico City's often hostile traffic.

In England, thanks to Malcolm Murray-Clark, Director of Congestion Charging in London, and Phil Davis, at Transport for London's London Traffic Control Centre. Peter Weeden of the Royal Kensington Borough Council graciously offered his time and expertise. John Adams, professor emeritus at University College London, offered his always trenchant thoughts on risk. At the Transport Research Laboratory in Wokingham, Janet Kennedy shared her expertise and the lab's driving simulator. Thanks also to John Groeger at the University of Surrey, Jake Desyllas at Intelligent Space, and Bill Hillier and Alain Chiaradia at Space Syntax. In Germany, Michael Schreckenberg at the University of Duisburg-Essen's Physics of Transport and Traffic department held a wide-ranging and illuminating symposium for me on the personal and system-wide physics of traffic. At the Bundesanstalt für Straßenwesen (Federal Highway Research Unit) in Bergisch Gladbach, Germany, Karl-Josef Höhnscheid and Kerstin Lemke answered my questions about the autobahn and other topics. Thanks also to Juergen Berlitz at the ADAC (Allgemeiner Deutscher Automobil-Club). In Copenhagen, thanks are due to the esteemed traffic guru Jan Gehl, at Jan Gehl Associates; and Steffen Rasmussen, of the city's Traffic and Planning Office. In Italy, many thanks to Paolo Borgognone and Giuseppe Cesaro of the Automobile Club d'Italia for the traffic knowledge and the excellent *cacio e pepe*. Thanks also to Andrea del Martino at the Laboratory of Complex Systems at "La Sapienza," and Max Hall, physics teacher and Roman Vespa rider.

In Beijing, thanks to Wang Shuling, Xian Kai, and Zhang Dexin at the Beijing Transportation Research Center for explaining the evolving complexities of traffic in the capital. Thanks also to Professors Rong Jian and Chen Yanyan at the Beijing University of Technology, as well as Dehui Lee. Thanks also to Lui Shinan at the *China Daily*; and Scott Kronick, Jonathan Landreth, and Alex Pasternak. In Shanghai, thanks to Jian Shuo Wang, and Zhongyin Guo of Tongji University; thanks also to Dan Washburn for hospitality and advice. In Japan, thanks to Paul Nolasco, Imai Tomomi, and James Corbett for arranging the tour of Toyota's Integrated System Engineering Division in Nagoya. In Hanoi, Vietnam, thanks to Walter Molt and Grieg Craft, who are, in their own different ways, trying to make the city's transportation better and safer. In Delhi, thanks to Maxwell Pereira; Geetam Tiwari and Dinesh Mohan at the Indian Institute of Technology; and Joint Commissioner of Police Qamar Ahmed. Thanks also to Rohit Baluja, Girish Chandra Kukreti, and Amandeep Singh Bedi of the Institute for Road Traffic Education.

Thanks must also go to a number of people, across the globe, who discussed their research, showed the way, corrected my mistakes. Again, in no order: Per Garder at the University of Maine; Eric Dumbaugh at Texas A&M University; Ezra Hauer, professor emeritus, University of Toronto; Walter Kulash, Dan Burden, and Ian Lockwood of Glatting Jackson in Orlando, Florida; Allan Williams and Kim Hazelbaker of the Insurers' Institute for Highway Safety; Sheila "Charlie" Klauer and Suzie Lee of the Virginia Tech Transportation Institute; Charles Zegeer at the Highway Safety Research Center; Erik Olson at the National Institute of Child Health and Human Development; Del Lisk, Bruce Moeller, and Rusty Weiss of DriveCam in San Diego; Christopher Patten of the Swedish Road Administration; John Dawson at the European Road Assessment Program; Tom Bernthal of Kelton Research; Sandi Rosenbloom at the University of Arizona; Tova Rosenbloom of Bar-Ilan University in Israel; Heikki Summala of the Traffic Research Unit at the University of Finland; Oliver Downs and Michele Largé at INRIX; Hussein Dia at the University of Queensland Intelligent Transport Systems lab; Graham Coe at the Transport Research Laboratory; Nick Fenton at U.K Highways Agency; Robert Gray at Arizona State University; Norman Garrick at the University of Connecticut; James Cutting at Cornell University; Anna Hackett, Bob Bondurant, Les Betchner, and Mike McGovern at the Bondurant School of High Performance Driving; Judie Zimomra and Amanda Rutherford in Sanibel Island, Florida; Charles Spence at the University of Oxford; Eric Bonabeau at Icosystem; Antti Oulasvirta at the University of California, Berkeley; Stephen Lea at the University of Exeter; Denis Wood at the University of North Carolina; Eleanor Maguire at University College London; Dale Purves at Duke University; Michael Spivey at Cornell University; Kara

Kockelman at the University of Texas; Moshe Ben-Akiva at the Massachusetts Institute of Technology; Gary Evans at Cornell University; John Kobza at Texas Tech University; Timothy McNamara at Vanderbilt University; John Van Horn at *Parking Today*; Andrew Velkey at Christopher Newport University; Franco Servadei at Ospedale "M. Buttalini," Cesena, Italy; Gary Davis at the University of Minnesota; Robert Cialdini at Arizona State University; Marc Ross at the University of Michigan; Nicholas Garber at the University of Virginia; Tom Wenzel at Lawrence Berkeley National Laboratory; Phil Jones of Phil Jones Associates in the United Kingdom; Jake Desyllas of Intelligent Space in London; Sidney Nagel and Lior Strahilevetz at the University of Chicago; Frank McKenna at the University of Reading; Geoff Underwood at the University of Nottingham; Daniel Lieberman at Harvard University; Stephen Popiel at Synovate; Asha Weinstein Agrawal at San Jose State University; Jeffrey Brown at Florida State University; Gordy Pehrson at the Office of Traffic Safety in St. Paul, Minnesota; David Levinson at the University of Minnesota; Charles Komanoff at Komanoff Energy Associates; Giuseppe La Torre at the Catholic University in Rome; Eric Poehler at the University of Virginia; Mark Horswill at the University of Queensland; Michael Paine at Vehicle Design and Research in Australia; Joseph Barton at Northwestern University; Anna Nagurney at the University of Massachusetts; David Gerard and Paul Fischbeck at Carnegie Mellon University; Andy Wiley-Schwartz, then of the Project for Public Spaces; Craig Davis at the University of Michigan; Bruce Laval, formerly of Disney; and Richard Larson at the Massachusetts Institute of Technology.

A handful of people deserve even more emphatic thanks for going above and beyond in sharing their research, or reading drafts of chapters. Leonard Evans, the dean of traffic safety, was always there to offer his expertise. Jeffrey Muttart made time to talk on countless occasions and ran experiments on my behalf. Stephen Most at the University of Delaware and Daniel Simons at the University of Illinois read parts of the manuscript and offered useful commentary, as did Matthew Kitchen of the Puget Sound Regional Council. Benjamin Coifman at Ohio State University helped me through the complexities of traffic flow. Ian Walker at the University of Bath is a brilliant scholar and all-around mensch. Iain Couzin at Oxford and Princeton led me through the world of ant traffic. James Surowiecki and Matt Weiland read drafts and offered honest feedback. Peter Hall graciously chipped in with research help. Ben Hamilton-Baillie, impassioned "shared space" advocate and wizard of the slide show, led me on an eye-opening tour through Germany and the Netherlands, where he generously introduced me to Joost Váhl, one of the seminal forces in traffic calming and engineering with a human face, and Hans Monderman, whose words and spirit pervade this book. My time spent with Hans, and subsequent conversations, revealed a man brimming with passion, insight, sly wit, and a surprisingly capacious range of interests. Into his discussion of left-turn gap acceptance or roundabout capacity he would percolate ideas on how the geography of the Netherlands fostered Dutch innovation, or quote Proust on how the automobile changed our conception of time. Hans died on January 7, 2008, after a several-year fight with cancer. I only hope I can help Hans's legacy live on in these pages.

I am indebted to Andrew Miller at Alfred A. Knopf, who encouraged me early on when the book was nothing but the grain of an idea, and subsequently was a steadfast presence, offering judicious editorial counsel, moral support, and the occasional football result. Sara Sherbill at Knopf also contributed a number of good criticisms, most of which helped shape the final book. Bonnie Thompson corralled wayward grammar, exposed logical lacunae, and kept facts this side of veracity. Thanks to the Knopf publicity team, Paul Bogaards, Gabrielle Brooks, Erinn Hartman, Nicholas Latimer, and Jason Kincade. Will Goodlad at Penguin UK offered all of the above from across the Atlantic. Lastly, I am immensely obliged to my agent and longtime friend, Zoe Pagnamenta, at PFD New York. She has been a tireless and sagacious advocate for me and the book, and I never felt as if I were going it alone. I am also grateful to Simon Trewin at PFD in London.

And finally, this book is dedicated to my family, near and far, who were there from the beginning of the journey; especially my wife, Jancee Dunn, my beautiful, brilliant co-passenger in the car, and in life.

# Notes

## Prologue: Why I Became a Late Merger

5 in a business magazine: Matt Asay, "How Team Works." *Connect*, November 2003. Retrieved from http://www.connect-utah.com/article.asp?r=189&iid=17&sid=4.

6 mingle so freely: There are exceptions to this, of course, as in the case of the ban on women drivers in Saudi Arabia (which extends even to golf carts) or the segregated highways in Israel for Palestinians and Israelis. See Brian Whitaker, "Saudi Driving Ban on Women Extends to Golf Carts," *Guardian*, March 3, 2006, and Steven Erlanger, "A Segregated Road in an Already Driving Land," *New York Times*, August 11, 2007.

7 people and things became interchangeable: Sean Dockray, Steve Rowell, and Fiona Whitton point out that while terms like *computer* and *typewriter* used to refer to people (e.g., the profession of a typewriter), they now refer exclusively to the technologies themselves. We have become traffic, they argue, but we do not like to admit that in our language. See "Blocking All Lanes," *Cabinet*, no. 17 (Spring 2005).

8 on certain streets altogether: See Eric Poehler, "The Circulation of Traffic in Pompeii's Regio VI," *Journal of Roman Archaeology*, vol. 19 (2005), pp. 53–74.

8 no traffic or street signs: Poehler argues that given the level of preservation at Pompeii, had these signs existed there would likely be archeological evidence today. Drivers, he suggests, relied instead upon the cues of other drivers or design cues in the streetscape, while people looking for addresses relied more upon relative cues (e.g., turn left at the butcher shop or right at the shrine). Correspondence with Eric Poehler.

8 Vico di Mercurio: Poehler suggests that these changes must have been overseen by some kind of Department of Traffic Engineering. "The inescapable implication is that the traffic system was carefully managed by a central, executive individual or group at the municipal level." See Eric Poehler, "A Reexamination of Traffic in Pompeii's Regio VI: The Casa del Fauno, the Central Baths, and the Reversal of Vico di Mercurio," *Archaeological Institute of America* (2005).

8 In ancient Rome: The Roman traffic history comes from *The Roads of the Romans*, by Romolo August Staccioli (Rome: L'Erma di Bretschneider, 2003), in particular pp. 21–23.

9 "a devil-fish from sleeping": quoted in ibid, p. 23.

9 "of the Mayor": The English traffic history comes from the wonderful book *Street Life in Medieval England*, by G. T. Salusbury Jones (Oxford: Pen-in-Hand, 1939).

9 "contesting for the way": The information on traffic fatalities and the accounts of London drivers are taken from Emily Cockayne's exemplary study *Hubbub: Filth, Noise and Stench in England* (New Haven: Yale University Press, 2007), pp. 157–80.

9 "reckless drivers": The 1867 pedestrian fatality figure comes from *Ways of the World: A History of the World's Roads and of the Vehicles That Used Them* (New Brunswick, N.J.: Rutgers University Press, 1992), p. 132.

9 "wish to pass over": *New York Times*, April 9, 1888.

10 "to show illumination at night": "Our Unwary Pedestrians," *New York Times*, December 24, 1879.

10 right-of-way to women?: For a delightful account of the impact of the bicycle on American culture, see Sidney H. Aronson, "The Sociology of the Bicycle," *Social Forces*, vol. 30, no. 3 (March 1952), pp. 305–12. Aronson noted, "Thus it can be concluded that the bicycle provided a preview on a miniature scale of much of the social phenomena which the automobile enlarged upon."

10 "good roads": For more on the history on the bicycle, including the Good Roads Movement, see David Herlihy's comprehensive *Bicycle: The History* (New Haven: Yale University Press, 2005), p. 5. Bicycle manufacturing, Herlihy notes, was the forerunner of the mass assembling of automobiles, and many bicycle-repair shops were converted into gas stations.

11 "social or business prominence": *New York Times*, September 15, 1903.

11 "right way to turn a corner": "Proposed Street Traffic Reforms," *New York Times Magazine* supplement, February 23, 1902.

11 "special indications meant": from Gordon M. Sessions, *Traffic Devices: Historical Aspects Thereof* (Washington, D.C.: Institute of Traffic Engineers, 1971), p. 63.

12 "red" time remained: The Wilshire and Western traffic light information comes from Sessions, *Traffic Devices*, ibid., p. 45.

12 learned red and green?: The story about color blindness and traffic signals comes from Clay McShane, "The Origins and Globalization of Traffic Control Signals," *Journal of Urban History*, March 1999. p. 396.

12 roles of city streets: Jeffrey Brown, "From Traffic Regulation to Limited Ways: The Effort to Build a Science of Transportation Planning," *Journal of Planning History,* vol. 5, no. 1 (February 2006), pp. 3–34.

12 collapse of the Berlin Wall: For a fascinating discussion of how German Democratic Republic traffic engineering was affected by the reunification of Germany, and the cultural underpinnings and consequences of those decisions, see Mark Duckenfield and Noel Calhoun, "Invasion of the Western *Ampelmännchen,*" *German Politics,* vol. 6, no. 3 (December 1997), pp. 54–69.

13 offers no improvement at all: As I was succinctly told by Michael Primeggia, deputy director of operations at New York City's Department of Transportation, "People have argued that the countdown signal gives more information to peds to make intelligent choices. Why would I think more info would be better, when right now I provide them good information and they choose to ignore it?" Some studies have found that pedestrians were *less* compliant with countdown signals; see, for example, H. Huang and C. Zegeer, "The Effects of Pedestrian Countdown Signals in Lake Buena Vista," University of North Carolina Highway Safety Research Center for Florida Department of Transportation, November 2000. Accessible via www.dot.state.fl.us/safety/ped_bike/handbooks_and_research/research/CNT-REPT.pdf. This could be an artifact, of course, of pedestrians rationally analyzing the situation and deciding that they have plenty of time to cross the street before their signal has expired. While they are technically "violating" the signal, they are also using the information smartly.

13 gradually rolled back: For a discussion of differential speed limits and their effects on safety, see "Safety Effects of Differential Speed Limits on Rural Interstate Highways," Federal Highway Administration, Washington, D.C., October 2005, FHWA-HRT-05-042.

13 "become more surrealistic": Henry Barnes, *The Man with the Red and Green Eyes* (New York: Dutton, 1965), p. 218.

14 "things you can do": Ralph Vartabadian, "Your Wheels," *Los Angeles Times,* May 14, 2003.

14 "explicit argument": The quote about convex mirrors comes from a telephone interview with Michael Flannagan.

15 insurance company surveys: A 2002 survey by Progressive Insurance, for example, which queried more than eleven thousand drivers who had filed a claim for a crash in 2001, found that 52 percent of the accidents occurred within five miles of the driver's home, and 77 percent occurred within fifteen miles of the home. Retrieved on October 3, 2007, from http://newsroom.progressive.com/2002/May/fivemiles.aspx.

15 A study by: See, for example, Tova Rosenbloom, Amotz Perlmana, and Amit Shahara, "Women Drivers' Behavior in Well-known Versus Less Familiar Locations," *Journal of Safety Research,* vol. 38, issue 3, 2007, pp. 283–88. Studies have also shown drivers are less likely to wear seat belts on shorter trips, which would seem to indicate a feeling of greater safety close to home. See, for one, David W. Eby, Lisa J. Molnar, Lidia P. Kostyniuk, Jean T. Shope, and Linda L.

Miller, "Developing an Optimal In-Vehicle Safety Belt Promotion System" (Ann Arbor: University of Michigan Transportation Research Institute, 2004).

15 food or health care: *Driven to Spend* (Surface Transportation Policy Project, 2001).

15 more own three than own one: Alan Pisarski, *Commuting in America III* (Washington, D.C.,: Transportation Research Board, 2006), p. 38.

16 has a three-car garage: Amy Orndorff, "Garages Go Gigantic: Car Buffs Opt for Bigger Spaces," *Washington Post*, September 13, 2006.

16 thirty-eight hours annually: See Tim Lomax and David Schrank, *2007 Annual Urban Mobility Report*, compiled for the Texas Transportation Institute (College Station: Texas A&M University, 2007).

16 by nearly half: Surface Transportation Policy Partnership, *Mean Streets 2002*, chapter 2. Retrieved at http://www.transact.org/report.asp?id=159.

16 "food and beverage venue": This phrase comes from *Food and Drug Packaging*, March 2002.

16 there were 504: *Frozen Food Age*, vol. 54, no. 1 (August 2005), p. 38.

16 84.4 billion in 2008: On-the-go eating figures come from the market research firm Datamonitor.

16 gentler, slower age: Drive-through sales figure comes from the *Wall Street Journal*, May 21, 2000.

16 through a car window: *Chicago Sun-Times*, October 7, 2005.

16 at least once per week: According to a survey commissioned by the Food Strategy Implementation Partnership (FSIP), Bord Bia, and Intertrade Ireland, and carried out by Invest NI, as quoted in *Checkout*, February 2006.

16 in order to speed traffic: Julie Jargon, "McD's Aims for the Fast Lane." *Crain's Chicago Business*, June 27, 2005, p. 3. The article does note that the two ordering lanes must merge into one paying lane; there is no word of any reported merging difficulties.

16 burgeoning drive-through customers: Geoffrey Fowler, "Drive-Through Tips for China," *Wall Street Journal*, June 20, 2006.

16 company-owned stores: Elizabeth M. Gillespie, "Starbucks Bows to Customer Demand," *Toronto Star*, December 27, 2005.

17 "handle well in the car": This quote comes from a press release accessed through Business Wire, retrieved at http://www.hispanicprwire.com/news.php?l=in&id=4394&cha=4. The dashboard-dining test was performed by Kelton Research; it was the firm's CEO, Tom Bernthal, whom I met with to discuss the test.

17 drive-through window: Carole Paquette, "Drive-Throughs Move Beyond Banks and Fast Food," *New York Times*, April 8, 2001.

17 Audio Publishers Association: Information on audiobooks comes from documents provided by the Audio Publishers Association.

17 to bear in L.A. traffic: Idan Ivri, "Gridlock: How Traffic Has Rerouted Jewish Life," *Jewish Journal*, July 9, 2004. The political scientist Robert Putnam suggests that every ten minutes of commute time "*cuts involvement in community affairs by 10 percent*" (Putnam's italics); from Putnam, *Bowling Alone: The Col-*

*lapse and Revival of American Community* (New York: Simon and Schuster, 2001), p. 213.

17 on their left sides: Based on research by Scott Fosko, Saint Louis University School of Medicine. Article retrieved from: http://www.aad.org/aad/Newsroom/Driving+An+Automobile.htm.

17 "toward the same horizon": From Alexis de Tocqueville, *Democracy in America* (repr.; London: Penguin, 2003), p. 328.

18 double since 1990: Elizabeth Rosenthal, "Car Boom Puts Europe on Road to a Smoggy Future," *New York Times*, January 7, 2007.

18 underground parking garages: "Car Ownership Boom Means Traffic Jams in Once-Tranquil Tibet," *International Herald Tribune*, November 7, 2007.

18 Caracas: Rory Carroll, "Carbon Leaves Caracas in One Big Jam," *Guardian*, November 23, 2006. The "seven cents" gas figure comes from Simon Romero, "Venezuela Hands Narrow Defeat to Chavez Plans," *New York Times*, October 30, 2007.

18 the legendary traffic: In a 2004 estimate, São Paulo was said to have just under four miles of freeways to accommodate more than five million vehicles. Los Angeles, by contrast, had nine hundred miles to handle about seven million vehicles. See Henry Chu, "São Paulo Seeks Road Map to Life Without Traffic Jams," *Los Angeles Times*, November 9, 2004. In 2007, a rising number of fatal helicopter crashes was prompting calls to limit the growing airborne traffic. See Cristina Christiano, "SP quer limitar tráfego de helicópteros," *O Globo*, September 24, 2007.

18 faster car-pool lanes: Matthew Moore, "Car Jockeys Often in for Rough Ride from Traffic Police," *Sydney Morning Herald*, December 26, 2002.

18 human "nav system": This information came via an e-mail correspondence with Jian Shou Wang.

18 cause of death: World Health Organization. Retrived from: http://www who int/world-health day/2004/infomaterials/en/brochure_jan04_en.pdf.

## Chapter One: Why Does the Other Lane Always Seem Faster?

20 "modal bias": This term was suggested to me in a conversation with Aaron Naparstek.

20 "change of mode": Hélène Fontaine and Yves Gourlet, "Fatal Pedestrian Accidents in France: A Typological Analysis," *Accident Analysis and Prevention*, vol. 39, no. 3 (1997), pp. 303–12.

20 "drives as he lives": W. A. Tillman and G. E. Hobbs, "The Accident-Prone Automobile Driver: A Study of the Psychiatric and Social Background," *American Journal of Psychiatry*, vol. 106 (November 1949), pp. 321–31. Many of us may think of "road rage" as a rather new concept, like "air rage" or "surfing rage," but it is really as old as the automobile itself. The year 1968, for example, might have been marked by violent social upheaval in metropolises from Paris to Mexico City, but there was another form of violence in the air: That year,

Mayer H. Parry published *Aggression on the Road,* while the *New York Times* reported on government testimony about "uncontrollable violent behavior" on the nation's roads. (Three years later, F. A. Whitlock followed up with his book *Death on the Road: A Study in Social Violence.*) See John D. Morris, "Driver Violence Tied to Crashes," *New York Times,* March 2, 1968.

20 risks on the road: For a discussion, see Patrick L. Brockett and Linda L. Golden, "Biological and Psychobehavioral Correlates of Credit Scores and Automobile Insurance Losses: Toward an Explication of Why Credit Scoring Works," *Journal of Risk and Insurance,* vol. 1, no. 74 (March 2007), pp. 23–63.

20 typically involve questionnaires: See, for example, David L. Van Rooy, James Rotton, and Tina M. Burns, "Convergent, Discriminant, and Predictive Validity of Aggressive Driving Inventories: They Drive as They Live," *Aggressive Behavior,* vol. 3, no. 2 (February 2006), pp. 89–98.

21 more aggressive manner: This is a virtual consensus in the field, as demonstrated by a survey of the scholarly literature in B. A. Jonah, "Sensation Seeking and Risky Driving: A Review and Synthesis of the Literature," *Accident Analysis and Prevention,* vol. 29, no. 5 (1997), pp. 651–65.

21 "Traffic tantrums": Thanks to Ian Walker for this phrase.

21 especially by novice drivers: Kazumi Renge, "Effect of Driving Experience on Drivers' Decoding Process of Roadway Interpersonal Communication," *Ergonomics,* vol. 43, no. 1 (1 January 2000), pp. 27–39.

21 Green Day bumper sticker: This brings up the point of whether there should really be any nonessential communication in traffic at all. As the German sociologist Norbert Schmidt-Relenberg has observed, "It could be said that cooperation in traffic is not a means to attain something positive, but to avoid something negative: every participant in the system attempts to attain his destination without friction. Hence traffic is a system all its own; the less its participants come into contact with each other and are compelled to interaction, the better it works: a system defined and approved in the reality by a principle of minimized contact." In other words, not only should we not honk at people with Green Day stickers, we should not put the sticker there in the first place. Norbert Schmidt-Relenberg, "On the Sociology of Car Traffic in Towns," in *Transport Sociology: Social Aspects of Transport Planning,* ed. Enne de Boer (Oxford, New York: Pergamon Press, 1986), p. 122.

22 violated traffic laws: María Cristina Caballero, "Academic Turns City into a Social Experiment," *Harvard University Gazette,* March 11, 2004.

22 associated with subordination: Katz suggests this may be why we so often call other drivers "assholes" and give the "up yours" finger.

22 by the honker: Andrew R. McGarva and Michelle Steiner, "Provoked Driver Aggression and Status: A Field Study," *Transportation Research F: Psychology and Behavior,* vol. 167 (2000), pp. 167–179.

22 anything but rude or hostile: What if our signals were more meaningful? A few years ago, before the Tokyo Motor Show, Simon Humphries, a designer for Lexus in Japan, told me in an e-mail exchange that the Toyota Motor Company had proposed a car—nicknamed POD—that would contain a "vehicle

expression operation control system." Accompanying the usual lights and arrows would be a new range of signals. The headlights would be "anthropomorphized" with "eyes" and "eyebrows," the antenna would "wag," and different colors would be used to show emotion. "As traffic grows heavier and vehicle use increases," reads the U.S. patent application, "vehicles having expression functions, such as crying or laughing, like people and other animals do, could create a joyful, organic atmosphere rather than the simple comings and goings of inorganic vehicles." Indeed, a German company even released an aftermarket version of this system, called Flashbox, that uses a series of blinks to signify things like "apology," "annoyed," and "stop for more?" Adding signals, however, creates many new problems. Everyone has to learn the new signals. More information in traffic means more time to process. The receiver of a "smile," moreover, may not understand why they have received it any more so than a honk. And flashing "angry" signals may provoke rather than defuse violence.

23 deficient male anatomy: One male Australian driver was actually fined because when a woman wagged her pinkie at him, he responded by hurling a plastic bottle at her windshield. The man claimed that the gesture was akin to a "sexual assault," a worse insult than the traditional finger. "The 'finger,' it's so common now, that we're over it, but this finger is a whole new thing and it's been promoted so much everybody knows it and you just get offended," he said. David Brouithwaite, "Driver Points to Ad Campaign for His Digitally Enhanced Road Rage," *The Sydney Morning Herald*, November 1, 2007.

23 "constructing moral dramas": For a more detailed discussion of Katz's investigation of anger in traffic, see Jack Katz, *How Emotions Work* (Chicago: University of Chicago Press, 1999), in particular the first chapter, "Pissed Off in L.A."

23 "the angry driver": Jack Katz, *How Emotions Work*, p. 48.

23 "actor-observer effect": See L. D. Ross, "The Intuitive Psychologist and His Shortcomings: Distortions in the Attribution Process," in *Advances in Experimental Social Psychology*, vol. 10, ed. L. Berkowitz (New York: Random House, 1977), pp. 173–220.

23 feel more in control: As Thomas Britt and Michael Garrity write, "individuals will probably err in the direction of assuming an internal locus of causality for the offending driver's behavior in order to feel some sense of control over events when driving." "Attributions and Personality as Predictors of the Road Rage Response," *British Journal of Social Psychology*, vol. 45 (2006), pp. 127–47.

24 required by the circumstances: This was the finding arrived at when a group of researchers for England's Transport Research Laboratory conducted a series of interviews with drivers, part of which included assessments of cyclist and driver behavior in traffic scenarios. They concluded, "The underlying unpredictability of cyclists' behavior was seen by drivers as stemming from the attitudes and limited competence of the cyclists themselves, rather than from the difficulty of the situations that cyclists are often forced to face on the road (i.e., drivers made a dispositional rather than a situational attribution). Despite their own evident difficulties in knowing how to respond, drivers never attributed these difficul-

ties to their own attitudes or competencies, nor did they do so in relation to other drivers (i.e. they made a situational attribution about their own and other drivers' behavior). This pattern of assignment of responsibility is characteristic of how people perceive the behavior of those they consider to be part of the same social group as themselves, versus those seen as part of a different social group." L. Basford, D. Davies, J. A. Thomson, and A. K. Tolmie, "Drivers' Perception of Cyclists," in *TRL Report 549: Phase I—a Qualitative Study* (Crowthorne: Transport Research Laboratory, 2002).

24 shares their birth date: See D. T. Miller, J. S. Downs, and D. A. Prentice, "Minimal Conditions for the Creation of a Unit Relationship: The Social Bond Between Birthday Mates," *European Journal of Social Psychology*, vol. 28 (1998), pp. 475–81. This idea was raised in an interesting paper by James W. Jenness, "Supporting Highway Safety Culture by Addressing Anonymity," AAA *Foundation for Traffic Safety*, 2007.

24 Katz says, cyborgs: This point was made as early as 1930, by a city planner in California who suggested that "Southern Californians have added wheels to their anatomy." The quote comes from J. Flink, *The Automobile Age* (Cambridge, Mass: MIT Press, 1988), p. 143, via an excellent article by John Urry, a sociologist at Lancaster University. See John Urry, "Inhabiting the Car," published by the Department of Sociology, Lancaster University, Lancaster, United Kingdom, available at http://www.comp.lancs.ac.uk/sociology/papers/Urry-Inhabiting-the-Car.pdf.

25 different people: See Henrik Walter, Sandra C. Vetter, Jo Grothe, Arthur P. Wunderlich, Stefan Hahn, and Manfred Spitzer, "The Neural Correlates of Driving," *Brain Imaging*, vol. 12, no. 8 (June 13, 2001), pp. 1763–67.

25 and following distance: See David Shinar and Richard Compton, "Aggressive Driving: An Observational Study of Driver, Vehicle and Situational Variables," *Accident Analysis & Prevention*, vol. 36 (2004), pp. 429–37.

25 give themselves over to the car: Research also suggests that single drivers are more susceptible to fatigue and being involved in crashes, and it is not difficult to speculate why. Passengers provide another "set of eyes" to warn of potential hazards and can aid in keeping the driver engaged. For the increased risk factors to single drivers see, for example, Vicki L. Neale, Thomas A. Dingus, Jeremy Sudweeks, and Michael Goodman, "An Overview of the 100-Car Naturalistic Study and Findings." National Highway Traffic Safety Administration, Paper Number 05-0400.

25 thirty-three citations: See F. K. Heussenstamm, "Bumper Stickers and the Cops," *Trans-Action (Society)*, vol. 8 (February 1971), pp. 32 and 33. The author acknowledged that the subjects' driving may have been affected by the experiment itself but argued that "it is statistically unlikely that this number of previously 'safe' drivers could amass such a collection of tickets without assuming real bias by police against drivers with Black Panther bumper stickers." The information about specially designated license plates comes from "New 'Scarlet Letter' for Predators in Ohio," Associated Press, March 1, 2007. The license plates raise, ironically, a problem similar to "Children at Play"

signs: They signify that a car without such plates is somehow safe for children to approach, just as the "Children at Play" signs suggest that drivers can act less cautiously in areas *without* the signs.

25 aggressive driving on her part: Women driving SUVs, as at least one set of very limited observational studies found, drove faster in 20-mile-per-hour school zones, parked more often in restricted shopping mall fire zones, came to a stop less frequently at stop signs, and were slower to move through an intersection when the light turned green, as compared to other drivers in other types of vehicles. As the author himself admits, the sample sizes were small, and the higher rates of women SUV drivers may simply have reflected the fact that the study took place in a setting where there happened to be a higher than average number of women driving SUVs. See John Trinkaus, "Shopping Center Fire Zone Parking Violators: An Informal Look," *Perceptual and Motor Skills*, vol. 95 (2002), pp. 1215–16; John Trinkaus, "School Zone Speed Limit Dissenters: An Informal Look," *Perceptual and Motor Skills*, vol. 88 (1999), pp. 1057–58.

26 at greater risk: See, for example, Devon E. Lefler and Hampton C. Gabler, "The Fatality and Injury Risk of Light Truck Impacts with Pedestrians in the United States," *Accident Analysis & Prevention*, vol. 36 (2004), pp. 295–304.

26 "grieving while driving": Paul C. Rosenblatt, "Grieving While Driving," *Death Studies*, vol. 28, (2004), pp. 679–86.

26 including nasal probing: Thanks to Daniel McGehee for this story.

26 not wearing hoods: Philip Zimbardo. "The Human Choice: Individuation, Reason, and Order vs. Deindividuation, Impulse, and Chaos." In *Nebraska Symposium on Motivation*, ed. W. J. Arnold and D. Levine (Lincoln: University of Nebraska Press, 1970). Zimbardo's description of the conditions that contribute to the sense of "deindividuation" are worth noting in light of traffic. He writes: "Anonymity, diffused responsibility, group activity, altered temporal perspective, emotional arousal, and sensory overload are some of the input variables that can generate deindividuated reactions." Arguably, *all* of Zimbardo's "input variables" can routinely be found in traffic situations. The quote comes from Zimbardo's "Depersonalization" entry in *International Encyclopedia of Psychiatry, Psychology, Psychoanalysis, and Neurology*, vol. 4, ed. B. B. Wolman (New York: Human Sciences Press, 1978), p. 52.

26 to the executioners: The hostage and firing squad information comes from David Grossman, *On Killing: The Psychological Cost of Learning to Kill in War and Society* (Boston: Back Bay Books, 1996), p. 128.

27 with the tops up: Patricia A. Ellison, John M. Govern, Herbert L. Petri, Michael H. Figler, "Anonymity and Aggressive Driving Behavior: A Field Study," *Journal of Social Behavior and Personality*, vol. 10, no. 1 (1995), pp. 265–72.

27 "online disinhibition effect": See J. Suler, "The Online Disinhibation Effect," *CyberPsychology and Behavior*, vol. 7 (2004), pp. 321–26.

28 relatively large social networks: See, for example, R. I. M. Dunbar, "Neocortical Size as a Constraint on Group Size in Primates," *Journal of Human Evolution*, vol. 22 (1993), pp. 469–93.

29 higher testosterone levels: Roxanne Khamsi, "Hormones Affect Men's Sense of Fair Play," *New Scientist*, July 4, 2007.

29 "strong reciprocity": See Ernst Fehr, Urs Fischbacher, and Simon Gächter, "Strong Reciprocity, Human Cooperation and the Enforcement of Social Norms," *Human Nature*, vol. 13 (2002), pp. 1–25.

30 altruistic all the same: The comments on road rage from Herbert Gintis come from an interview posted at www.innoarticles.com. The example of the bird signaling a predator's approach comes from Olivia Judson, "The Selfish Gene," *Atlantic Monthly*, October 2007, p. 92. It has also been speculated that animals raising an alarm against a predator are actually sending a signal to the *predator* that it has been spotted. For an interesting theoretical discussion, see C. T. Bergstrom and M. Lachmann, "Alarm Calls as Costly Signals of Antipredator Vigilance: The Watchful Babbler Game," *Animal Behaviour*, vol. 61 (2001), pp. 535–43.

30 around 20 miles per hour: Ben Hamilton-Baillie, "Improving Traffic Behaviour and Safety Through Urban Design," *Civil Engineering*, vol. 158 (May 2005), pp. 39–47.

31 result was the same: P. C. Ellsworth, J. M. Carlsmith, and A. Henson, "The Stare as a Stimulus to Flight in Human Subjects: A Series of Field Experiments," *Journal of Personality and Social Psychology*, vol. 21 (1972), pp. 302–11.

31 were not present: Kevin J. Haley and Daniel M. T. Fessler, "Nobody's Watching? Subtle Cues Affect Generosity in an Anonymous Economic Game," *Evolution and Human Behavior*, vol. 26 (2005), pp. 245–56.

31 university break room: See Melissa Bateson, Daniel Nettle, and Gilbert Roberts, "Cues of Being Watched Enhance Cooperation in a Real-World Setting," *Biology Letters*, June 2, 2006.

31 cooperation in humans: Michael Tomasello, Brian Hare, Hagen Lehmann, and Josep Call, "Reliance on Head Versus Eyes in the Gaze Following of Great Apes and Human Infants: The Cooperative Eye Hypothesis," *Journal of Human Evolution*, vol. 52 (2007), pp. 314–20.

31 direction of one's gaze: Smiling might help as well, at least if you're female and the person you're smiling at is male, a French study showed. The study had male and female subjects try to hitch rides by smiling or not smiling at passing drivers. More women received rides when they smiled, but for men, alas, this did not work. Also, every driver that stopped was male. Nicolas Guegue and Jacques Fischer-Lokou, "Hitchhikers' Smiles and Receipt of Help," *Psychological Reports*, vol. 94, (2004), pp. 756–60.

31 tilt your head up: Michael Tomasello, Brian Harea, Hagen Lehmanna, and Josep Call, "Reliance on Head Versus Eyes in the Gaze Following of Great Apes and Human Infants: The Cooperative Eye Hypothesis," *Journal of Human Evolution*, vol. 52, no. 3 (March 2007), pp. 314–20.

33 if one does not make eye contact: Robert Wright explains this phenomenon succinctly: "When we pass a homeless person, we may feel uncomfortable about failing to help. But what really gets the conscience twinging is making

eye contact and still failing to help. We don't seem to mind not giving nearly so much as we mind being seen not giving." (As for why we should care about the opinion of someone we'll never encounter again: Perhaps in our ancestral environment, just about everyone encountered was someone we might well encounter again.) From *The Moral Animal* (New York: Alfred A. Knopf, 1994), p. 206.

33 "if there are more than two": Thomas Schelling, *Choice and Consequence* (Cambridge, Mass.: Harvard University Press, 1984), p. 214.

33 first through the intersection: Schelling also suggested throwing one's steering wheel out the window as a sign that one has committed to one's course of action.

33 at the oncoming car: A. Katz, D. Zaidel, and A. Elgrishi. "An Experimental Study of Driver and Pedestrian Interaction During the Crossing Conflict," *Human Factors*, vol. 17, no. 5 (1975), pp. 514–27.

34 Or was it just chivalry?: Jeffrey Z. Rubin, Bruce D. Steinberg, and John R. Gerrein, "How to Obtain the Right of Way: An Experimental Analysis of Behavior at Intersections," *Perceptual and Motor Skills*, vol. 34 (1974), pp. 1263–74.

34 in New York City: Of course, the faster pace of life in New York City also has an affect on the traffic culture. Michael Primeggia, the deputy director of the New York City Department of Transportation, told me the following joke: "What's the shortest amount of recorded time? The time between the light turning green in New York and the horn sounding."

34 visibly express anger: Andrew R. McGarva, Matthew Ramsey, and Suzannah A. Shear, "Effects of Driver Cell-Phone Use on Driver Aggression," *Journal of Social Psychology*, vol. 146, no. 2 (2006), pp. 133–46.

35 replicated in Australia: S. Bochner, "Inhibition of Horn-Sounding as a Function of Frustrator's Status and Sex: An Australian Replication and Extension of Doob and Gross," *Australian Journal of Psychology*, vol. 6 (1968), pp. 194–99.

35 doing the blocking: A. N. Doob and A. E. Gross, "Status of Frustrator as an Inhibitor of Horn-Honking Responses," *Journal of Social Psychology*, vol. 76 (1968), pp. 213–18.

35 you guessed right: Andreas Diekmann, Monika Jungbauer-Gans, Heinz Krassnig, Heinz Lorenz, and Sigrid Lorenz, "Social Status and Aggression: A Field Study Analyzed by Survival Analysis," *Journal of Social Psychology*; vol. 136, no. 6 (December 1996), pp. 761–68.

35 been at work: See Ben Jann, "Driver Aggression as a Function of Status Concurrence: An Analysis of Horn-Honking Responses," Bern, Switzerland, 2002; available at www.socio.ethz.ch/de/jann. Interestingly, this study found, as in the earlier mentioned birthday study, that drivers of a vehicle were less likely to honk at a vehicle when it was of the *same* status. The researcher noted, however, that "our data do not reveal whether it is actually *similarity* which *reduces* aggression or rather *difference* that *increases* it."

35 when it was a man: Kay Deux, "Honking at the Intersection: A Replication and Extension," *Journal of Social Psychology*, vol. 84 (1971), pp. 159–60.

35 a driving "lesson": H. Yazawa, "Effects of Inferred Social Status and a Begin-

ning Driver's Sticker upon Aggression of Drivers in Japan," *Psychological Reports*, vol. 94 (2004), pp. 1215–20.

35 from another country: The study, interestingly, found that French, Spanish, and Italian drivers were faster to the horn than German drivers (Italians were the fastest). Drivers also honked more when the visible sticker was German and not the less recognizable Australian identity sticker. See Joseph P. Forgas, "An Unobtrusive Study of Reactions to National Stereotypes in Four European Countries," *Journal of Social Psychology*, vol. 99 (1976), pp. 37–42.

35 suspected these things: Drivers, of course, may simply be honking in a "nonaggressive" way simply to let the driver ahead know that the light has changed. But as Dwight Hennessey has pointed out, the frequency and latency of honks indicates that more than just polite signaling is at work. See Dwight Hennessey, "The Interaction of Person and Situation Within the Driving Environment: Daily Hassles, Traffic Congestion, Driver Stress, Aggression, Vengeance and Past Performance" (Ph.D. dissertation, York University, Toronto, Ontario, April 1999).

36 In another study: Ian Walker, "Signals Are Informative but Slow Down Responses When Drivers Meet Bicyclists at Road Junctions," *Accident Analysis & Prevention*, vol. 37 (2005), pp. 1074–85.

36 In a previous study: Ian Walker, "Road Users' Perceptions of Other Road Users: Do Different Transport Modes Invoke Qualitatively Different Concepts in Observers?" *Advances in Transportation Studies*, section A, no. 6 (2005), pp. 25–32.

37 rendered invisible by the car: Perhaps the subjects were distracted by simply recognizing the make and model of car. Psychologists at Vanderbilt University have shown in clinical tests that car aficionados shown pictures of cars were less able to identify faces at the same time. Car fanciers were looking at cars *as if they were* faces, causing a "perceptual traffic jam" in a part of the brain implicated in the "holistic" visual processes of facial recognition. See Isabel Gauthier and Kim M. Curby, "A Perceptual Traffic Jam on Highway N170: Interference Between Face and Car Expertise," *Current Directions in Psychological Science*, vol. 14, no. 1 (February 2005), pp. 30–33.

37 people's eyes meet: See, for example, A. Gale, G. Spratt, AJ Chapman, and A. Smallbone, "EEG correlates of eye contact and interpersonal distance," *Biological Psychology*, vol. 3, no. 4 (December 1975), pp. 237–45.

38 to the actual road: For further details on the study, see Ian Walker, "Drivers Overtaking Bicyclists: Objective Data on the Effects of Riding Position, Helmet Use, Vehicle Types and Apparent Gender," *Accident Analysis & Prevention*, vol. 39 (2007), pp. 417–25.

39 the driver slows: There is conceivably no limit to the number and variety of stereotypes drivers possess about other vehicles and the people driving them—for example, BMW drivers are aggressive, minivan drivers are slow. How all these secret interactions all play out in traffic is virtually beyond study. Do certain car drivers act a certain way, and do we act differently toward certain cars or drivers? Do you get the finger in a Hummer and a cute smile in a Mini, and

does this then affect the way you drive, which then reinforces the stereotype? Research has suggested one drawback of these stereotypes: When subjects were read the description of a crash between two cars in which the actual facts were unknown, they estimated that the speed of one car was higher when the driver was younger and in a stereotypical "boy racer" car. (The effect was even stronger when the color was red!) See Graham M. Davies and Darshana Patel, "The Influence of Car and Driver Stereotypes on Attributions of Vehicle Speed, Position on the Road and Culpability in a Road Accident Scenario," *Legal and Criminal Psychology*, vol. 10, (2005), pp. 45–62.

39 automatic reponses: Irene V. Blaire and Mahzarin R. Banaji, "Automatic and Controlled Processes in Stereotype Priming," *Journal of Personality and Social Psychology*, vol. 70, no. 6 (1996), pp. 1142–63.

41 waiting in line: See David Maister, "The Psychology of Waiting in Line," available at http://davidmaister.com/articles/1/52/.

41 on the highway itself: L., Zhang, F. Xie, and D. Levinson, "Variation of the Subjective Value of Travel Time Under Different Driving Conditions." Paper presented at the Eighty-four Transportation Research Board Annual Meeting, January 9–13, 2005, Washington, D.C.

41 groups often move faster: See David A. Hensher, "Influence of Vehicle Occupancy on the Valuation of Car Driver's Travel Time Savings: Identifying Important Behavioural Segments," Working Paper ITLS-WP-06-011, May 2006, Institute of Transport and Logistics Studies, University of Sydney.

41 with our perception of time: A curious example of this are the new "smart" elevator systems being installed in high-rise buildings around the world. Instead of simply calling an elevator, users are grouped according to which floor they want. In theory, this speeds up the average journey by 50 percent, but it also prompts impatience in people who see elevators bound for other floors arriving and leaving before theirs; they think they are actually waiting longer. See Clive Thompson, "Smart Elevators," *New York Times*, December 10, 2006.

42 "At least I'm better off than you": See Rongrong Zhou and Dilip Soman, "Looking Back: Exploring the Psychology of Queuing and the Effect of the Number of People Behind," *Journal of Consumer Research*, vol. 29 (March 2003).

42 "irritated with that": On the differences in queue systems between Wendy's and McDonald's, there is another factor to consider: customers' perceptions of the *length* of the line. McDonald's says that people will renege on a line that looks longer; hence it prefers shorter multiple lines, despite Wendy's claims that a single line is faster. See "Merchants Mull the Long and Short of Lines," *Wall Street Journal*, September 3, 1998.

43 *an eighty-minute drive*: The lane-changing experiment was conducted by the CBC's *Fifth Estate*. Details are available at http://www.cbc.ca/fifth/road warriors/research.html.

43 did passing them: Donald A. Redelmeier and Robert J. Tibshirani, "Why Cars in the Next Lane Seem to Go Faster," *Nature*, vol. 35, September 2, 1999.

43 at the forward roadway: See, for example, Alexei R. Tsyganov, Randy B.

Machemehl, Nicholas M. Warrenchuk, and Yue Wang, "Before-After Comparison of Edgeline Effects on Rural Two-Lane Highways," Report No. FHWA/TX-07/0-50902 (Austin: Center for Transportation Research, University of Texas at Austin, 2006).

43 stay in our lane: See, for example, D. Salvucci, A. Liu, and E. R. Boer, "Control and Monitoring During Lane Changes," in *Vision in Vehicles: 9*, conference proceedings (Brisbane, Australia, 2001).

43 looking in the rearview mirror: The forward and rearview percentages are drawn from M. A. Brackstone and B. J. Waterson, "Are We Looking Where We Are Going? An Exploratory Examination of Eye Movement in High Speed Driving." Paper 04-2602, *Proceedings of the 83rd Annual Meeting of the Transportation Research Board* (Washington D.C., January 2004).

44 "loss aversion": The notion of loss aversion was first hypothesized by Daniel Kahneman and Amos Tversky, "Prospect Theory: An Analysis of Decision Under Risk," *Econometrica*, vol. 47 (1979), pp. 263–91.

44 sensitive to loss: See Sabrina M. Tom, Craig R. Fox, Christopher Trepel, and Russell A. Poldrack, "The Neural Basis of Loss Aversion in Decision-Making Under Risk," *Science*, vol. 315, no. 5811 (26 January 2007), pp. 515–18. See also William J. Gehring and Adrian R. Willoughby, "The Medial Frontal Cortex and the Rapid Processing of Monetary Gains and Losses," *Science*, vol. 295, no. 5563 (2002), pp. 2279–82.

44 "endowment effect": D. Kanheman, J. L. Knetsch, and R. H. Thaler, "Experimental Tests of the Endowment Effect and the Coase Theorem," *Journal of Political Economy*, vol. 98 (1990) pp. 1325–48.

44 to the person leaving it: The parking lot studies were chronicled in R. Barry Ruback and Daniel Juieng, "Territorial Defense in Parking Lots: Retaliation Against Waiting Drivers," *Journal of Applied Social Psychology*, vol. 27, no. 9 (1997), pp. 821–34. The authors suggest another theory: that fighting for the "symbolic value" of the parking space when it is threatened by an intruder helps give the parking spot owner a feeling of heightened control over the situation. This is why, they suggest, people will take even longer to vacate a spot when the waiting driver honks. It is a threat to their "sense of freedom," and the best response is to simply stay longer in the parking space, thus asserting that sense of freedom.

44 involved lane changes: Basav Sen, John D. Smith, and Wassim G. Najm, "Analysis of Lane Change Crashes," DOT-VNTSC-NHTSA-02-03, National Highway Traffic Safety Administration, March 2003.

44 how many were discretionary?: One study that compared crashes to traffic volume (obtained via loop-inductor data) found that most lane-change crashes occurred, perhaps not surprisingly, when the variability of highway speeds across lanes was highest—in other words, the time when most people would find it advantageous to change lanes. See Thomas F. Golob, Wilfred W. Recker, and Veronica M. Alvarez, "Freeway Safety as a Function of Traffic Flow," *Accident Analysis & Prevention*, vol. 36 (2004), pp. 933–46.

44 decisions we make while driving: At Cooper University Hospital in New Jersey,

for example, doctors estimate that 60 percent of the trauma intensive care unit patients are the victims of car crashes; see Geoff Mulvihill, "In Corzine's Hospital Unit, Handling Terrible Accidents Routine," *Newsday*, April 23, 2007.

45 work zones: The work-zone fatality statistic comes from the U.S. Federal Highway Administration (http://safety.fhwa.dot.gov/wz/wz_facts.htm).

46 "merging difficulties": From *Understanding Road Rage: Implementation Plan for Promising Mitigation Measures*, by Carol H. Walters and Scott A. Cooner (Texas Transportation Institute, November 2001).

47 lane that will close: Information on work-zone merge strategies was drawn from a number of useful sources, including "Dynamic Late Merge Control Concept for Work Zones on Rural Freeways," by Patrick T. McCoy and Geza Pesti, Department of Civil Engineering, University of Nebraska.

48 smoothly through the work zone: The TRL data comes from a report by G. A. Coe, I. J. Burrow, and J. E. Collins, "Trials of 'Merge in Turn' Signs at Major Roadworks." Unpublished project report, PR/TT/043/95, N207, October 30, 1997.

48 exactly where to merge: For a sample discussion of U.K. merging ambiguity, see http://www.pistonheads.com/gassing/topic.asp?f=154&h=&t=256729. Retrieved on December 1, 2007.

48 which is also safer: See Federal Highway Administration, U.S. Department of Transportation, "Methods and Procedures to Reduce Motorist Delays in European Work Zones," FHWA-PL-00-001, October 2000.

48 One important caveat: Another simulation study showed that the Late Merge strategy was more effective when two lanes narrowed to one than when three narrowed to two. According to one report, "A possible explanation may be evident in the way vehicles appeared to be behaving in the simulations. When simulation animations of the 3-to-2 lane configurations of the late merge control were viewed, it appeared that vehicles driving in the middle lane would move to the far left lane to avoid merging from the closing lane. This interaction slowed vehicles in the far left lane enough that throughput may have been significantly reduced." *Evaluation of the Late Merge Workzone Traffic Control Strategy*, by Andrew G. Beacher, Michael D. Fontaine, and Nicholas J. Garber. Virginia Transportation Research Council, August 2004, VTRC 05—R6.

49 summer of 2003: The Minnesota Dynamic Late Merge information was drawn from two reports, "Dynamic Late Merge System Evaluation: Initial Deployment on I-10,"prepared by URS for the Minnesota Department of Transportation," and a follow-up study, "Evaluation of 2004 Dynamic Late Merge System for the Minnesota Department of Transportation," also prepared by URS.

49 blocked by trucks: Garber, in a telephone conversation, also noted the particular tendency of trucks to perform blocking maneuvers. He found that Late Merge worked best when the total volume of heavy vehicles in the traffic stream was less than 20 percent.

## Chapter Two: You're Not as Good a Driver as You Think You Are

51 fifteen hundred "subskills": This estimate comes from A. J. McKnight and B. Adams, *Driver Education Task Analysis*, vol. 1, *Task Descriptions*, Washington D.C.: National Highway Traffic Safety Administration, 1970.

51 twenty per mile: Leslie George Norman, "Road Traffic Accidents: Epidemiology, Control and Prevention" (World Health Organization, Public Health Papers no. 12, 1962), p. 51.

51 440 words, per minute: This figure comes from William Ewald, *Street Graphics* (Washington, D.C.: American Society of Landscape Architects Foundation), p. 32.

53 "avoiding obstacles": See *Urban Challenge Rules* (Arlington, Va.: Defense Advanced Research Projects Agency, July 10, 2007).

53 in the future: The cognitive scientist Donald D. Hoffman points out that an average traffic scene of a tree-lined street with cars creates a multitude of problems for computer intelligence, as analysis by researcher Scott Richman has revealed. Hoffman notes, "Several problems that Richman faced are evident from this picture: clutter, trees moving in the wind, shadows dancing on the road, cars in front hiding cars behind. A sophisticated analysis of motion, using several frames of motion at once, allows Richman's system to distinguish the motion of cars from that of trees and shadows. . . . [Richman's] system can track cars through shadows, a feat that is trivial for our visual intelligence but, heretofore, quite difficult for computer vision systems. It's easy to underestimate our sophistication at constructing visual motion. That is, until we try to duplicate that sophistication on a computer. Then it seems impossible to overestimate it." From Donald D. Hoffman, *Visual Intelligence* (New York: W. W. Nortion, 1998), p. 170.

54 "caution for the caution": See, for example, Don Leavitt, "Insights at the Intersection," *Traffic Management and Engineering*, October 2003.

55 *sooner* than necessary: H. Kölla, M. Badera, and K. W. Axhausen, "Driver Behavior During Flashing Green Before Amber: A Comparative Study," *Accident Analysis & Prevention*, vol. 36, no. 2 (March 2004), pp. 273–80.

55 without the flashing green: D. Mahalel and D. M. Zaidel, "Safety Evaluation of a Flashing Green Light in a Traffic Signal," *Traffic Engineering and Control*, vol. 26, no. 2 (1985), pp. 79–81.

55 chances to crash: This point is made in L. Staplin, K. W. Gish, L. E. Decina, K. H. Lococo, D. L. Harkey, M. S. Tarawneh, R. Lyles, D. Mace, and P. Garvey in *Synthesis of Human Factors Research on Older Drivers and Highway Safety*, vol. 2, Publication No. FHWA-RD-97-095, 1997. Available at http://www.fhwa.dot.gov/tfhrc/safety/pubs/97094/97094.html.

56 "bump itself up the queue": One might think that robot drivers would be free from the complicated psychological dynamics that trouble humans at intersections; yet, perhaps like humans, it all depends on how they are wired. "Robots can be more aggressive or more conservative," Montemerlo told me. You might, for example, "program your robot to always ignore the queuing order

and always go first, to be a pushy robot." But whether or not this strategy works depends on how the other robots have been programmed. Four pushy robots at a four-way stop could get ugly quickly.

57 "They slow everyone down": This recalls a comment from T. C. Willet's *Criminal on the Road: A Study of Serious Motoring Offences and Those Who Commit Them* (London: Tavistock Publications, 1964). As Willet noted: "Some years ago a contest was arranged between two cars to be driven across a city area. One driver had to observe all signs, traffic lights, and speed regulations. The other was allowed to ignore all three if he could do so without endangering the lives of other road users. The law-breaking motorist arrived at this destination just—and only just—ahead of his law-abiding antagonist" (p. 129).

58 "without a hitch": The eBay quote comes from Theresa Howard, "Ads Pump up eBay Community with Good Feelings," *USA Today*, October 17, 2004.

58 more in revenue: Paul Resnick, Richard Zeckhauser, John Swanson, and Kate Lockwood, "The Value of Reputation on eBay: A Controlled Experiment." John F. Kennedy School of Government, Harvard University; Working Paper No. RWP03-007.

58 (provided it's authentic): See, for example, John Morgan and Jennifer Brown, "Reputation in Online Auctions: The Market for Trust," *California Management Review*, Fall 2006. About 98 percent of feedback on eBay is positive, which has led economist Axel Ockenfels of the University of Cologne in Germany to suspect that people may be afraid of negative retaliatory feedback. Ockenfels has worked with eBay to introduce mechanisms that allow users to post honest, negative feedback with less fear of reprisal. See Christoph Uhlhaas, "Is Greed Good?" *Scientific American Mind*, August–September 2007, p. 67.

58 "rising insurance premiums": Lior J. Strahilevitz, "How's My Driving? For Everyone (and Everything?)," Public Law and Legal Theory Working Paper No. 125, Law School, University of Chicago. Accessed from http://ssrn.com/abstract_id=899144.

59 have been tried: The Web site uncivilservants.org, for example, posts pictures of New York City cars with various official parking permits that are nonetheless parked illegally (many cars also have bootleg parking permits).

60 actual driving record: C. E. Preston and S. Harris, "Psychology of Drivers in Traffic Accidents," *Journal of Applied Psychology*, vol. 49 (1965), pp. 284–88.

60 they were "better": For a good summary of these studies, see D. Walton and J. Bathurst, "An Exploration of the Perceptions of the Average Driver's Speed Compared with Perceived Driver Safety and Driving Skill," *Accident Analysis & Prevention*, vol. 30 (1998), 821–30.

60 most dangerous thing: John Groeger, a psychologist at the University of Surrey in England, points out that this behavior may be a way to "protect ourselves from the anxieties involved in constantly placing ourselves at risk by developing confidence in our ability which we are rarely likely to be forced to realize is misplaced." See Groeger, *Understanding Driving* (East Sussex: Psychology Press, 2001), p. 163.

60 *smallest* returns: Brad M. Barber and Terrance Odean, "Trading Is Hazardous

to Your Wealth: The Common Stock Investment Performance of Individual Investors," *Journal of Finance*, vol. 55, no. 2 (2000).

60 car accident: Julie M. Kos and Valerie A. Clarke, "Is Optimistic Bias Influenced by Control or Delay?" *Health Education Research: Theory and Practice*, vol. 16, no. 5 (2001), pp 533–40.

60 have done it: The texting while driving poll comes from Reuters, August 7, 2007. Retrieved from http://www.reuters.com/article/idUSN0640649920070807.

60 underestimate our own risk: For an interesting discussion of this phenomenon in the context of seat-belt usage, see "Unconscious Motivators and Situational Safety Belt Use," *Traffic Safety Facts: Traffic Tech*, No. 315 (Washington, D.C.: National Highway Traffic Safety Administration, 2007).

60 social mores and traffic laws: For a seminal discussion of these problems, see H. Laurence Ross, "Traffic Law Violation: A Folk Crime," *Social Problems*, vol. 8, no. 3 (1960–61) pp. 231–41.

61 in question are ambiguous: See R. B. Felson, "Ambiguity and Bias in the Self-Concept," *Social Psychology Quarterly*, vol. 44 (March 1981), pp. 64–69.

61 "unskilled and unaware of it": Justin Kruger and David Dunning, "Unskilled and Unaware of It: How Difficulties in Recognizing One's Own Incompetence Lead to Inflated Self-Assessments," *Journal of Personality and Social Pscyhology*, vol. 77, no. 6, (1999), pp. 1121–34.

61 "better" (i.e., safer) drivers: E. Kunkel, "On the Relationship Between Estimate of Ability and Driver Qualification," *Psychologie und Praxis*, vol. 15 (1971), pp. 73–80.

61 (particularly men): See Frank P. McKenna, Robert A. Stanier, and Clive Lewis, "Factors Underlying Illusory Self-Assessment of Driving Skill in Males and Females," *Accident Analysis & Prevention*, vol. 23, no. 1 (1991), pp. 45–52.

61 outnumbered the courteous: *New Jersey Star-Ledger*, September 28, 1998.

61 by low self-esteem: Mayer Perry writes, for example, that "if an individual lacks 'personal drive' or dominance, either is easily afforded in the driving situation, and in compensating for this lack, he frequently over-compensates." Perry, *Aggression on the Road* (London: Tavistock, 1968), p. 7.

61 promotes aggressive driving: George E. Schreer, "Narcissism and Aggression: Is Inflated Self-Esteem Related to Aggressive Driving?" *North American Journal of Psychology*, vol. 4, no. 3 (2002), pp. 333–42.

61 claim to have had: See Gina Kolata, "The Median, the Math, and the Sex," *New York Times*, August 19, 2007.

62 than doing it: See "Aggravating Circumstances," a report produced by Public Agenda (available at http://www.publicagenda.com). It could be, of course, that the people in the sample (maybe the kind of people who answer surveys) just happened to be an extraordinarily well-behaved group of drivers who really were subject to an inordinate number of louts (the sort who do not answer surveys). There could also be recall bias at work; it is far easier to remember the isolated aggressive acts of others than the uneventful stream of well-behaved driving. This in itself, however, would not explain why people's perceptions would have changed over time.

62 "view of the self": J. M. Twenge, S. Konrath, J. D. Foster, W. K. Campbell, and B. J. Bushman, *Egos Inflating over Time: A Test of Two Generational Theories of Narcissism*, 2006. Cited in "Primary Sources," *Atlantic*, July–August 2007.

62 attributes to police officers: Still, getting a ticket may be a form of at least temporarily effective feedback: One study, looking at ten million Ontario drivers for more than a decade, found that each conviction for a traffic offense led to a 35 percent decrease in relative risk of death over the next month for that driver and others. See Donald A. Redelmeier, Robert J. Tibsharani, and Leonard Evans, "Traffic-Law Enforcement and Risk of Death from Motor-Vehicle Crashes: Case-Crossover Study," *Lancet*, vol. 361, no. 9376 (2003), pp. 2177–82.

62 "experience is a mixed blessing": James Reason, *Human Error* (Cambridge: Cambridge University Press, 1990), p. 86.

63 environment for workers: *Census of Fatal Occupational Injuries* (Bureau of Labor Statistics, 2006). Available at http://www.bls.gov. See also P. Lynn and C. R. Lockwood, *The Accident Liability of Company Car Drivers*, TRL Report 317 (Crowthorne: Transport Research Laboratory, 1998). This study found that company car drivers were 49 percent more likely to be involved in a crash, even after accounting for higher mileage and other factors.

64 at the bottom: Heinrich's safety philosophies have proved controversial over the years, but the idea that near misses are scaleable to more serious incidents remains powerful, particularly in traffic, where "human factors," it is commonly believed, are responsible for up to 90 percent of all crashes. Indeed, a large-scale study of "naturalistic driving behavior" in 2006, which for the first time was able to reliably estimate the near-miss incidents, reported the following distribution after a year's worth of study: 69 crashes, 761 near crashes, and 8,295 "incidents." This means, roughly, that for every 120 incidents, there were 11 minor-injury crashes and 1 serious crash—a more robust frequency than that proposed by Heinrich. See, for example, the work of Fred Manuele, such as *On the Practice of Safety* (New York: Wiley Interscience, 2003).

65 Investigators learned: Associated Press, May 5, 2007.

65 killed a motorcyclist: Information on the Janklow case comes from the *Argus Leader*, August 31, 2003.

66 "more unintentional than others": See Teresa L. Kramer, Brenda M. Booth, Han Xiaotong, and Keith D. Williams, "Some Crashes Are More Unintentional Than Others: A Reply to Blanchard, Hicking, and Kuhn," *Journal of Traumatic Stress*, vol. 16, no. 5 (October 2003), pp. 529–30.

66 "hindsight bias": For a seminal account, see Baruch Fischoff, "Hindsight Is Not Equal to Foresight: The Effect of Outcome Knowledge on Judgment Under Uncertainty," *Journal of Experimental Psychology: Human Perception and Performance*, vol. 1, no. 2 (1975), pp. 288–99.

66 intentional or not: In 1958, this number was said to be 88 out of 100. This figure, taken from a National Safety Council study, comes from H. Laurence Ross, "Traffic Law Violation: A Folk Crime," *Social Problems*, vol. 8, no. 3 (1960–61), pp. 231–41.

66 "then it's an accident": Shamus Toumey, "Ryan Crash Kills Man Who Had Just Arrived from Mexico," *Chicago Sun-Times*, October 6, 2006.

66 over the limit and kills someone: In an excellent survey of the legal penalties for drivers who kill "vulnerable road users" (pedestrians and cyclists), Jake Voelcker notes that juries have long been unwilling to levy the most serious charges of manslaughter against even negligent drivers because, as drivers themselves, they expressed a feeling of "there, but for the grace of God, go I." He cites, as well, examples of subtle bias among judges who imply that accidents are themselves unavoidable—for example, the "accident happened because the appellant was driving rather too fast, as young men will." The "genuine accident" involving a sober driver, he notes, tends to be avoided by legislation. "Is this simply an unfortunate fact of modern life for which no one is to blame?" he asks. "Or is the driver responsible for the very presence of his car?" Voelcker points to a number of other unresolved legal issues. What should the appropriate penalty be for dangerous driving that narrowly misses killing someone? Why are convicted criminals given harsher sentences for driving-related offenses than others, even when the standard of driving involved in the offense is the same? Should drivers be held to a certain level of causality simply by choosing to operate a machine that is known to be dangerous, thus imposing potential risk on others? See Jake Voelcker, "A Critical Review of the Legal Penalties for Drivers Who Kill Cyclists or Pedestrians," April 2007. Retrieved from www.jake-v.co.uk/cycling.

67 majority were men: Phillip C. Shin, David Hallett, Mary L. Chipman, Charles Tator, and John T. Granton, "Unsafe Driving in North American Automobile Commercials," *Journal of Public Health*, vol. 27, no. 4 (December 2005), pp. 318–25.

67 themselves as "unlucky"): See Richard Wiseman, *The Luck Factor* (New York: Miramax Books, 2003).

67 back in time they happened: See, for example, J. Maycock, C. Lockwood, and J. F. Lester, *The Accident Liability of Car Drivers*, Research Report No. 315 (Crowthorne: Transport and Road Research Laboratory, 1991).

68 end of their trip: G. Underwood, P. Chapman, Z. Berger and D. Crundall, "Driving Experience, Attentional Focusing, and the Recall of Recently Inspected Events," *Transportation Research F: Psychology and Behaviour*, vol. 6 (2003), pp. 289–304.

68 more experienced drivers: P. Chapman, D. Crundall, N. Phelps, and G. Underwood, "The Effects of Driving Experience on Visual Search and Subsequent Memory for Hazardous Driving Situations," in *Behavioural Research in Road Safety: Thirteenth Seminar* (London: Department for Transport, 2003), pp. 253–66.

68 experience and expertise: When expert chess players are given a short glimpse of a chess board, for example, they can remember almost twice as much of the board's positions as novices can. For a discussion of this see Groeger, *Understanding Driving*, p. 101.

68 scan the whole picture): See Stine Vogt and Svein Magnussen, "Expertise in

Pictorial Perception: Eye-Movement Patterns and Visual Memory in Artists and Laymen," *Perception*, vol. 36, no. 1, 2007, pp. 91–100.

69 "right above the threshold": For a more detailed account of McGehee's study, see Daniel V. McGehee, Mireille Raby, Cher Carney, John D. Lee, and Michelle L. Reyes, "Extending Parental Mentoring Using an Event-Triggered Video Intervention in Rural Teen Drivers," *Journal of Safety Research*, vol. 38, no. 2 (2007), pp. 215–27.

70 was not the case: "Vehicle Monitoring Systems Please Providers and Patients," *EMS Insider*, August 2004, p. 7.

71 in the "wrong" lanes: Mohamed Abdel-Aty and J. G. Klodzinski, "Safety Considerations in Designing Electronic Toll Plazas: Case Study," *ITE Journal*, March 2001.

72 when it is minor: E. Walster, "Assignment of Responsibility for an Accident," *Journal of Personality and Social Psychology*, vol. 3 (1966), pp. 73–79.

72 no glass was broken: Elizabeth F. Loftus and John C. Palmer, "Reconstruction of Automobile Destruction: An Example of the Interaction Between Language and Memory," *Journal of Verbal Learning and Verbal Behavior*, 1974. This study has been questioned for its "ecological validity" because it took place in a laboratory setting and not in the traumatic, unexpected real-life environment of actually witnessing a car crash and then testifying in court. In those situations, even more distortion could be expected.

72 "tend to explain": J. Stannand Baker, "Single Vehicle Accidents on Route 66," The Journal of Criminal Law, Criminology, and Police Science, vol. 58, no. 4 (December 1967), pp. 58–95.

## Chapter Three: How Our Eyes and Minds Betray Us on the Road

74 "the attention it deserves": Thanks to Leonard Evans for this quote.

74 people who study driving: See, for example, Walter Miles, "Sleeping with the Eyes Open," *Scientific American*, June 1929, pp. 489-92.

75 one-third of the time: K. Karrer, S. Briest, T. Vohringer-Kuhnt, T. Baumgarten, and R. Schleicher, "Driving Without Awareness," Unpublished paper, Center of Human-Machine-Systems, Berlin University of Technology, Germany.

75 become fully automatic: See John Groeger, *Understanding Driving* (East Sussex: Psychology Press, 2001), p. 69.

76 does not seem overly taxing: Studies have suggested that merely *changing* one's speed occasionally can help keep a driver more alert. See Pilar Tejero and Mariano Choliz, "Driving on the Motorway: the Effect of Alternating Speed on Drivers' Activation Level and Mental Effort," *Ergonomics*, vol. 45, no. 9 (2002), pp. 605–18.

76 than on a highway: L. Harms, "Drivers' Attention Responses to Environmental Variation: A Dual-Task Real Traffic Study," in *Vision in Vehicles*, ed. A. G. Gale et al. (Amsterdam: Elsevier Science Publishers, 1986), pp. 131–38.

77 the less we actually remember: These findings were reported in L. Bergen,

T. Grimes, and D. Potter, "How Attention Partitions Itself During Simultaneous Message Presentations," *Human Communication Research*, vol. 31, no. 3 (2005), pp. 311–36. See also C. Blain and R. Meeds, "Effects of Television News Crawls on Viewers' Memory for Audio Information in Newscasts" (unpublished manuscript, Kansas State University, Manhattan, 2004).

78 10.8 times per hour: See J. C. Stutts, J. R. Feaganes, E. A. Rodgman, C. Hamlett, T. Meadows, D. W. Reinfurt, K. Gish, M. Mercadante, and L. Staplin, *Distractions in Everyday Driving* (Washington, D.C.: AAA Foundation for Traffic Safety, 2003). Available at: http://www.aaafoundation.org/pdf/DistractionsInEverydayDriving.pdf.

78 for 0.6 seconds: L. Tijerina, "Driver Eye Glance Behavior During Car Following on the Road," Society of Automotive Engineers Paper 1999-01-1300, 1999.

78 skipping a song: Susan L. Chisholm, Jeff K. Caird, Julie Lockhart, Lisa Fern, and Elise Teteris, "Driving Performance While Engaged in MP-3 Player Interaction: Effects of Practice and Task Difficulty on PRT and Eye Movements," *Proceedings of the Fourth International Driving Symposium on Human Factors in Driver Assessment, Training and Vehicle Design* (Iowa City, 2007).

78 "fifteen-second rule": See, for example, Paul Green, "The 15-Second Rule for Driver Information Systems," *ITS America Ninth Annual Meeting Conference Proceedings* (Washington, D.C.: Intelligent Transportation Society of America, 1999).

80 "and they're in trouble": This raises the interesting question of why people who were following *closer* than two seconds did not account for the majority of rear-end crashes, as one might suspect. Klauer suggested that when people "are aggressively tailgating, or trying to maintain their position against all vehicles in their surrounding environment, they're paying very close attention." Does that mean we should all tailgate? "It's an interesting finding," Klauer said. "We tried to be very careful in the way that we reported that, because that's exactly what we did not want people to take away from this—'Oh, tailgating is a perfectly safe thing to do.' All we're saying is we didn't see a whole lot of crashes that were a result of it." This raises the question of which demon you would rather face: the driver hanging far back but talking on a cell phone or the frenetically attentive tailgater.

80 like changing lanes: A simulator study by William Horrey and Daniel Simons found that drivers under "single" and "dual" task conditions did not change the spacing they allowed during lane changing, unlike the greater headway drivers tend to allow when following a car and talking on a cell phone. The authors suggest that "dual-task interference might be more dangerous when drivers must actively decide how to interact with traffic than when their decisions are constrained by the driving context." W. J. Horrey and D. J. Simons, "Examining Cognitive Interference and Adaptive Safety Behaviors in Tactical Vehicle Control," *Ergonomics*, vol. 50, no. 8 (August 2007), pp. 1340–50.

80 to their speed: See James Reason, *Human Error* (Cambridge: Cambridge Univeristy Press, 1990), p. 81.

80 task got harder: J. Verghese, G. Kuslansky, R. Holtzer, M. Katz, X. Xue, H. Buschke, and M. Pahor, "Walking While Talking: Effect of Task Prioritiza-

tion in the Elderly," *Archives of Physical Medicine and Rehabilitation*, vol. 88, no. 1 (2006), pp. 50–53.

80 "sample the environment": A. Oulasvirta, S. Tamminen, V. Roto, and J. Kuorelahti, "Interaction in 4-Second Bursts: The Fragmented Nature of Attentional Resources in Mobile HCI," *Proceedings of CHI 2005* (New York: ACM Press, 2005), pp. 919–28. See also V. Lantz, J. Marila, T. Nyyssönen, and H. Summala, "Mobile Measurements of Mobile Users," in Lucas Noldus and Fabrizio Grieco, *Proceedings of Measuring Behavior 2005: Fifth International Conference on Methods and Techniques in Behavioral Research*, ed. (Wageningen, Netherlands, 2005).

80 longer to do so: J. Hatfield and S. Murphy, "The Effects of Mobile Phone Use on Pedestrian Crossing Behaviour at Signalised and Unsignalised Intersections," *Accident Analysis & Prevention*, vol. 39, no. 1 (2006), pp. 197–205.

80 suffers from a bottleneck: Mei-Ching Lien, Eric Ruthruff, and James C Johnston, "Attentional Limitations in Doing Two Tasks at Once: The Search for Exceptions," *Current Directions in Psychological Science*, vol. 15, no. 2 (2006), pp. 89–93.

81 forgot most of them: For a good summary of this research, see David Shinar, *Psychology on the Road: The Human Factor in Traffic Safety* (New York: Wiley, 1978), p. 27.

81 is not useful to our lives: Indeed, scientists have demonstrated, neurologically, how forgetting things helps us in the process of remembering. See Brice A. Kuhl, Nicole M. Dudukovic, Itamar Kahn, and Anthony D. Wagner, "Decreased Demands on Cognitive Control Reveal the Neural Processing Benefits of Forgetting," *Nature Neuroscience*, vol. 10 (2007), pp. 908–14.

82 again for "confirmation": See Helmut T. Zwahlen and Thomas Schnell, "Driver Eye Scan Behavior When Reading Symbolic Warning Signs," in *Vision in Vehicles VI*, ed. A. Gale, I. D. Brown, C. M. Haslegrave, and S. P. Taylor (Amsterdam: Elsevier Science, 1998), p. 3.

82 ("effectively blind"): See Graham Hole's concise and authorative study, *The Psychology of Driving* (Mahwah, New Jersey: Lawrence Erlbaum Associations, 2007), p. 60.

82 had already been made: H. Shinoda, M. Hayhoe, and A. Shrivastava, "What Controls Attention in Natural Environments?" *Vision Research*, vol. 41 (2001), pp. 3535–46.

83 basketball players: Daniel J. Simons and Christopher F. Chabris, "Gorillas in Our Midst: Sustained Inattentional Blindness for Dynamic Events," *Perception*, vol. 28 (1999), pp. 1059–74.

83 locked on the video screen: One of Simon's key findings is that subjects were less likely to see the gorilla when they were asked to count the number of passes made by the team wearing white T-shirts. This meant, according to Simons, that people did not see the gorilla because it did not look like what they were looking for—or because it *did* look like what they were ignoring (the team wearing black shirts). As Simons put it, "The more you're focused on what you expect to see, the less likely you are to see unexpected stuff."

83 "failure to see": The role that a car driver's vision (or lack thereof) plays in car-

motorcycle crashes is huge: For nine of the ten types of leading car-motorcycle crashes, the motorcycle is proceeding in a straight line (with the most common being the car turning left across the path of an approaching motorcycle). See P. A. Hancock, G. Wulf, D. R. Thom, and P. Fassnacht, "Contrasting Driver Behavior During Turns and Straight Driving," paper presented at the 33rd Annual Meeting of the Human Factors Society, Denver, Colorado, October 1989.

83 on the road: Another response, of course, is the "loud pipes save lives" approach, by which motorcyclists insist that an ear-shattering exhaust system will surely alert drivers of their presence. The problem is that drivers are often unaware of the *direction* of such sounds. Another problem is that for the people who have to listen to the loud pipes, the issue of saving motorcyclists' lives might not exactly be a pressing agenda.

84 change lanes or slow down: *USA Today*, July 4, 2007.

84 moths to a flame: At a meeting I attended in Los Angeles, for example, the California Highway patrol was concerned with a recent spate of these crashes, which had claimed the lives of six officers over just a few months. "For whatever reason they tend to find us on the side of the road," a CHP officer said at a traffic reporters' meeting one morning. "It's just a dangerous place to be."

84 we see something interesting: Driving simulator studies have suggested that drivers have a tendency to at least momentarily steer in the direction of their gaze, "in many cases without the conscious awareness of doing so at all." W. O. Readinger, A. Chatziastros, D. W. Cunningham, H. H. Bülthoff, and J. E. Cutting, "Gaze-Eccentricity Effects on Road Position and Steering," *Journal of Experimental Psychology: Applied*, vol. 8, no. 4 (Dec. 2002), pp. 247–58. In an e-mail correspondence, James Cutting made the further point that he thought the reason drivers were not constantly driving off the road when they looked at something had to do with balance: "The 'looking where you are going' phenomenon is, I think, strongly related to balance. This is why it is a problem with novice motorcycle drivers, and can have a small effect in walking. Balance is not much an issue in driving (although people do tilt their heads when going into a turn, and they obviously don't need to). Usually, when driving, one maintains direction while shifting gaze simply because the arm motions to make a turn are not reflexively relative to gaze direction. Balance is."

84 their position in the lane: For a concise roundup of moth effect research, see Marc Green, "Is the Moth Effect Real?" Accessed from http://www.visualexpert.com/Resources/motheffect.html.

84 while we are moving: See Mark Nawrot, Benita Nordenstrom, and Amy Olson, "Disruption of Eye Movements by Ethanol Intoxication Affects Perception of Depth from Motion Parallax," *Psychological Science*, vol. 15, no. 12 (2004), pp. 858–65.

84 for both cars: Martin Langham, Graham Hole, Jacqueline Edwards, and Colin O'Neil, "An Analysis of 'Looked but Failed to See' Accidents Involving Parked Police Vehicles," *Ergonomics*, vol. 45, no. 3 (2002), pp. 167–85. Another study found that police cars equipped with the more visible roof-top bar lights were struck just as often as cars with the less visible rear-deck lights, also suggesting

that visibility per se may not be the most important factor in these crashes. See Lieutenant James D. Wells Jr., "Patrol-Car Crashes: Rear-End Collision Study—1999," Florida Highway Patrol, 1999.

85 typically takes longer: Interestingly, a French study had subjects first take a Stroop test and then participate in a driving test on a closed course that required an unexpected evasive maneuver. Subjects who did poorly on the Stroop test tended to also do less well during the driving exercise. Christian Collet, Claire Petit, Alain Priez, and Andre Dittmar, "Stroop Color-Word Test, Arousal, Electrodermal Activity and Performance in a Critical Driving Situation," *Biological Psychology*, vol. 69 (2005), pp. 195–203.

85 in the way of the less automatic: See Colin M. McLeod, "Half a Century of Research on the Stroop Effect: An Integrative Review," *Psychological Bulletin*, vol. 109, no. 2 (1991), pp. 163–201.

85 (i.e., the word itself): This idea comes from Jennifer J. Freyd, Susan R. Martorello, Jessica S. Alvardo, Amy E. Hayes, and Jill C. Christman, "Cognitive Environments and Dissociative Tendencies: Performance on the Standard Stroop Task for High Versus Low Dissociators," *Applied Cognitive Psychology*, vol. 12 (1998), pp. 91–103.

85 than the arrow: S. B Most and R. S. Astur, "Feature-Based Attentional Set as a Cause of Traffic Accidents," *Visual Cognition*, vol. 15 (2007), pp. 125–32.

85 consultant in California: P. L. Jacobsen, "Safety in Numbers: More Walkers and Bicyclists, Safer Walking and Bicycling," *Injury Prevention*, vol. 9 (2003), pp. 205–09. The "safety in numbers" effect has been found in many other studies as well. For example, Noah Radford and David Ragland of the University of California at Berkeley looked at the city of Oakland, California. They found that nearly all of the city's most dangerous intersections were on the city's east side, an area with low pedestrian volumes. Only one of the most dangerous intersections for pedestrians was downtown. Noah Radford and David R. Ragland, "Space Syntax: An Innovative Pedestrian Volume Modeling Tool for Pedestrian Safety," U.C. Berkeley Traffic Safety Center, Paper UCB-TSC-RR-2003-11, December 11, 2003. Available at http://www.repositories.cdlib.org/its/tsc/UCB-TSC-RR-2003-11.

85 the slower they drive: See Kenneth Todd, "Pedestrian Regulations in the United States: A Critical Review," *Transportation Quarterly*, vol. 46, no. 4 (October 1992), pp. 541–59.

86 for a longer period: The Danish transportation planner Jan Gehl makes this point in his seminal book *Life Between Buildings* (New York: Van Nostrand Reinhold, 1986), p. 79.

86 safest place to be a cyclist: Conversation with Dan Burden.

87 asked to remember something: A. M. Glenberg, J. L. Schroeder, and D. A. Robertson, "Averting the Gaze Disengages the Environment and Facilitates Remembering," *Memory & Cognition*, vol. 26 (July 1998), pp. 651–58.

87 thought to aid memory: See A. Parker and N. Dagnall, "Effects of Bilateral Eye movements on Gist Based False Recognition in the DRM Paradigm," *Brain and Cognition*, vol. 63, no. 3 (April 2007), pp. 221–25.

87 other things, like driving: M. A. Recarte and L. M. Nunes, "Effects of Verbal

and Spatial-Imagery Tasks on Eye Fixations While Driving," *Journal of Experimental Psychology: Applied*, vol. 6, no.1 (2000), pp. 31–43.

87 on our mental workload: See, for example, M. C. Lien, E. Ruthruff, and D. Kuhns, "On the Difficulty of Task Switching: Assessing the Role of Task-Set Inhibition," *Psychonomic Bulletin & Review*, vol. 13 (2006), pp. 530–35.

87 for us to process things: C. Spence and L. Read, "Speech Shadowing While Driving: On the Difficulty of Splitting Attention Between Eye and Ear," *Psychological Science*, vol. 14 (2003), pp. 251–56.

87 consumes even more effort: Curiously, this has not been studied extensively per se in driving simulator studies, but the work of Nilli Lavie, at the Institute of Cognitive Neuroscience at University College London, and her colleagues hints at the problem. In a study, subjects were asked to perform a "linguistic task" that was either "high load" or "low load"; when the task was low load, they were more likely to notice an irrelevant display of motion than when it was high load. Her finding—that people are unable to ignore *irrelevant* stimuli when their "perceptual load" is not fully taxed, carries, as she mentioned to me in a conversation, the reverse implication that relevant stimuli will be less likely to be noticed under high-load conditions. See G. Rees, C. D. Frith, and N. Lavie, "Modulating Irrelevant Motion Perception by Varying Attentional Load in an Unrelated Task," *Science*, vol. 278 (1997), pp. 1616–19. In more recent research, Lavie found that people engaged in intensive visual tasks were less likely to notice sounds at a low volume. It is not difficult to extrapolate from this that an intensive auditory task—for example, straining to hear a voice at low volume on a cell phone—will exact more pressure on the "perceptual load" and thus reduce performance in performing visual tasks.

88 they still remembered fewer: David L. Strayer and Frank A. Drews, "Multitasking in the Automobile," in *Attention: From Theory to Practice*, ed. A. Kramer, D. Wiegmann, and A. Kirlik (New York: Oxford University Press, 2006).

89 When first asked this question: This is drawn from a conversation with Benjamin Coifman.

90 100 miles per hour: Robert Winkler, "The Need for Speed," *New York Times*, November 13, 2005.

91 sequential "frames": Tim Andrews and Dale Purves, "The Wagon Wheel Illusion in Continuous Light, *Trends in Cognitive Neuroscience*, vol. 9, no. 6 (2005), pp. 261–63.

91 demonstration of motion parallax): Mark Nawrot provided me with a simple exercise in "seeing" motion parallax at work: "For example, pick out two objects, near and far, on your desktop. Hold up your two index fingers, near to your face, one below the objects pointing up, one above pointing down. Hold your fingers stationary, fixate on the 'near' object, close one eye, and move your head side to side. Easy. Now do the same as you move your top finger along with your head movement so that it 'matches' the distant object. If you had to guess, you'd now say your top finger is farther away than your lower finger." For further interesting research on the mechanics of motion parallax, see Mark Nawrot, "Eye Movements Provide the Extra-retinal Signal Required for the

Perception of Depth from Motion Parallax," *Vision Research*, vol. 43 (2003), pp. 1553–62.

92 more realistic: In the scene in *The Lord of the Rings* in which the "beacons" are being lit to sound the alarm for the impending danger to Rohan, the aerial camera sweeps across the landscape, but the beacon remains in the center of the shot as the background sweeps by. Nawrot suggests that the motion might trigger an involuntary "optokinetic response." To prevent us from simply being visually swept up in that background movement, however, the eye responds with a "smooth pursuit" movement to effectively countermand the motion and maintain the fixation on the lit beacon. This, Nawrot posits, mimics the series of compensatory eye movements we are constantly making in real life. Mark Nawrot and Chad Stockert, "Motion Parallax in Motion Pictures: The Role of Background Motion and Eye Movements" (unpublished paper, Department of Psychology, North Dakota State University). For a further fascinating discussion on human vision and the movies, see James E. Cutting, "Perceiving Scenes in Film and in the World," in *Moving Image Theory: Ecological considerations*, ed. J. D. Anderson and B. F. Anderson (Carbondale: Southern Illinois University Press, 2005), pp. 9–27.

92 "illusory pavement markings": Actually, *any* pavement marking is rather illusory.

92 reduced their speed: "Evaluation of the Converging Chevron Pavement Marking Pattern," AAA Foundation for Traffic Safety (Washington, D.C.), July 2003.

92 have been mixed: "A Review of Two Innovative Pavement Marking Patterns That Have Been Developed to Reduce Traffic Speeds and Crashes," AAA Foundation for Traffic Safety (Washington, D.C.), August 1995.

93 at the higher speed: G. G. Denton, "The Influence of Adaptation on Subjective Velocity for an Observer in Simulated Rectilinear Motion," *Ergonomics*, vol. 19 (1976), pp. 409–30.

93 sensation of moving backward: In a study by Stuart Anstis, subjects asked to jog on a treadmill for as little as a minute experienced this aftereffect. Once the treadmill was stopped, subjects asked to jog in place actually jogged, on average, 162 centimeters *forward*. Anstis notes, "The backward motion of the treadmill produces an artificial mismatch between motor output and normal postural feedback, for which the adaptation compensates or nulls out by adjusting internal gain parameters to bring output and feedback back into line. But once the runner steps on to solid ground these newly adjusted parameters are now inappropriate and manifest themselves as an aftereffect, which dissipates as the parameters automatically update to match the solid ground. So these new aftereffects reveal the continuous neural recalibration of the gait control system." See Stuart Anstis, "Aftereffects from Jogging," *Experimental Brain Research*, vol. 103 (1995), pp. 476–78.

93 when asked to speed up: For an excellent discussion of this issue see John Groeger, *Understanding Driving* (East Sussex, Psychology Press: 2001), p. 14.

93 largely, it is thought: This theory is credited to the pioneering work of J. J. Gibson, who wrote: "The aiming point of any locomotion is the center of the cen-

trifugal flow of the ambient optic array." Gibson, *The Ecological Approach to Visual Perception* (Boston: Houghton Mifflin, 1979), p. 182. The "steering" process is much more complicated than this, as we must somehow compensate, like Steadicams, for the fact that our eyes and heads are also moving as we move. For a good discussion of some of these complexities, see William H. Warren, "Perception of Heading Is a Brain in the Neck," *Nature Neuroscience*, vol. 1, no. 8 (1998), pp. 647–49. Warren also provides the example of the *Millennium Falcon* in hyperspace to describe the radial pattern away from the focus of expansion.

93 "global optical flow": Not all of our sense of motion comes from visual inputs, of course. The reason I, like many other people, experienced bouts of "simulator sickness" in the various driving simulators in which I drove is that the picture of the moving road I was looking at did not correspond to what my vestibular system (the "balance" system of the inner ear) was experiencing.

93 our "target": In an interesting experiment at Brown University, researchers used virtual reality to create an optically impossible situation in which subjects had to walk toward something without the use of optical flow, instead of merely walking toward the object via its egocentric direction (its direction in space relative to the subject). Subjects were less accurate in their approach without the optic flow. See W. H. Warren, Bruce Kay, Wendy Zosh, Andrew Duchon, and Stephanie Sahue, "Optic Flow Is Used to Control Human Walking," *Nature Neuroscience*, vol. 4, no. 2 (2001), pp. 213–16.

93 a kind of radial pattern: This is not an entirely resolved issue and is still being debated. Gibson, for example, observed: "The behavior involved in steering an automobile, for instance, has usually been misunderstood. It is less a matter of aligning the car with the road than it is a matter of keeping the focus of expansion in the direction one must go." But as vision researcher Michael Land has pointed out, this argument may not account for a driver's behavior around curves: "On a curved trajectory the locations of the stationary points in the flow-field vary with distance, generating a curved line across the ground plane, not a single focus of expansion." Land notes that we rely instead of the inner edge of the road in driving around curves, with some 80 percent of driver's glances being directed in that region. See Michael F. Land, "Does Steering a Car Involve Perception of the Velocity Flow Field?" in *Motion Vision–Computational, Neural, and Ecological Constraints*, ed. Johannes M. Zanker and Jochen Zeil (New York: Springer Verlag, 2001).

93 our sense of speed: It has also been argued that optic flow influences our estimates of distance while driving as well. See M. Lappe, A. Grigo, F. Bremmer, H. Frenz, R. J. V. Bertin, and I. Israel, "Perception of Heading and Driving Distance from Optic Flow," *Driving Simulation Conference 2000* (Paris), pp. 25–31.

93 tree-lined roads: This information comes from T. Triggs, "Speed Estimation," in *Automotive Engineering and Litigations*, vol. 2, ed. G. A. Peters and B. Peters (New York: Garland Law Publishing), pp. 569–98.

94 flow at the same speed: Christopher Wickens, *Engineering Psychology and Human Performance* (Upper Saddle River, N.J.: Prentice Hall, 2000), p. 162.

94 those at lower heights: See, for example, Christina M. Rudin-Brown, "The Effect of Driver Eye Height on Speed Choice, Lane-Keeping, and Car-Following Behavior: Results of Two Driving Simulator Studies," *Traffic Injury Prevention*, vol. 7, no. 4 (December 2006), pp. 365–72; or B. R. Fajen and R. S. David, "Speed Information and the Visual Control of Braking to Avoid a Collision," *Journal of Vision*, vol. 3, no. 9 (2003), pp. 555–555a.

94 than they intend to: See C. M. Rudin-Brown, "Vehicle Height Affects Drivers' Speed Perception: Implications for Rollover Risk," *Transportation Research Record No. 1899: Driver and Vehicle Simulation, Human Performance, and Information Systems for Highways; Railroad Safety; and Visualization in Transportation* (Washington, D.C.: National Research Council, 2004), pp. 84–89.

94 speed more than others: See, for example, Allan F. Williams, Sergey Y. Kyrchenko, and Richard A. Retting, "Characteristics of Speeders," *Journal of Safety Research*, vol. 37 (2006), pp. 227–32. Of course, any findings that drivers of SUVs and pickups drove faster than other vehicles brings up other "confounding" factors, such as a higher rate of male drivers for those vehicle categories, or the idea that people who choose to drive SUVs and pickups may be more prone to speeding or feel safer and thus are more likely to drive at a higher speed—instead of the vehicle making them more prone to speeding.

94 slowly than they really were: N Harré, "Discrepancy Between Actual and Estimated Speeds of Drivers in the Presence of Child Pedestrians," *Injury Prevention*, vol. 9 (2003), pp. 38–41.

94 slow down slightly: See "Research Shows Speed Trailers Improve Safety in Temporary Work Zones," *Texas Transportation Researcher*, vol. 36, no. 3 (2000).

94 Some highway agencies: *Minnesota Tailgating Pilot Project* (St. Paul, Mn: Department of Public Safety, 2006). The Pac-Man information comes from the *Star Tribune*, December 20, 2006.

95 how fast they're going: For a good roundup of research, see Leonard Evans, *Traffic Safety* (Bloomfield Hills, Mich.: Science Serving Society, 2004), p. 173.

95 279 feet: I am using the example provided by crash investigator and human factors researcher Marc Green, available at http://www.visualexpert.com/Resources/reactiontime.html.

95 directly at a fielder: For a fascinating discussion of the complexities of catching a ball, among other things, see Mike Stadler, *The Psychology of Baseball* (New York: Gotham Books, 2007).

96 as much as several seconds: Robert Dewar and Paul Olson note that drivers "often perceive a stationary vehicle as moving, even with five seconds' viewing." Dewar and Olson, *Human Factors in Traffic Safety* (Tuscon: Lawyers and Judges Publishing, 2002), p. 23.

96 no idea of the rate: For a good discussion of this, see Olson and Farber, *Forensic Aspects of Driver Perception and Response* (Tucson: Lawyers and Judges Publishing Co., 2003), p. 112.

96 overtaking crashes: The psychologists Rob Gray and David Regan suggest that what is going on here is that as we stare for a while at things like the white stripes on the road, or trees on the side of the road, our brains quickly adapt;

they compare the effect to the well-known "waterfall effect": You stare at water rushing down a waterfall for a while, and then look at a nearby rock—it will seem to be moving upward. When we come off the highway, something similar happens, and it may look to us as if the stop sign at the end of the ramp is farther away than it really is, which is why engineers have tested chevrons and other patterns on off-ramps: to break up the illusion of those white stripes. Rob Gray and David Regan, "Risky Driving Behavior: A Consequence of Motion Adaptation for Visually Guided Motor Action," *Journal of Experimental Psychology: Human Perception and Performance*, vol. 26, no. 6 (2000), pp. 1721–32.

96 really tell the difference: This has long been known to people who study driving. In *Human Limitations in Automobile Driving* (Garden City: Doubleday, Doran & Company, 1938), authors J. R. Hamilton and Louis L. Thurstone (psychologists at Harvard University) observed: "From eight hundred feet right down to where the other car is almost on top of you, the average eye will not have any idea of the *rapidity* of motion, or speed, of the oncoming car. It will perceive motion, and that is all. The distance at which motion is first perceived, as we have said above, does not depend very much on the speed of either car. *But the distance at which rapidity of motion is perceived depends entirely upon the speed of each car.* [italics in original] With two cars traveling 40 miles an hour, that distance where the average eye suddenly perceives rapidity of motion is about 145 feet between cars. When two cars are traveling at 50 miles an hour, that distance is about 70 feet. Now we begin to have some understanding of the reason for the frightful collision accidents on the highway."

96 speed of the opposing car: See D. A. Gordon and T. M. Mast, "Driver's Decisions in Overtaking and Passing," *Highway Research Record*, no. 247, Highway Research Board, 1968.

97 your attempted passing: One study remarked on a "conundrum" about passing difficulty and passing risk, noting that drivers were found "to be somewhat poor at making the judgments required for passing maneuvers, particularly judgments about opposing vehicle speed, but the safety record of passing maneuvers is very good. This suggests that passing maneuvers occur in a relatively forgiving environment. First, while drivers are relatively poor in making passing judgments, many drivers may inherently understand this and make very conservative decisions about passing. Second, the buffer area provided downstream of each passing zone provides a margin of safety against collisions resulting from poor driver judgments." From "Passing Sight Distance Criteria," NCHRP Project 15-26, MRI Project 110348, prepared for the National Cooperative Highway Research Program, Transportation Research Board National Research Council, Midwest Research Institute, March 2000.

97 up by only 30 percent: L. Staplin, "Simulator and Field Measure of Driver Age Differences in Left-Turn Gap Judgments," *Transportation Research Board Record*, no. 1485, Transportation Research Board, National Research Council, 1995.

97 to actually see: R. E. Eberts and A. G. MacMillan, "Misperception of Small Cars," in *Trends in Ergonomics/Human Factors*, vol 2, ed. R. E. Ebert and C. G. Eberts (North Holland: Elsevier Science Publishers, 1985).

98 slower the object seems: H. W. Leibowitz, "Grade Crossing Accidents and Human Factors Engineering," *American Scientist*, vol. 73, no. 6 (November–December 1985), pp. 558–62. Leibowitz also noted another potential reason—the "deceptive geometry of collisions"—for overestimating the distance of an approaching train, similar to the problem mentioned with drivers trying to judge the distance of an approaching car. A car and a train that are approaching each other will retain consistent positions. He wrote, "There is no lateral motion, and thus the principal cue to velocity is the increase in size of the visual angle subtended or the expansion pattern. . . . The rate of increases of the expansion pattern is not linear but rather is described by a hyperbolic function. For distant objects, the rate of change in the expansion is low. As the distance decreases, the visual angle subtended increases at an accelerated rate." This is somewhat similar to a phenomenon known as "motion camouflage," which has been observed in the natural world—male hoverflies, for example, move in a way to conceal the fact that they are moving when they are tracking female hoverflies. They do so, it has been argued, by "approaching along a path such that its image projected onto the prey's eye emulates that of a distant stationary object (a *fixed point*). During its attack, the predator must ensure that it is always positioned directly between the current position of the prey and this fixed point." Humans, research has suggested, are also susceptible to this effect. See Andrew James Anderson and Peter William McOwan, "Humans Deceived by Predatory Stealth Strategy Camouflaging Motion," *Proceedings of the Royal Society B: Biological Sciences*, vol. 270, Supp. 1 (August 7, 2003), pp. S18–S20.

98 latter was moving faster: Joseph E. Barton and Theodore E. Cohn, "A 3D Computer Simulation Test of the Leibowitz Hypothesis," U.C. Berkeley Traffic Safety Center, Paper UCB-TSC-TR-2007-10, April 1, 2007; http://repositories.cdlib.org/its/tsc/UCB-TSC-TR-2007-10.

98 human vision is an illusion: See Sandra J. Ackerman, "Optical Illusions: Why Do We See the Way We Do?" *HHMI Bulletin*, June 2003, p. 37.

98 (much more at night): Dewar and Olson, *Human Factors in Traffic Safety*, p. 88.

98 remember more at night): D. Shinar and A. Drory, "Sign Registration in Daytime and Night Time Driving," *Human Factors*, vol. 25 (1983), pp. 117–22.

99 blind to our blindness: See H. W. Leibowitz, "Nighttime Driving Accidents and Selective Visual Degradation," *Science*, vol. 197 (July 29, 1977), pp. 422–23.

99 as drivers actually do: M. J. Allen, R. D. Hazlett, H. L. Tacker, and B. L. Graham, "Actual Pedestrian Visibility and the Pedestrian's Estimate of His Own Visibility," *American Journal of Optometry and Archives of the American Academy of Optometry*, vol. 47 (1970), pp. 44–49, and David Shinar, "Actual Versus Estimated Night-time Pedestrian Visibility," *Ergonomics*, vol. 27, no. 8 (1984), pp. 863–71, and Richard Tyrrel, Joanne Wood, and Trent Carberry, "On-road Measures of Pedestrians' Estimates of Their Own Nighttime Conspicuity," *Journal of Safety Research*, vol. 35, no. 5 (December 2004), pp. 483–90.

99 drive 20 miles per hour: See Olsen, *Forensic Aspects of Driver Perception and Response*, p. 157.

99 through the landscape: The contrast experiment discussed can be viewed at http://www.psy.ucsd.edu/~sanstis/Foot.html. For an interesting discussion of the experiment and the traffic implications, see Stuart Anstis, "Moving in a Fog: Contrast Affects the Perceived Speed and Direction of Motion," *Proceedings of the Conference on Neural Networks*, Portland, Ore., 2003.

99 signs have been set up: See C. Arthur MacCarley, Christopher Ackles, and Tabber Watts, "A Study of the Response of Highway Traffic to Dynamic Fog Warning and Speed Advisory Messages," TRB 06-3086, Transportation Research Record, National Research Council, Washington, D.C., February 2007.

99 not brake accordingly: For an excellent discussion of snowplow visibility, see Albert Yonas and Lee Zimmerman, "Improving the Ability of Drivers to Avoid Collisions with Snowplows in Fog and Snow," Minnesota Department of Transportation, St. Paul, Minn., July 2006.

100 glances over the shoulder: The rearview mirror information is drawn from Thomas Ayres, Li Li, Doris Trachtman, and Douglas Young, "Passenger-Side Rear-View Mirrors: Driver Behavior and Safety," *International Journal of Industrial Ergonomics*, vol. 35 (2005), pp. 157–62.

100 actually it is *half*: This example was proposed by the art historian E. H. Gombrich in *Art and Illusion* (Oxford: Phaidon Press, 1961) and was later confirmed and studied further by Marco Bertamini and Theodore E. Parks in "On What People Know About Images on Mirrors," *Cognition*, vol. 98 (2005), pp. 85–104. Their use of the phrase "on mirrors" immediately reveals one of the disconnects we tend to have with mirrors, as we tend to say "in mirrors," as if the image lurked behind the glass. The authors note, "Both the fact that our image is half the physical size, and the fact that this relationship is independent of how far we are from the mirror, are counterintuitive. However, they become clearer as soon as we realize that a mirror is always located halfway between oneself and our virtual self."

101 "they ought to be": For details on Flannagan's work with rearview mirrors, see M. J. Flannagan, M. Sivak, J. Schumann, S. Kojima, and E. Traube, "Distance Perception in Driver-Side and Passenger-Side Convex Rearview Mirrors: Objects in Mirror are More Complicated Than They Appear," Report No. UMTRI-97-32, July 1997.

## Chapter Four: Why Ants Don't Get into Traffic Jams

102 "cricket war": William G. Harley, "Mormons, Crickets, and Gulls: A New Look at an Old Story," *Utah Historical Quarterly*, vol. 38 (Summer 1970), pp. 224–39.

103 "black carpet": From Peter Calamai, "Crickets March with Religious Fervor," *Toronto Star*, August 2, 2003.

104 as a tight swarm: A good way to think about this in human terms, as complex-systems theorist Eric Bonabeau has cleverly done, is to imagine a cocktail party.

Each person in the room is given a command: Pick two people at random, A and B, and then place yourself so that A is constantly between B and you. In a room of people, this results in a loose crowd always on the move, shifting to stay in the right position, some people at times drifting around the periphery like timid wallflowers. Now change the rules, however, so that *you* are always between A and B. Instead of milling, the crowd will clump into a "single, almost stationary cluster." A seemingly minor change in the way each person acts completely alters the group. Could you have predicted that? From Eric Bonabeau, "Predicting the Unpredictable," *Harvard Business Review*, vol. 80, no. 3 (March 2002). For a more in-depth discussion of the dynamics involved, see Bonabeau, Pablo Funes, and Belinda Orme, "Exploratory Design of Swarms," *Proceedings of the Second International Workshop on the Mathematics and Algorithms of Social Insects* (Atlanta, GA: Georgia Institute of Technology, 2003), pp. 17–24.

104 to play by the rules: Matt Steinglass made an important point while writing about a collision that Seymour Papert, the founder of MIT's Artificial Intelligence Lab, suffered with a motorbike while crossing the street in Hanoi, Vietnam, a city where the traffic behavior is as much explained by "emergent behavior" as it is by formal traffic rules (if not more so): "One thing about emergent phenomena that the pioneers of the field tended not to emphasize is that they are often unkind to their constituent agents: Ant colonies are not very solicitous of the lives of individual ants. Hanoi traffic is a fascinating emergent phenomenon, but it didn't take good care of Seymour Papert when he became one of its constituent agents." Steinglass, "Caught in the Swarm," *Boston Globe*, December 17, 2006.

104 the "wrong" direction: For a fascinating discussion of the dynamics of the wave, see I. Farkas, D. Helbing, and T. Vicsek, "Mexican Waves in an Excitable Medium," *Nature*, vol. 419 (2002), pp. 131–32. For a simulation and videos, see http://angel.elte.hu/wave/.

104 none died: Gregory A. Sword, Patrick D. Lorch, and Darryl T. Gwynne, "Migratory Bands Give Crickets Protection," *Nature*, vol. 433 (February 17, 2005).

105 a congested mess: This recalls a number of studies of how animal behavior changes under increasingly crowded conditions. A study that looked at cats found results that sound a lot like rush-hour freeways: "The more crowded the cage is, the less relative hierarchy there is. Eventually a despot emerges, 'pariahs' appear, driven to frenzy and all kinds of neurotic behavior by continuous and pitiless attack by all others; the community turns into a spiteful mob. They all seldom relax, they never look at ease, and there is a continuous hissing, growling, and even fighting. Play stops altogether and locomotion and exercises are reduced to a minimum." Quoted in E. O. Wilson, *Sociobiology: The New Synthesis* (Cambridge, Mass.: Harvard University Press, 1995), p. 255.

106 difference in the number of cars: David Shinar and Richard Compton, "Aggressive Driving: An Observational Study of Driver, Vehicle, and Situational Factors," *Accident Analysis & Prevention*, vol. 36 (2004), pp. 429–37.

106 road signs and white stripes: The biologist E. O. Wilson notes that "in general, it appears that the typical ant colony operates with somewhere between 10 and 20 signals, and most of these are chemical in nature." E. O. Wilson and Bert Holldöbler, *The Ants* (Cambridge, Mass.: Havard University Press, 1990), p. 227.

106 army ant trail in Panama: I. D. Couzin and N. R. Franks, "Self-organized Lane Formation and Optimized Traffic Flow in Army Ants," *Proceedings of the Royal Society: Biological Science*, v. 270 (1511), January 22, 2003, pp. 139–46.

107 "pinnacle of traffic organization": Ant foraging models have been deployed in the human world to improve the routing performance of trucking and other companies. For a good account see Peter Miller, "Swarm Theory," *National Geographic*, July 2007.

110 ongoing labor dispute: Sharon Bernstein and Andrew Blankstein, "2 Deny Hacking Into L.A.'s Traffic Light System," *Los Angeles Times*, January 9, 2007.

111 feel their neighbors' presence: Stephen Johnson writes that "the problem with all car-centric cities is that the potential for local interaction is so limited by the speed and the distance of the automobile that no higher-level order can emerge. . . . There has to be feedback between agents, cells that change in response to the changes in other cells. At sixty-five miles an hour, the information transmitted between agents is too limited for such subtle interactions, just as it would be in the ant world if a worker ant suddenly began to hurtle across the desert floor at ten times the speed of her neighbors." See Johnson, *Emergence* (New York: Scribner, 2001), p. 96.

112 even ATSAC's computers: John Fisher would point this fact out again later in a newspaper story announcing the state of California's $150 million plan to synchronize all the city's signals, which, officials announced, could shave commutes by "up to 16%." *Los Angeles Times*, October 17, 2007.

112 more people die in cars each year: Gerald Wilde pointed this out to me.

112 "pedestrian interference": See, for example, N. M. Rouphail and B. S. Eads, "Pedestrian Impedance of Turning-Movement Saturation Flow Rates: Comparison of Simulation, Analytical, and Field Observations," *Transportation Research Record*, No. 76, Annual Meeting of the Transportation Research Board, Washington, D.C., 1997, pp. 56–63.

112 to help move the fewer cars: The city of Amsterdam, for example, has instituted a "green wave" for cyclists, so that cyclists moving at 15 to 18 kilometers per hour get a succession of green lights. (Cars, which tend to move more quickly than that, will find themselves seeing more red.) From "News from Amsterdam," retrieved from http://www.nieuwsuitamsterdam.nl/English/2007/11/green_wave.htm.

113 green wave for walking?: Indeed, as the urbanist William H. Whyte pointed out, the signals on Fifth Avenue seem designed to *thwart* the pedestrian: "Traffic signals are a particular vexation. They are, for one thing, timed to benefit cars rather than pedestrians. Take Fifth Avenue. You want to make time going north. At the turn of the light to green you start walking briskly. You have about 240 feet to go to reach the next light. You will reach it just as the light turns red.

Only by going at flank speed, say 310 feet per minute, will you beat the light." From William H. Whyte, *City* (New York: Doubleday, 1988), p. 61.

113 even higher authorities: This is not such a far-fetched premise. A study by a team of researchers at Bar-Ilan University in Israel examined pedestrian behavior in two cities: the "ultra-Orthodox" Bnei Brak and the "secular" Ramat-Gan. While traffic and infrastructure conditions were essentially the same in both locations, pedestrians in Bnei Brak were three times more likely to commit what the researchers judged "unsafe" pedestrian behaviors. This may be a function of the fact that fewer residents of Bnei Brak own cars; thus they're less cognizant of drivers' abilities or less willing to consider them. But the researchers suggested another reason, citing studies that note "a strong connection between the belief in supremacy of other laws (i.e. religious laws) over state laws, and a readiness to violate the law." See Tova Rosenbloom, Dan Nemrodova, and Hadar Barkana, "For Heaven's Sake Follow the Rules: Pedestrians' Behavior in an Ultra-Orthodox and a Non-Orthodox City," *Transportation Research Part F: Traffic Psychology and Behaviour*, vol. 7, no. 6 (November 2004), pp. 395–404. For more on the link between religious belief and compliance with laws, see A. Ratner, D. Yagil, and A. Pedahzur, "Not Bound by the Law: Legal Disobedience in Israeli Society," *Behavioral Sciences and the Law*, vol. 19 (2001), pp. 265–83.

113 "crosswalk on the Sabbath": Letter from the Rabbinical Council of California to John Fisher, August 9, 2004.

114 stops by 31 percent: F. Banerjee, "Preliminary Evaluation Study of Adaptive Traffic Control System (ATCS)," City of Los Angeles Department of Transportation, July 2001.

117 previous night's fireworks: In 2005, the CHP reported, there were thirty-four Code 1125-A incidents on Tuesday, July 5, roughly 50 percent more than the previous or following Tuesday. Data provided by Joe Zizi of the CHP.

118 "driving on ice, literally": The link between precipitation intervals and crash risk is well-known driver lore, and studies back it up. See Daniel Eisenberg, "The Mixed Effects of Precipitation on Traffic Crashes," *Accident Analysis & Prevention*, vol. 36 (2004), pp. 637–47.

119 for many decades: G. F. Newell, a researcher at the University of California at Berkeley, observed that "in later years, indeed even to the present time, some researchers try to associate with vehicular traffic all sorts of phantom phenomena analogous to the effects in gases. They don't exist." He also argued that traffic is not "like any of the idealized models that the mathematical statisticians theorize about. It is messy and can be analyzed only by crude approximations." G. F. Newell, "Memoirs on Highway Traffic Flow Theory in the 1950s," *Operations Research*, vol. 50, no. 1 (January–February 2002), pp. 173–78.

120 "puzzles remain unsolved": See Carlos Daganzo, "A Behavioral Theory of Multi-lane Traffic Flow, Part I: Long Homogeneous Freeway Sections," *Transportation Research Part B: Methodological*, vol. 36, no. 2 (February 2002), pp. 131–58.

120 "heterogeneity of driver behavior": In his superb book *Critical Mass*, Philip

Ball, noting the increasing inclusion of "psychological" and other such factors in traffic modeling, points out a conundrum: "The more complex the model, the harder it becomes to know what outcomes are in any sense 'fundamental' aspects of traffic flow and which follow from the details of the rules." See Philip Ball, *Critical Mass* (New York: Farrar, Straus and Giroux, 2004), p. 160.

120 when they followed passenger cars: The researchers who conducted the study speculated that following drivers may believe that SUVs, like tractor-trailers, take longer to stop than a car, and thus it is safer to follow at a closer distance. Another theory is that "ignorance is bliss"—that is, drivers worry less about what they cannot see than what they can (or they merely focus on the vehicle immediately in front of them, rather than a stream of several vehicles, because it seems easier). See James R. Sayer, Mary Lynn Mefford, and Ritchie W. Huang, "The Effects of Lead-Vehicle Size on Driver-Following Behavior: Is Ignorance Truly Bliss?" Report No. UMTRI-2000-15, University of Michigan, Transportation Research Institute, June 2000.

120 Los Gatos effect: Carlos F. Daganzo, "A Behavioral Theory of Multi-Lane Traffic Flow," Part I: Long Homogeneous Freeway Sections." Transportation Research Part B: Methodological, vol. 36, no 2 (Febryary 2002), pp 131–58.

121 traveling at 55 miles per hour: In 1985, the *Highway Capacity Manual*, the bible of highway engineers, put maximum capacity at 2,000 vehicles per lane per hour. That was raised to 2,300 in 1994 and raised again in 1998 to its current figure. Drivers, it seems, are willing to drive at a closer distance to the car ahead of them and to do so at higher speeds in the past. Why are drivers willing to take on more risk? It may be because vehicles have better handling, or because drivers are finding themselves having to cover more distance in a commute, and are thus willing to drive more aggressively to reduce the time. See Federal Highway Administration, *2004 Status of the Nation's Highways, Bridges and Transit: Conditions and Performance* (Washington, D.C.: 2004), U.S. Department of Transportation, pp. 4–16. Similarly, where previous estimates calculated that maximum flow occurred at 45 miles per hour, research by Pravin Varaiya in California, drawn from inductor-loop figures, now puts that figure at 60 miles per hour. See Z. Jia, P. Varaiya, C. Chen, K. Petty, and A. Skabardonis, "Maximum Throughput in L.A. Freeways Occurs at 60 MPH," University of California, Berkeley, PeMS Development Group, January 16, 2001.

121 that it is being underused: As with many things in traffic, there is a debate as to the actual efficacy of HOV lanes from a traffic point of view (and not a social perspective). Do they improve the total flow of the highway or, more narrowly, simply give HOV drivers a faster trip? Or do they actually accomplish neither? In one study, by University of California researchers Pravin Varaiya and Jaimyoung Kwon, based on loop-detector data taken from freeways in the San Francisco area, the HOV lane, it was argued, not only increased congestion in the other lanes (as one might expect if only a minority of drivers are using the HOV lane), but itself suffered from a 20 percent "capacity penalty." The reason? As it was a single lane, any driver stuck behind a "snail"—in California, driving 60

miles per hour earns you this characterization—in the HOV lane had to travel the speed of the snail (as the other lanes were even slower, it would not do to try to pass the HOV snail). An additional potential complication that has emerged is that in California, cars bearing a hybrid fuel sticker (85,000 of the most recent version were issued) are legally permitted to drive in HOV lanes. Those drivers may indeed wish to travel around 60 miles per hour, as that will produce higher fuel efficiency (as indicated by the in-car displays). In a later study by fellow University of California researchers Michael J. Cassidy, Carlos F. Daganzo, Kitae Jang, and Koohong Chung (of the California Department of Transportation), the authors reexamined Varaiya and Kwon's data and came to the conclusion that while overall traffic speeds did drop concurrently with the time the HOV lane was actuated (which, it must be pointed out, is precisely when the roads begin to get crowded; hence the HOV lane), they could not attribute this decline to the HOV lanes themselves, and in some cases, the HOV lanes actually *enhanced* the flow of traffic through troublesome bottlenecks. See J. Kwon and P. Varaiya, "Effectiveness of High-Occupancy Vehicle (HOV) Lanes in the San Francisco Bay Area," July 2006, available at http://www.sci.csuhayward.edu/~jkwon/, and Michael J. Cassidy, Carlos F. Daganzo, Kitae Jung, and Koohong Chung, "Empirical Reassessment of Traffic Operations: Freeway Bottlenecks and the Case for HOV Lanes," Research Report UCB-ITS-RR-2006-6, December 2006.

122 nowhere near critical density: "Possible Explanations of Phase Transitions in Highway Traffic," C. F. Daganzo, M. J. Cassidy, and R. L. Bertini, Department of Civil and Environmental Engineering and Institute of Transportation Studies, University of California, Berkeley, May 25, 1998.

122 If done properly: This is not to say that ramp meters always work perfectly, because nothing in traffic is ever so easy. Timing patterns may be skewed (although this is being addressed with real-time, system-wide adaptive ramp meters). Ramp metering done without carefully studying the traffic terrain can lead to "perverse outcomes," one study suggests, as in the case of metered on-ramp drivers being held hostage by a "downstream" off-ramp they will not even use (congestion caused "not by too many cars getting on the freeway but by too many cars trying to get off"). Too many cars held on the ramp, no matter how desirable for the freeway, can back up into local streets, triggering other jams. Needless to say, for metering to work properly, people actually need to obey the signals. There is a fairness issue as well, as the authors of the Minnesota study pointed out: Ramp metering favors those making longer trips and actually hurts those traveling only a few exits. See Michael Cassidy, "Complications at Off-Ramps," *Access* magazine, January 2003, pp. 27–31.

122 one-third less time: The rice experiment (proposed by Paul Haase) was the winning entry in a contest sponsored by the Washington DOT for the best way to visualize "throughput maximization"; Susan Gilmore, "Rice Is Nice When Trying to Visualize Highway Traffic," *Seattle Times*, December 29, 2006.

123 "like cars on the highway": To wit: "Traffic flow resembles granular flow nowhere more closely than on the highway. Here the individual behavior of

the drivers forms a relatively small statistical perturbation on the deterministic part of the collective motion, and hence the cars can be treated as physical particles. Both are many particle systems far from equilibrium, in which the constant competition between driving forces and dissipative interactions leads to self-organized structures: Indeed, there is a strong analogy between the formation of traffic jams on the highway and the formation of particle clusters in a granular gas." From K. van der Weele, W. Spit, T. Mekkes, and D. van der Meer, "From Granular Flux Model to Traffic Flow Description," in *Traffic and Granular Flow 2003*, eds. S. P. Hoogendoorn, S. Luding, P. H. L. Bovy, M. Schreckenberg, and D. E. Wolf (Berlin: Springer, 2005), pp. 569–78. On the other hand, G. F. Newell, a seminal traffic flow researcher, once cautioned that "some researchers try to associate with vehicular traffic all sorts of phantom phenomena analogous to the effects in gases. They don't exist." G. F. Newell, "Memoirs on Highway Traffic Flow Theory in the 1950s," *Operations Research*, vol. 50, no. 1 (January–February 2002), pp. 173–78.

123 "through the hopper": Rice is not a perfect metaphor for traffic either. As Benjamin Coifman points out, "Traffic is mostly a one-dimensional system within the lane, with occasional coupling to adjacent lanes. Traditional granular flow is three-dimensional. And then in traffic you are dealing with smart particles." (Author interview.)

123 between the grains: The German physicist and traffic researcher Dirk Helbing has observed a similar phenomenon at work in the "outflow" of people from crowded rooms. "Panicking pedestrians often come so close to each other, that their physical contacts lead to the buildup of pressure and obstructing friction effects." This can occur even when the exits are fairly wide. Why? "This comes from disturbances due to pedestrians, who expand in the wide area because of their repulsive interactions or try to overtake one another." His simulations have found that columns placed asymmetrically in front of door openings can help "reduce the pressure at the door." As with rice, when you organize the flow, slower is faster. See Dirk Helbing, "Traffic and Related Self-Driven Many-Particle Systems," *Reviews of Modern Physics*, vol. 73, no. 4 (2001), pp. 1067–1141.

123 with ramp meters than without: See David Levinson and Lei Zhang, "Ramp Meters on Trial: Evidence from the Twin Cities Metering Holiday," Department of Civil Engineering, University of Minnesota, May 30, 2002; see also Cambridge Systematics, "Twin Cities Ramp Meter Evaluation," prepared for Minnesota Department of Transportation, February 1, 2001.

124 rarely have to stop: Jerry Champa, "Roundabout Intersections: How Slower Can Be Faster," *California Department of Transportation Journal*, vol. 2 (May–June 2002), pp. 42–47.

124 1,320 vehicles per hour: Robert Herman and Keith Gardels, "Vehicular Traffic Flow," *Scientific American*, vol. 209, no. 8 (December 1963).

125 more lost time: According to one study, SUVs reduce traffic flow in another way as well, by blocking the view of following drivers, who tend to leave more headway as their sight distance drops and they are less sure of traffic conditions

ahead. This, of course, diverges from the findings of another study, cited above in the note for the phrase "when they followed passenger cars." The difference in results may be due to the different types of roads on which the two studies were conducted or some other unidentified artifact. Kara M. Kockelman and Raheel A. Shabih, "Effect of Vehicle Type on the Capacity of Signalized Intersections: The Case of Light-Duty Trucks," *Journal of Transportation Engineering*, vol. 126, no. 6 (1999), pp. 506–12.

125 stop on red: See, for example, Matt Helms, "Wait Just Two Seconds Before You Start," *Free Press*, June 18, 2007.

125 Drivers talking on cell phones: University of Utah psychology professor David Strayer found in one driving-simulator experiment that subjects talking on a cell phone tended to drive more slowly and make fewer lane changes to avoid slower moving traffic (which may be read as a surrogate for a delayed ability to react). The total of this activity, Strayer estimates, adds 5 to 10 percent to total commuting times (then again, driving more slowly has safety and environmental benefits). See Joel M. Cooper, Ivana Vladisavljevic, David L. Strayer, and Peter T. Martin, "Drivers' Lane-Changing Behavior While Conversing on Cell Phone in Variable-Density Simulated Highway Environment," paper submitted to 87th Transportation Research Board meeting, Washington, D.C., 2008.

125 about 12 miles per hour: Robert L. Bertini and Monica T. Leal, "Empirical Study of Traffic Features at a Freeway Lane Drop," *Journal of Transportation Engineering*, vol. 131, no. 6 (2005), pp. 397–407.

126 wreak progressive havoc: See Philip Ball, "Slow, Slow, Quick, Quick, Slow," *Nature*, April 17, 2000. For the original research, see T. Nagatani, "Traffic Jams Induced by Fluctuation of a Leading Car," *Physical Review E*, vol. 61 (2000), pp. 3534–40.

128 effects of a shock wave: See P. Breton, A. Hegy, B. De Schutter, and H. Hellendoorn, "Shock Wave Elimination/Reduction by Optimal Coordination of Variable Speed Limits," *Proceedings of the IEEE Fifth International Conference on Intelligent Transportation* Systems (ITSC '02), Singapore, pp. 225–30, September 2002.

128 trip times declined: Highways Agency, *M25 Controlled Motorways: Summary Report*, November, 2004.

128 slower can be faster: These systems require careful planning, however, to avoid unintended effects. The speed-limit step-down cannot be too sudden, or that itself could cause a shock wave. The ideal system would be coordinated along the length of the highway, to avoid simply sending one well-coordinated group of drivers smack into another jam farther down the road—and inadvertently helping to extend that jam or cause another one. See, for example, P. Breton et al., "Shock Wave Elimination/Reduction by Optimal Coordination of Variable Speed Limits."

128 or the opposite: Boris Kerner notes, "The traffic flow instability is related to a finite reaction time of drivers. This reaction time is responsible for the vehicle over-deceleration effect: if the preceding vehicle begins to decelerate unexpectedly, a driver decelerates stronger than is needed to avoid collisions." From

Boris Kerner, *The Physics of Traffic: Empirical Freeway Pattern Features, Engineering Applications, and Theory* (Berlin: Springer, 2004), p. 69.

128 each car behind it will stop: One simulation compared the "oscillations" and "amplifications" found in stop-and-go traffic to those found in queues. "Perturbations" in the queue, or the way people stopped and started, were often observed to grow larger from the front to the back of the queue in simulators using cellular automata. See Bongsoo Son, Tawan Kim, and Yongjae Lee, "A Simulation Model of Congested Traffic in the Waiting Line," *Computational Science and Its Applications: ICCSA 2005*, vol. 3481 (2005), pp. 863–69.

128 the harder it is to predict: An interesting parallel has been drawn between the way nonlinear traffic flows behave and the way supply chains work in the world of business. Supply chains suffer from what has been called the "bullwhip effect"—the farther a supplier is from the consumer, the higher the potential for variability (e.g., oversupply or undersupply). For example, when a person orders a beer in a bar, there is direct communication between the patron and the bartender. The order is placed and then filled. But this immediacy becomes increasingly more difficult moving out along the supply chain. If there is a sudden surge in demand for a type of beer at a bar, the bartender will be instantly aware of this; it will take longer for the brewer of the beer to realize this, and even longer for the grower of the hops (and by the time they react to the changed demand, it may have changed again). In traffic, Carlos Daganzo has pointed out, cars flow through a bottleneck rather smoothly; cars far upstream of the bottleneck, however, experience wide "oscillations" in speed. They are less aware of the actual conditions of supply and demand than those cars moving through the bottleneck. See "The Beer Game and the Bullwhip," *ITS Berkeley Online Magazine*, vol. 1, no. 2 (Winter 2005).

129 *by the car following them:* Gary A. Davis and Tait Swenson, "Identification and Simulation of a Common Freeway Accident Mechanism: Collective Responsibility in Freeway Rear-End Collisions," CTS 06-02. Intelligent Transportation Systems Institute, Center for Transportation Studies, University of Minnesota, April 2006.

129 car was given ACC: The ACC study results are described in L. C. Davis, "Effect of Adaptive Cruise Control Systems on Traffic Flow," *Physical Review E*, vol. 69 (2004).

## Chapter Five: Why Women Cause More Congestion Than Men

131 1.1 hours: Andreas Schafer and David Victor, "The Past and Future of Global Mobility," *Scientific American*, October 1997, pp. 58–63.

131 made more frequent, shorter trips: Yacov Zahavi, "The 'UMOT' Project," August 1979, prepared for the U.S. Department of Transportation and the Ministry of Transport, Federal Republic of Germany, Bonn.

132 in one hour: Cesare Marchetti, "Anthropological Invariants in Travel Behavior," *Technological Forecasting and Social Change*, vol. 47 (1994), pp. 75–88.

132 thirty minutes each way: M. Wachs, B. D. Taylor, N. Levine, and P. Ong, "The Changing Commute: A Case-study of the Jobs-Housing Relationship over Time," *Urban Studies*, vol. 30, no. 10 (1993), pp. 1711–29.

133 jobs were located: See David Levinson and Ajay Kumar, "The Rational Locator," *Journal of the American Planning Association*, vol. 60, no. 3 (1994), pp. 319–43. Similar trends have been observed in the Portland area, as described in Robert L. Bertini, "You Are the Traffic Jam: An Examination of Congestion Measures," paper submitted to Eighty-fifth Annual Meeting of the Transportation Research Board, January 2006, Washington, D.C.

133 jacking up the numbers: D. Levinson and Y. Wu, "The Rational Locator Reexamined," *Transportation*, vol. 32 (2005), pp. 187–202.

134 prompts more driving: See Nancy McGuckin, Susan Liss, and Bryant Gross, "Do More Vehicles Make More Miles?" *National Household Travel Survey* (Washington, D.C.: Federal Highway Administration, 2001).

134 the worse the traffic congestion: Anthony Downs, "Why Traffic Congestion Is Here to Stay . . . and Will Get Worse," *Access Magazine*, no. 25 (Fall 2004). See also Scott F. Festin, *Summary of National and Regional Travel Trends: 1970–1995* (Washington, D.C.: U.S. Department of Transportation, Federal Highway Administration, 1996).

134 figure is 48 percent: Figures supplied by Alan Pisarski.

135 roughly 16 percent: Alan Pisarski, *Commuting in America III* (Washington, D.C.: Transportation Research Board, 2006), p. 2.

135 over 32 miles: Susan Handy, Andrew DeGarmo, and Kelly Clifton, *Understanding the Growth in Non-Work VMT*, Research Report SWUTC/02/167222 (Austin, Texas: Southwest Region University Transportation Center, University of Texas, February 2002), p. 6.

135 whole day to complete: For a good discussion of recent changes in women's travel behavior, see Rachel Gossen and Charles Purvis, "Activities, Time, and Travel: Changes in Women's Travel Time Expenditures, 1990–2000," *Research on Women's Issues in Transportation, Report of a Conference, Vol. 2* (Washington, D.C.: Transportation Research Board, 2004).

136 are now fam-pools: Nancy McGuckin and Nandu Srinivasan, "The Journey-to-Work in the Context of Daily Travel," paper presented at the Transportation Research Board meeting, Washington, D.C., 2005.

136 statistically driving more miles: Survey data in the United States indicates what seems like an intuitive fact: The more members in a household, the more miles it drives. "Travel within households increases by household size and income," as Nancy McGuckin put it to me in an e-mail correspondence.

136 precocious car poolers: See, for example, Christina Sidecius, "Car Pool Lane Not for Dummies," *Seattle Times*, August 2, 2007.

136 more often than men do: See *Research on Women's Issues in Transportation: Report of a Conference* (Washington, D.C.: Transportation Research Board, National Research Council, 2005), p. 30.

136 about 15 percent do: Jane Brody, "Turning the Ride to School into a Walk," *New York Times*, September 11, 2007.

136 by some 30 percent: See U.S. Environmental Protection Agency, *Travel and Environmental Implications of School Siting*, EPA 231-R-03-004, October 2003, and Department of Environment, Transport and the Regions, London, Greater Vancouver Regional District, *Morning Peak Trip by Purpose*, 1999.

137 sports in America *doubled*: Charles Fishman, "The Smorgasbord Generation," *American Demographics*, May 1999.

137 trips are getting longer: Handy, DeGarmo, and Clifton, *Understanding the Growth in Non-Work VMT*.

137 typical rush hours: See *Highway Statistics 2005* (Washington, D.C.: Office of Highway Policy Information, Federal Highway Administration).

137 closest to their home: Susan L. Handy and Kelly J. Clifton, "Local Shopping as a Strategy for Reducing Automobile Dependence," *Transportation*, vol. 28, no. 4 (2001), pp. 317–46.

137 did a few decades ago: Handy, DeGarmo, and Clifton, p. 31.

138 it was .79 miles: Handy, DeGarmo, and Clifton, p. 29.

138 was completely alien: See the report by the Technical Committee of the Colorado-Wyoming Section of the Institute for Transportation Engineers, "Trip Generation of Coffee Shops with Combination Drive-Through and Sit-Down Facilities"; retrieved from http://www.cowyite.org/technical/.

138 left turn during rush hour: Starbucks also anticipates traffic flow in another way: It likes to locate stores near dry cleaners and video rental shops in order to capture the "dropping off" and "picking up" traffic flows (two chances to sell that double latte). See Taylor Clark, *Starbucked* (New York: Little, Brown, 2007).

139 stalled queues of cars: Andrew Downie, "Postcard: Brazil," *Time*, September 27, 2007. The author drily notes: "Motorbikes account for 9% of the city's vehicles but they cause more accidents than all the rest combined, according to city traffic officials. That means moto-medics also come with a dose of irony."

139 all other travel methods: Pisarski, *Commuting in America III*, p. 109.

139 those without one: "Poverty and Mobility in America," *NPTS Brief* (Washington, D.C.: U.S. Department of Transportation, Federal Highway Administration, December 2005).

139 than public transit: See Brian D. Taylor, "Putting a Price on Mobility: Cars and Contradictions in Planning," *Journal of the American Planning Association*, vol. 72, no. 3 (Summer 2006), pp. 279–84.

139 near the top: Daniel Kahneman, Alan Krueger, Norbert Schwarz, and Arthur Stone, "A Survey Method for Characterizing Daily Life Experience: The Day Reconstruction Method," *Science*, vol. 306, no. 5702 (December 2004), pp. 1776–78.

139 but sixteen minutes: Mokhtarian raises the point that people in such surveys may be confusing the idea of "ideal commute" with what commute they would be *willing* to make; she also notes that they might be giving what they consider to be a "realistic" ideal and not, say zero minutes. See Patricia L. Mokhtariand and Lothlorien S. Redmond, "The Positive Utility of the Commute: Modeling Ideal Commute Time and Relative Desired Commute Amount," Berkeley: University of California Transportation Center, Reprint UCTC No. 526.

140 figuring out alternatives: S. Handy, L. Weston, and Patricia L. Mokhtarian, "Driving by Choice or Necessity?" *Transportation Research Part A: Policy and Practice*, vol. 39, nos. 2–3 (2005), pp. 183–203.

140 rational perspective: Alois Stutzer and Bruno S. Frey, "Stress That Doesn't Pay Off: The Commuting Paradox" (September 2004), IZA Discussion Paper No. 1278, Zurich IEER Working Paper No. 151. Available at SSRN: http://ssrn.com/abstract=408220.

140 grown the most: Robert H. Frank, *Falling Behind* (Berkeley: Univ. of California Press, 2007), p. 82.

141 "hedonic adaptation": See S. Frederick and G. Loewenstein, "Hedonic Adaptation," in *Scientific Perspectives on Enjoyment, Suffering, and Well-Being*, ed. D. Kahneman, E. Diener, and N. Schwartz (New York: Russell Sage Foundation, 1999), pp. 303–29.

141 more prone it is to variability: Nancy McGuckin and Nandu Srinivasan, "The Journey-to-Work in the Context of Daily Travel," paper presented at the Transportation Research Board meeting, 2005. Washington, D.C.

141 actual time itself: See, for example, Harry Cohen and Frank Southworth, "On the Measurement and Valulation of Travel Time Variability Due to Incidents on Freeways," *Journal of Transportation and Statistics*, vol. 2, no. 2 (Dec. 1999), as well as David Brownstone and Kenneth A. Small, "Valuing Time and Reliability: Assessing the Evidence from Road Pricing Demonstrations," *Transportation Research Part A: Policy and Practice*, vol. 39, no. 4 (2005), pp. 279–93.

141 "hell every day": Jonathan Clements, "Money and Happiness? Here's Why You Won't Laugh," *Wall Street Journal*, August 16, 2006.

141 higher rate than are passenger cars: T. Cohn, "On the Back of the Bus," *Access*, vol. 21 (1999), pp. 17–21.

141 into early retirement: The information on urban bus drivers comes primarily from the work of Gary Evans, a professor of human ecology at Cornell University. See, for example, Gary Evans, "Working on the Hot Seat: Urban Bus Drivers," *Accident Analysis & Prevention*, vol. 26 (1994), pp. 181–93; G. Evans, M. Palsane, and S. Carrere, "Type A Behavior and Occupational Stress: A Cross-cultural Study of Blue-Collar Workers," *Journal of Personality and Social Psychology*, vol. 52 (1987), pp. 1002–07; and Gary W. Evans and S. Carrere, "Traffic Congestion, Perceived Control, and Psychophysiological Stress Among Urban Bus Drivers," *Journal of Applied Psychology*, vol. 76 (1991), pp. 658–63.

141 how much they're dating: F. Strack, L. L. Martin, and N. Schwarz, "Priming and Communication: The Social Determinants of Information Use in Judgments of Life-Satisfaction," *European Journal of Social Psychology*, vol. 18, 1988, pp. 429–42.

142 "focusing illusion": Daniel Kahneman, Alan B. Krueger, David Schkade, Norbert Schwarz, and Arthur A. Stone, "Would You Be Happier If You Were Richer? A Focusing Illusion," *Science*, vol. 312, no. 5782 (June 30, 2006), pp. 1908–10.

142 makes them think it is: We are also quite capable of changing the way we feel about something—or the way we *think* we feel about something—simply by

subtly changing our definitions of what is important. A fascinating example of this was seen when a group of psychologists from various countries decided to interview solo drivers before and after a car-pool lane was built on a highway in the Netherlands. They conducted similar interviews on a "control" highway that was not getting a new car-pool lane. When the car-pool lane was added, saving about twenty minutes for those in it, solo drivers' attitudes seemed to change. It was not as if they suddenly had a more positive opinion of driving alone and a more negative opinion of carpooling, per se. What did change was how important they felt certain aspects of their commute were. Suddenly, "flexibility" ranked as more important, and saving money or travel time less so. On the highway without a car-pool lane, drivers' attitudes remained the same. But on the highway where the new car-pool lane appeared, teasing solo drivers with its uncongested pleasures, they suddenly had less of a preference for carpooling than when it had not been there. Rather than change their behavior or be haunted every day by not "doing the right thing," they were suddenly telling themselves new stories about what was important to them. (Interestingly, they did not change their attitudes toward what was best for the environment, even if their own behavior did not follow suit.) They were justifying their actions to themselves—that is, making themselves feel better. It could be that rounding up the car pool would take longer than the lane would save (even if a car pool would still be better for the environment and traffic congestion). It could also be that many people, as mentioned above, simply cannot carpool. But it also seems that people, when actually shown an alternative that would be better for society at large, are good at finding ways to explain why it would not be good for them. A driver stuck in traffic watching a commuter train speed by does not necessarily think, "I wish I were on that train," but instead tries to console himself with the reasons he cannot be on that train. And so the roads are filled with people wondering why there are so many other people on the roads, all of them convinced of the reasons they need to be there. See Mark Van Vugt, Paul A. M. Van Lange, Ree Meertens, and Jeffrey Joireman, "How a Structural Solution to a Real-World Social Dilemma Failed: A Field Experiment on the First Carpool Lane in Europe," *Social Psychology Quarterly*, vol. 59 (1996), pp. 364–74.

142 less than 15 percent: Brian Taylor, "Rethinking Traffic Congestion," *Access*, Fall 2002, pp. 8-16.

143 like a bell: There are interesting regional variations on this. In Arizona, for example, it has been observed that parking spaces *closest* to the store are often empty, as cars gravitate first toward the perimeter of the lot, where trees might provide some shade. As one article put it, "A long walk to the store is far better than driving home in a car that has baked for hours in the desert heat." From Diane Boudreau, "Urban Ecology: A Shady Situation," *Chain Reaction*, vol. 4 (2003), pp. 18–19. For more on the microclimate differences between tree-shaded parking lots and those without, see Klaus I. Scott, James R. Simpson, and E. Gregory McPherson, "Effects of Tree Cover on Parking Lot Microclimate and Vehicle Emissions," *Journal of Arboriculture*, vol. 25, no. 3 (May 1999), pp. 129–41.

143 bell-curve arrangement: This idea was first suggested, as far as I can discern, at the following Web site: http://vandersluys.ca/?p=7914.

144 not necessarily being chosen: Velkey's findings matched those predicated by two engineering professors in a "probabilistic model." See C. Richard Cassady and John E. Kobza, "A Probabilistic Approach to Evaluate Strategies for Selecting a Parking Space," *Transportation Science*, vol. 32, no. 1 (January 1998), pp. 30–42.

144 to walk somewhere: *Travel Behaviour Research Baseline Survey 2004: Sustainable Travel Demonstration Towns* (SUSTRANS and Socialdata, 2004). Retrieved from http://www.sustrans.org.uk/webfiles/travelsmart/STDT%20Research%20FINAL.pdf.

144 was at work: The "availability heuristic" is credited to Daniel Kahneman and Amos Tversky. (*Heuristic* is a sophisticated-sounding word that really just means "mental shortcut.") When people are asked to imagine how often something happens, they tend to overestimate the probability of things that can be more easily recalled from memory—that is, that are "available"—or that loom more vividly in the imagination.

144 mixed conclusions on this: See, for example, R. G. Golledge, K. L. Lovelace, D. R. Montello, and C. M. Self, "Sex-Related Differences and Similarities in Geographic and Environmental Spatial Abilities," *Annals of the Association of American Geographers*, vol. 89 (1999), pp. 515–34.

144 as the distance did: A. J. Velkey, C. Laboda, S. Parada, M. L. McNeil, and R. Otts, "Sex Differences in the Estimation of Foot Travel Time," paper presented at the annual meeting of the Eastern Psychological Association, Boston, March 2002. One factor that might lead women to overestimate distances is that, as previous studies have shown, distance estimations tend to be skewed in unpleasant or stressful surroundings. Women may not feel safe in large parking lots, which may help distort the sensation of how close or far a potential parking space is. See Sigrid Schmitz, "Gender Differences in Acquisition of Environmental Knowledge Related to Wayfinding Behavior, Spatial Anxiety and Self-Estimated Environmental Competencies," *Sex Roles: A Journal of Research*, July 1999.

145 "optimal foraging": For a good introduction to optimal foraging, see T. Schoener, "A Brief History of Optimal Foraging Ecology," in *Foraging Behavior*, ed. A. C. Kamil, J. R. Krebs, and H. R. Pulliam (New York: Plenum Press, 1987), pp. 5–67. See also Jeffrey A. Kurland and Stephen J. Beckerman, "Optimal Foraging and Hominid Evolution: Labor and Reciprocity," *American Anthropologist*, vol. 87, no. 1 (March 1985), pp. 73–93.

146 the effort of looking: This example is given in an interesting paper by Elizabeth Newell, a biologist at Hobart and William Smith Colleges, titled "The Energetics of Bee Foraging." Retrieved from http://www.life.umd.edu/Faculty/inouye/Pollination%20Exercises/Beth's.html.

146 is the better option: Esa Ranta, Hannu Rita, and Kai Lindstrom, "Competition Versus Cooperation: Success of Individuals Foraging Alone and in Groups," *American Naturalist*, vol. 142, no. 1 (July 1993), pp. 42–58.

147 spot to a destination: Mark Schlueb, "To Get to Game or Show, Parking May Be Tricky," *Orlando Sentinel*, December 1, 2006.

147 destination is in sight: See Daniel R. Montello, "The Perception and Cognition of Environmental Distance: Direct Sources of Information," in *Spatial Information Theory: A Theoretical Basis for GIS* (Berlin: Springer, 1997), pp. 297–311, and Lorin J. Staplin and Edward K. Sadalla, "Distance Cognition in Urban Environments," *Professional Geographer*, vol. 33 (1981), pp. 302–10.

147 is "good enough": See Herbert Simon, *Administrative Behavior*, 4th ed. (New York: Free Press, 1997).

148 of their time parked: Donald Shoup, *The High Cost of Free Parking* (Chicago: American Planning Association, 2005), p. 6.

148 subsidized parking spots: Bruce Schaller, "Free Parking, Congested Streets," March 1, 2007; available at http://www.schallerconsult.com/pub/index.html.

148 "as has cycle parking space": City of Copenhagen, *Traffic and Environmental Plan 2004*, p. 16.

149 "that will avoid shortages": Shoup, *The High Cost of Free Parking*, p. 303.

149 metered street spots: Donald C. Shoup, "Cruising for Parking," *Transport Policy*, vol. 13 (2006), pp. 479–86.

149 to thirteen minutes: Shoup, *The High Cost of Free Parking*, p. 279.

150 "vehicle per block was enough": William Whyte, *City* (New York: Doubleday, 1988), p. 72.

150 all urban traffic collisions: See Paul C. Box, "Curb Parking Findings Revisited," *Transportation Research Circular 501* (Washington, D.C.: Transportation Research Board, 2000).

150 8 miles per hour: This estimate, for streets with both parking and trees, comes from Dan Burden, "22 Benefits of Street Trees," Glatting Jackson/Walkable Communities, Summer 2006.

151 shiny black sealcoat: See Peter C. Van Metre, Barbara J. Mahler, Mateo Scoggins, and Pixie A. Hamilton, "Parking Lot Sealcoat: A Major Source of Polycyclic Aromatic Hydrocarbons (PAHs) in Urban and Suburban Environments," *Fact Sheet 2005–3147* (Austin: U.S. Geological Survey, January 2006). Not surprisingly, the authors report that PAHs seem to be on the rise: "USGS findings show that concentrations of total PAHs in the majority of lakes and reservoirs in urban and suburban areas across the nation increased significantly from 1970 to 2001. The increases were greatest in lakes with rapidly urbanizing watersheds (urban sprawl); for example, over the last 10 years, the concentrations of PAHs in Lake in the Hills (suburban Chicago, Illinois) increased tenfold as the watershed was rapidly developed."

151 *three to one:* Douglas M. Main, "Parking Spaces Outnumber Drivers 3-to-1, Drive Pollution and Warming," Purdue University News Service, September 11, 2007.

## Chapter Six: Why More Roads Lead to More Traffic

154 during the shutdown: See Jon D. Haveman and David Hummels, *California's Global Gateway: Trends and Issues* (San Francisco: Public Policy Institute of California, 2004), p. 62.

154 "all right by Friday": See Richard Clegg, "It'll Be Alright by Friday: Traffic Response to Capacity Reduction," Department of Mathematics, University of York.

154 "based on those changes": This equilibrium effect seems to happen even in extreme cases, like the 2005 transit strike in New York City. Suddenly, private vehicles, the only way to get into the city, needed to carry at least four passengers to enter during the peak hours of five a.m. to eleven a.m. The world was basically turned upside down. On the first day of the strike, the number of vehicles entering the Central Business District was down 24 percent. People were no doubt confused, unsure of what traffic would be like, or hoping for a quick end to the strike. By the second day, 21 percent fewer vehicles than normal entered. People began testing the waters or could not stay home from work any longer. And on the third day, the number was down to 13 percent fewer vehicles. The strike ended that day, so there is no way to know if traffic would have returned to normal; but clearly, people were adapting, either coming in much earlier (traffic levels at four a.m. tripled) or later than normal, or suddenly becoming believers in car pools. The numbers come from "2005 Transit Strike: Summary Report," New York City Department of Transportation, February 2006.

155 "for other lines": This line was quoted in the PBS documentary *New York Underground* (*American Experience*).

155 like population growth: See Lewis M. Fulton, Robert B. Noland, Daniel J. Meszler, and John V. Thomas, "A Statistical Analysis of Induced Travel Effects in the U.S. Mid-Atlantic Region," *Journal of Transportation and Statistics*, vol. 3, no. 1 (2000), pp. 1–14. A study in California found that a 1 percent increase in lane-miles creates an immediate increase in vehicle-miles traveled of 0.2 percent. See Mark Hansen and Huang Yuanlin, "Road Supply and Traffic in California Urban Areas," *Transportation Research* A, vol. 31 (1997), pp. 205–18. Robert B. Noland, a scientist at Imperial College London, has compiled an extensive bibliography of "induced demand" research; it's available at http://www.vtpi.org/induced_bib.htm.

156 on the affected roads: See S. Cairns, S. Atkins, and P. Goodwin, "Disappearing Traffic? The Story So Far," *Municipal Engineer*, vol. 151, no. 1 (March 2002), pp. 13–22. There was an interesting example of this phenomenon in New York City. When Christo's *The Gates* was on display in New York's Central Park and the roads that crisscross the park were closed to traffic, the city's transportation department did find local streets more crowded, for the short time that the art was installed. But commute *speeds* were not hugely affected, largely, according to the DOT, because of special preparations. It is not difficult to imagine that

the DOT could also make preparations for closing the park drive to vehicles permanently. *The Gates* was a huge draw, of course, so we need to factor in how much of the new traffic volume was from people coming to see the art.

157 congestion itself as an evil: Asha Weinstein Agrawal, a professor of urban planning at San Jose State University, has shown, using Boston as a case study, that the notion of exactly *why* congestion is bad is quite fluid, often depending on the needs of a political class. At the turn of the century, safety and personal travel time were often invoked as reasons to cure what the mayor called the "evils of congestion," but by the 1920s, arguments usually tended to focus on the negative economic consequences of congestion, including a rise in the cost of living. Why? "The growing emphasis on congestion and the cost of living was most likely a political effort to convince the larger population that congestion-generated delay was a problem for them, too, even if they didn't directly experience it as auto drivers," she writes. "Once the subway eliminated the congestion-induced delay experienced by people traveling downtown on the streetcars, proponents of expensive and controversial congestion relief projects like the loop highway needed a new argument to convince the general public that they should support these policies, and the cost-of-living argument filled that role." See Agrawal, "Congestion as a Cultural Construct: The 'Congestion Evil' in Boston in the 1890s and 1920s," *Journal of Transport History*, vol. 27, no. 2 (September 2006), pp. 97–113.

157 "less crowded roads elsewhere": Brian D. Taylor, "Rethinking Traffic Congestion," *Access* (October, 2002), pp 8–16.

157 boosts productivity: Timothy F. Harris and Yannis M. Ioannides, "Productivity and Metropolitan Density," Dept. of Economics, Tufts University, 2000, http://ase.tufts.edu/econ/papers/200016.pdf.

157 the hassles of congestion: Helena Oliviero, "Looking for Love in All the Close Places," *Atlanta Journal Constitution*, October 15, 2002, and Katherine Shaver, "On Congested Roads, Love Runs Out of Gas," *Washington Post*, June 3, 2002. These citations come from Ted Balaker, *Why Mobility Matters to Personal Life*, Policy Brief 62 (Washington, D.C.: Reason Foundation, July 2007).

157 Brookings Institution: See Anthony Downs, *Still Stuck in Traffic: Coping with Peak-Hour Traffic Congestion* (Washington, D.C.: Brookings Institution, 2004), p. 27.

157 close to $12 billion: This, and the $108 billion figure, come from Gabriel Roth, ed., *Street Smart: Competition, Entrepreneurship, and the Future of Roads* (New Brunswick: Transaction Publishers, 2006), p. 7.

157 since Juvenal's Rome: Asha Weinstein Agrawal argues that "the essential challenges of traffic congestion are fundamental to urban life, and therefore unlikely to disappear as long as people choose to base their social and economic institutions around the free and frequent interaction that becomes possible in cities and towns." From "Congestion as a Cultural Construct."

158 Dietrich Braess: Dietrich Braess (translated from the orginal German by A. Nagurney and T. Wakolbinger), "On a Paradox of Traffic Planning," *Transportation Science*, vol. 39 (2005), pp. 446–50.

158 J. G. Wardrop: J. G. Wardrop, "Some Theoretical Aspects of Road Traffic Research," *Proceedings of the Institute of Civil Engineers, Part II* (1952) pp. 325–78.

159 total travel time would *drop:* My example for traffic equilibrium and the Braess paradox was inspired by an article by Brian Hayes, "Coping with Selfishness," *American Scientist,* November 2005.

159 really makes the head spin: When I asked Anna Nagurney, an expert in networks at the University of Massachusetts at Amherst who helped translate Braess's paper into English, if Braess's paradox actually exists in the real world, she said that while he was treating the problem mathematically, there is no reason it could not; she also noted that "Braess even lucked out by picking that [traffic] demand because it lies within a range where the Braess paradox will occur."

159 "selfish routing": Tim Roughgarden, *Selfish Routing and the Price of Anarchy* (Cambridge, Mass.: MIT Press, 2005).

160 more than $2,000: Aaron Edlin and Pinar Karaca-Mandic, "The Accident Externality from Driving," U.C. Berkeley Public Law Research Paper No. 130; available at http://ssrn.com/abstract=424244.

160 2.3 cents per mile: The original estimate comes from Ken Small and Camilla Kazimi, "On the Costs of Air Pollution from Motor Vehicles," *Journal of Transport Economics and Policy,* January 1995, pp. 7–32. The updating to 2005 dollars is from Ian Parry, Margaret Walls, and Winston Harrington, "Automobile Externalities and Policies," Resources for the Future Discussion Paper No. 06-26, January 2007.

160 $10 billion per year: M. A. Delucchi and S.-L. Hsu, "The External Damage Cost of Noise from Motor Vehicles," *Journal of Transportation and Statistics,* vol. 1, no. 3 (October 1998), pp. 1–24.

160 rates and speeds: William T. Hughes Jr. and C. F. Sirmans, "Traffic Externalities and Single-Family House Prices," *Journal of Regional Science,* vol. 32, no. 4 (1992), pp. 487–500.

160 prices often rise: After Clematis Street in West Palm Beach, Florida, was narrowed and retrofitted with bulb-outs and other traffic-calming measures, property values doubled. See "The Economic Benefits of Walkable Communities," report published by the Local Government Commission Center for Livable Communities, Sacramento, California.

160 and coronary problems: There is a huge literature examining the potential links between traffic and health; for example, see A. J. Venn, S. A. Lewis, M. Cooper, et al., "Living Near a Main Road and the Risk of Wheezing Illness in Children," *American Journal of Respiratory and Critical Care Medicine,* vol. 164 (2001), pp. 2177–80. The fact that houses tend to be cheaper near heavy traffic introduces epidemiological uncertainty, however, because in general the lives of people near the road are not the same, in socioeconomic terms, as those of people living on estates well back from the road. Is it living near the road that gives a person health problems, or are the problems due to something else about the lives of people who dwell near the road?

160 tendency of birds to breed: Harvard University's Richard Forman, the dean of the "road ecology" movement, noted in a typical study that bobolinks and other grassland birds in Massachusetts do not breed when their nest sites are close to high-traffic streets (on streets with three thousand or fewer vehicles a day, they do breed). The suggested culprit is noise. See R. T. T. Forman, B. Reineking, and A. M. Hersperger, "Road Traffic and Nearby Grassland Bird Patterns in a Suburbanizing Landscape," *Environmental Management*, vol. 29 (2002), pp. 782–800, and R. T. T. Forman, et al., *Road Ecology: Science and Solutions* (Washington, D.C.: Island Press, 2003). See also J. A. Jaeger, L. Fahrig, and W. Haber, "Reducing Habitat Fragmentation by Roads: A Comparison of Measures and Scales, in *Proceedings of the 2005 International Conference on Ecology and Transportation*, eds. C. L. Irwin, P. Garrett, and K. P. McDermott (Raleigh: Center for Transportation and the Environment, North Carolina State University, 2006), pp. 13–17.

160 less able to afford cars: See Donald Appleyard, M. Sue Gerson, and Mark Lintell, *Livable Urban Streets: Managing Auto Traffic in Neighborhoods*, a report prepared for the Federal Highway Administration, 1976. Many of Appleyard's findings were reconfirmed in a study by the New York City group Transportation Alternatives, "Traffic's Human Toll," 2006; available at http://www.transalt.org/press/releases/061004trafficshumantoll.html.

160 were taxing the poor: It follows that poorer areas also suffer more exposure to the exhaust of passing traffic. Studies in Leeds, England, for example, found that economically disadvantaged areas had higher levels of nitrogen dioxide. See G. Parkhurst, G. Dudley, G. Lyons, E. Avineri, K. Chatterjee, and D. Holley, "Understanding the Distributional Impacts of Road Pricing," Department of Transport, United Kingdom, 2006.

161 by Garrett Hardin: See Garrett Hardin, The Tragedy of the Commons." *Science*, December 13, 1968.

161 oft-invoked "tragedy": Shi-Ling Hsu, "What *Is* a Tragedy of the Commons? Overfishing and the Campaign Spending Problem," February 21, 2005, bepress Legal Series, Working Paper 463; http://law.bepress.com/expresso/eps/463.

161 any traffic engineer: Gary Toth, a planner with the New Jersey Department of Transportation, told me in a conversation in early 2007: "We ran a calculation this week for the twenty congestion-related projects that I have in my division. Those twenty represent about ten percent of the congestion in New Jersey. The construction cost to fix those is $6.7 billion." Given that about $100 million of the department's $600 to $700 million budget can be spent on congestion projects, he said that "at the rate the public is providing funding for us," he could expect those congestion projects to be completed in 67 years.

161 build new ones: See, for example, Joel Kotkin, "Road Work," *Wall Street Journal*, August 28, 2007.

162 all those fuel taxes: Mark Delucchi of the Institute of Transportation Studies at UC-Davis estimates that current payments in the form of fees and taxes by car users to the federal government fall below the costs the federal government

pays for car use by some 20 to 70 cents per gallon of fuel. See Mark A. Delucchi, "Do Motor-Vehicle Users in the US Pay Their Way?" Institute of Transportation Studies, Research Report UCD-ITS-RP-07-17, University of California, Davis, 2007.

162 in the 1960s: See "The Gasoline Tax: Should It Rise?" *Wall Street Journal*, August 18–19, 2007.

162 "90 percent of the time": Martin Wachs, "Fighting Traffic Congestion with Information Technology," *Issues in Science and Technology*, vol. 19 (2002), pp. 43–50.

162 two Canadian researchers: See K. Mucsi and A. M. Khan, "Effectiveness of Additional Lanes at Signalized Intersections," *Institute of Transportation Engineers Journal*, January 2003, pp. 26–30. The authors also note that additions to larger intersections will become congested more quickly than additions to smaller crossroads. They write: "If a one-lane road (per direction) gets saturated at 1,000 vehicles per hour (vph) and annual growth is 3 percent, the additional lane will have an uncongested lifetime of approximately 24 years. If a three-lane road (per direction) gets saturated at 3,000 vph and annual growth again is 3 percent, the uncongested lifetime of the additional lane is only 10 years, even without factoring in the diminishing marginal capacity benefit of the additional lane. The diminishing capacity benefits of additional lanes only speed up the process."

163 the fourth just 385: Engineers, for their part, have responded to the problems of large intersections by building highway-style overpasses, which are not just expensive but can look rather freakish rising out of an otherwise flat suburban environment, or with the so-called continuous-flow intersection, a breathtakingly complex creature that removes the left-turn conflict from the main intersection by having drivers turn left *before* they get to the actual intersection; this is a bit unnerving for some drivers, as the design makes it seem as if they are headed into the oncoming lane. Early studies, however, have shown that these designs actually move more traffic more safely than conventional intersections. At an intersection in Baton Rouge, Louisiana, wait times were reduced from four minutes to one. For a good roundup of CFI intersections, with animations, visit AMBD Engineering's Web site at http://www.abmb.com/cfi.html.

163 an estimated 12.7 percent: This number is taken from H. Teng and J. P. Masinick, "An Analysis of the Impact of Rubbernecking on Urban Freeway Traffic," Center for Transportation Studies, University of Virginia, Report No. UVACTS-15-0-62, 2004, p. 47.

163 "it is a bad bargain": Thomas Schelling, *Micromotives and Macrobehavior* (New York: W.W. Norton, 2006), p. 125.

163 photos of incidents: Melissa Leong, "Best and Worst: Driving GTA's Highways with Sgt. Cam Woolley," *National Post*, July 18, 2007.

163 "or other vehicles": Andrea Glaze and James Ellis, "Pilot Study of Distracted Drivers," Center for Public Policy, Virginia Commonwealth University, January 2003.

164 would have gone up: As a thought experiment, consider that the salad bar was

actually free. What would happen? There would be huge queues of people lined up for the free food. As Tim Harford points out, "We recognize that food, clothes, and houses cannot be free or we would have quickly run out of them. It is because roads are free that we have run out of spare road space." From Harford, *The Undercover Economist* (Oxford: Oxford University Press, 2004), p. 88.

165 more people want to use them?: William Vickrey, "Pricing in Urban and Suburban Transport," *American Economic Review*, vol. 53 (1963). Reprinted in Richard Arnott, Kenneth Arrow, Anthony B. Atkinson, and Jacques H. Drèze., eds., *Public Economics: William Vickrey* (Cambridge: Cambridge University Press, 1994).

165 the results to friends: The Vickrey story is taken from a working paper by Ron Harstad at the University of Missouri, available at www. economics.missouri .edu/working-papers/2005/wp0519_harstad.pdf.

165 rationalize its loss: For an interesting discussion of these ideas based on laboratory experiments, see Erica Mina Okada and Stephen J. Hoch, "Spending Time Versus Spending Money," *Journal of Consumer Research*, vol. 31 (2004), pp. 313–23.

166 than on another day: Richard Clegg, "An Empirical Study of Day-to-Day Variability in Driver Travel Behavior," Department of Mathematics, University of York, Heslington. Retrieved at www.richardclegg.org/pubs/rgc_utsg2005.doc.

166 dropped by 13 percent: Kitchen, in an e-mail, pointed out that all results are "non-equilibrium." That is, if the roads were actually tolled, traffic speeds would improve, attracting additional users.

166 increase speeds by 50 percent: John D. McKinnon, "Bush Plays Traffic Cop in Budget Request," *Wall Street Journal*, February 5, 2007.

167 jump by 5 percent: Philip Bagwell, *The Transport Revolution* (London: Routledge, 1988), p. 375.

167 go into buses: As Puget Sound's Kitchen points out, the revenues generated from economically efficient tolling are greater than the total surplus that is gained through drivers' saved time, which makes the question of how revenues from pricing get redistributed an important, if often neglected, one.

167 thus more popular: For more on this "virtuous circle," see Kenneth A. Small, "Unnoticed Lessons from London: Road Pricing and Public Transit," *Access*, vol. 26 (2005), pp. 10–15.

168 show up so often in networks: An interesting example from the traffic world that recalls Laval's monorail case is Route 29 in Trenton, New Jersey. A product of the 1960s era in which cities elected to build massive high-speed highways through the middles of downtowns or alongside waterways, Route 29 is a dangerous road, with numerous crashes and some two fatalities over a fifteen-year period, as I was told by Gary Toth, an engineer with the New Jersey Department of Transportation. Part of the reason was that cars were "blitzing" down a road that was marked for 45 miles per hour but designed more like a 65-mile-per-hour freeway (with all the standard "safety" provisions of clear zones and the like). Drivers would then inevitably encounter the back of a queue of cars waiting at a signalized intersection; it was a classic "hurry up and wait" situa-

tion. Rather than have a bunch of high-speed cars encounter a single light with a long delay, Toth and his colleagues wondered what would happen if Route 29 was converted from a highway into a more aesthetically appropriate and pleasant "urban boulevard," with a lower speed limit and several more sets of signalized crossings. Wouldn't that just *cause* more congestion? Wouldn't it foist an unconscionable delay upon drivers? When they ran simulations, they found that the new system added only two minutes to the total trip during peak times. Instead of one large queue at a signal, the wait would be redistributed among a set of lights. Importantly, the new system carries the added benefit of being much safer as well, as it involves less sudden braking at high speed.

169 "for Easter Sunday": There are other strange dynamics at work; after running simulations, Laval rejected a plan to double the capacity of the Country Bear Jamboree. "People had the perception it was popular because it had such long lines," he said. "It was really just because it had limited capacity. It's a common misperception."

169 *because* it is expensive: See, for example, Daniel Machalaba, "Paying for VIP Treatment in a Traffic Jam," *Wall Street Journal*, June 21, 2007.

169 as the toll goes up: As Moshe Ben-Akiva, director of the Intelligent Transportation Systems program at the Massachusetts Institute of Technology, described it to me, the challenge with dynamic pricing is that the price changes depending on your objective: "You may want to charge people for time they actually save. That will mean if congestion builds up on the toll road, you reduce the price. On the other hand, you may want to maintain a certain level of speed on the toll road. If congestion builds up you may want to *increase* the toll so as to not have stop-and-go traffic on the toll road. There is some confusion going on right now as to what strategy is best."

170 by changing their plans: Ronald Koo and Younbin Yim, "Commuter Response to Traffic Information on an Incident," September 1, 1998, California Partners for Advanced Transit and Highways (PATH), Working Papers: Paper UCB ITS PWP-98-26; http://repositories.cdlib.org/its/path/papers/UCB-ITS-PWP-98-26.

171 have some information: In one experimental study, for example, eighteen subjects had to choose between two roads, one of which was faster only if an equal number of people chose the opposite road. The subjects would receive a higher payoff for successfully choosing the quickest route. As it happens, over the long run most people split evenly onto the two roads. But there were "daily" fluctuations, and, more important, these were still happening after two hundred trials. The reason is that people continually tried to outguess each other with better strategies (it turned out drivers did better when they simply chose the same route each time) or find out if the other road was in fact better. Interestingly, in a second trial, drivers were given information about the travel time of the route they did not take, meaning they did not have to change roads to know what the conditions were. The fact that drivers had this information had only a "small effect" on the fluctuation between the two roads from day to day. See Reinhard Selten, Michael Shreckenberg, Thomas Pitz, Thorsten Chmur, and Sebastian Kube, "Experiments and Simulations on Day-to-Day Route Choice-

Behaviour," April 2003, CESifo Working Paper Series No. 900; available at http://ssrn.com/abstract=393841.

171 flocked to the highway: Virginia Groark, "Dan Ryan Traffic Flow Changes by Minute—Like Chicago Weather," *Chicago Tribune*, April 5, 2006.

172 shown the best routes: See Moshe Ben-Akiva, Andre De Palma, and Isam Kays, "Dynamic Network Models and Driver Information Systems," *Transportation Research* A, vol. 25A, no. 5 (1991), pp. 251–66.

172 in two-way traffic: Sarah Murray, "The Green Way to Keep on Trucking," *Financial Times*, March 13, 2007.

173 no longer rise tomorrow: Tim Harford, *The Undercover Economist* (Oxford: Oxford University Press, 2005), p. 138.

173 to the same problem: There is an interesting analogy in all this between traffic and the stock market. In theory, as individual investors are able to more closely track the real-time fluctuation of stock prices via the Internet, having access to more and more bits of information about companies, they should be better able to make informed decisions that more quickly translate into stock prices, and market volatility should go down (see Daniel Gross, "Where Have All the Stock Bubbles Gone?" *Slate*, January 3, 2006). But Brad Barber and Terrance Odean have suggested several potential problems that may arise as a result of many more people having access to inexpensive, almost-instantaneous stock trading via the Internet, including the availability of "faster feedback" that may prompt investors to focus too much on recent performance. Those trying to profit from short-term "momentum cycles," they write, may actually increase volatility. (They note that individual stocks have increased in volatility over the past several decades, for reasons that they say are not well understood.) Brad M. Barber and Terrance Odean, "The Internet and the Investor," *The Journal of Economic Perspectives*, vol. 15, no. 7 (Winter 2001), pp. 41–54.

173 "once the prediction is broadcast": Inrix, for example, predicted, ahead of the big I-5 highway closure in Seattle, that traffic would not be as bad as people were making it out to be (for the "disappearing traffic" reasons already mentioned). And it was not. Not everyone heard Inrix's prediction, however, or at least they did not have enough faith in it against the wall of dire predictions of traffic mayhem. See Danny Westneat, "Math Whiz Had I-5's number," *Seattle Times*, August 22, 2007.

174 real-time, the better: See I. Kaysi, "Frameworks and Models for the Provision of Real-Time Driver Information" (Ph.D. thesis, Department of Civil Engineering, Massachusetts Institute of Technology,1992).

174 travel times and congestion: See, for example, Daniel Florian, "Simulation-Based Evaluation of Advanced Traveler Information Services (ATIS)" (dissertation, Massachusetts Institute of Technology, 2004). For a useful review of previous studies, see David Levinson, "The Value of Advanced Traveler Information Systems for Route Choice," *Transportation Research Part C*, vol. 11 (2003), pp. 75–87.

174 as more people have it: See Levinson, ibid.

174 for the savvy taxi driver: Other studies, however, have suggested that as more

people have information about traffic conditions, traffic can actually get *worse*. The reason goes back to the noncooperative nature of the traffic network. If everyone is told at once that route A is better than route B, and people self-interestedly and immediately all move to route A, it will no longer be good. People who study networks call these "concentration" and overreaction problems. This is where imperfect information can be worse than no information at all: If no one is told anything, the outcome will be random—each route might be good or bad. It all depends on how quickly people get the information and the choices they make. Ideally, the roads would then be like, for instance, the rows of customs inspectors' queues at an airport. Everyone can see how much each window is being used at once. If a new window opens up, people can exit every other queue and fill up the new one so that the new queue is as long as the others. The system is in equilibrium. Does it always work so well for the individual, however? You may have moved to the line a bit too slowly and found yourself farther back than you were in the queue you left. You had the information, but did you make the right decision? See H. S. Mahmassani and R. Jayakrishnan, "System Performance and User Response Under Real-Time Information in a Congested Traffic Corridor," *Transportation Research* A, vol. 25, no. 5 (1991), pp. 293–307. See also R. Arnott, A. de Palma, and R. Lindsey, "Does Providing Information to Drivers Reduce Traffic Congestion?" *Transportation Research* A, vol. 25, no. 5 (1991), 309–18, and A. M. Bell, W. A. Sethares, and J. A. Bucklew, "Coordination Failure as a Source of Congestion," *IEEE Transactions on Signal Processing*, vol. 51. no. 3, March 2003.

174 congestion has been passed: In a simulation by David Levinson, a professor of civil engineering at the University of Minnesota, travelers could save the most time through real-time information when traffic conditions were at 95 percent of the available capacity. This is the moment, he suggested, before queues have begun to form and the options begin to dwindle. From Levinson, "Value," op cit.

174 huge majority of the traffic: In the Puget Sound study, interestingly, it was found that 5 percent of the tolled networks generated 50 percent of the hypothetical revenue for the study. Data from an e-mail exchange with Matthew Kitchen.

175 10 percent of the roads: See S. Lammer, B. Gehlsen, and Dirk Helbing, "Scaling Laws in the Spatial Structure of Urban Road Networks," *Physica* A, vol. 363, no. 1 (2006), pp. 89–95.

175 because they are the fastest: A similar dynamic, interestingly, exists in ant-trail formation. As noted in the book *Self-Organization in Biological Systems*, ants tend to congregate on the paths that lead to the richest food sources or are the fastest: "The shortest path enables ants to minimize the time spent traveling between nest and food source, takes less time to complete, and therefore allows ants to consume their food more quickly, minimizing the risk that a good source of food will be discovered and monopolized by a larger or more aggressive neighboring colony. Shorter paths also mean lower transportation costs." Attractive trails are visited by more ants, who lay more pheromones, which

attracts even more ants, in a "feedback mechanism." When a trail branches, ants will choose the branch that has been chosen by more ants. See Scott Camazine, Jean-Louis Denéoubourg, Nigel R. Franks, et al., *Self-Organization in Biological Systems* (Princeton: Princeton University Press, 2001), particularly Chapter 13.

175 havoc with local roads: In England, for example, rural towns have seen traffic surge on roads that are essentially one-lane tracks, as SatNav-equipped drivers looking for shortcuts are sent on routes that "look good on paper," as it were, but are ill-prepared to deal with a large influx of new drivers. See David Millward, "End of the Road for Unreliable SatNavs," *Daily Telegraph,* June 11, 2006.

175 was still the best: I kept having this experience. In Phoenix, I tried repeatedly to find alternate routes when I ran into congestion, and the phone, always pleasant, kept advising, "No alternate routes available."

175 traffic in things: As pointed out by transportation researcher G. F. Newell, many people are resistant to treating vehicle transportation like any other good. "Economic theory is seriously flawed as applied to transportation," he wrote, "because most economists treat transportation like a consumer good that can be sold to the highest bidder, but they don't ask: 'What does society want?' " He added, "I don't know either." See G. F. Newell, "Memoirs on Highway Traffic Flow Theory in the 1950s," *Operations Research,* vol. 50, no. 1 (January–February 2002), pp. 173–78.

## Chapter Seven: Why Dangerous Roads Are Safer

177 *New York Times* observed darkly: Paul J. K. Friedlanden, "H-Day Is Coming to Sweden," *New York Times,* August 20, 1967. See also "Sweden May Shift Road Traffic to the Right to Curb Accidents," *New York Times,* November 12, 1961; "All Goes Right as Sweden Shifts Her Traffic Pattern," *New York Times,* September 4, 1967; "Swedes Face the Trauma of Shifting to Right Side," *New York Times,* April 10, 1966; and "Swedes Adjust, Some Grumpily, to Switching Traffic to the Right," *New York Times,* September 5, 1967.

177 year before the changeover: See R. Näätänen and H. Summala, *Road-User Behavior and Traffic Accidents* (New York: Elsevier, 1976), pp. 139–40.

178 *half* that of conventional intersections: The speed and conflict information for roundabouts comes from Timothy J. Gates and Robert E. Maki, "Converting Old Traffic Circles to Modern Roundabouts: Michigan State University Case Study," in *ITE Annual Meeting Compendium* (Washington, D.C.: Institue for Transportation Engines, 2000).

179 about 90 percent: R. A. Retting, B. N. Persaud, P. E. Garder, and D. Lord, "Crash and Injury Reduction Following Installation of Roundabouts in the United States," *American Journal of Public Health,* vol. 91, no. 4 (April 2001), pp. 628–31.

179 about to hit: See Kenneth Todd, "Traffic Control: An Exercise in Self-Defeat," *Regulation Magazine,* vol. 27, no. 3 (Fall 2004).

180 free of junctions): See "The Impact of Driver Inattention on Crash/Near-Crash Risk: An Analysis Using the 100-Car Naturalistic Driving Study Data," DOT HS 810-594, U.S. Department of Transportation, April 2006, p. 118.

180 "allow it on the roads": Jake Voelcker, in his article "A Critical Review of the Legal Penalties for Drivers Who Kill Cyclists or Pedestrians," makes the useful point that "Health and Safety regulations would not permit thousands of one-tonne steel and glass machines with exposed moving parts to repeatedly pass feet or inches away from unprotected workers on the shop floor at well over 10 m/s (HSE 1998, Sect. 11). Yet this is the situation in our towns and cities today. Why are drivers allowed to impose this danger on pedestrians without more strict prosecution of liability?" Retrieved from www.jake-v.co.uk/cycling.

181 no posts: V. P. Kallberg, "Reflector Posts—Signs of Danger?" *Transportation Research Record*, vol. 1403, pp. 57–66.

181 than when it is not: See, for example, S. Comte, A. Várhelyi, and J. Santos, "The Effects of ATT and Non-ATT Systems and Treatments on Driver Speed Behaviour," Working Paper R 3.1.1 in the MASTER project, VTT Communities & Infrastructure (VTT, Finland), August 1997.

182 it confuses traffic people too: See Raymond A. Krammes, Kay Fitzpatrick, Joseph D. Blaschke, and Daniel B. Fambro, *Speed: Understanding Design, Operating, and Posted Speed*, Report No. 1465–1 (Austin, TX: Texas Dept. of Transportation, March 1996).

183 "time-consuming effort": See David Shinar, *Psychology on the Road: The Human Factor in Traffic Safety* (New York: Wiley, 1978), p. 87.

184 in the period studied: Neal E. Wood, "Shoulder Rumble Strips: A Method to Alert 'Drifting' Drivers," Pennsylvania Turnpike Commission, Harrisburg, Pennsylvania, January 1994.

184 nor is it always easy to locate: Think for a moment about when you come across a hill on a freeway. It is rarely very steep, but then again, it is not flat. Notice how the road gently unspools before you as you near the crest of the hill. The road has been designed so that if there is an unexpected obstacle lurking over that hill, the average driver should be able to see it and have enough time to react and stop. This seems like a smart idea. But how high should the imaginary obstacle be? What would make the road "safe"? Ezra Hauer, a retired Canadian professor of engineering, has observed that early on, highway engineers settled on a four-inch obstacle—a hypothetical "dead dog." They did not know whether three-inch obstacles might also be dangerous, or even how many people were hitting four-inch roadkill as they came over a hill. All they really knew was that building the road so that drivers could stop in time for a three-inch obstacle would require more excavating, and thus more money. Little decisions like this may seem trivial, but in a larger sense they literally help shape the way our world looks to us (and how people behave in it). And so for every highway on a hill in America, the road was designed so the average driver could stop in time for a four-inch "dead dog." In the absence of real information about how, why, and when obstacles on the road lead to crashes, this was, at the very least, prudent engineering, Hauer argues, based on the most efficient construction costs. But over time, something strange happened. Cars

began to get lower. Suddenly, drivers could not see the four-inch obstacle in the given time. So the "dead dog" grew two inches taller—even though, Hauer says, "no link has been found between the risk of collisions with small fixed objects on crest curves and the available sight distance." New roads were built with the new standard (and on existing roads, the driver had just better pray there were no four-inch dogs lying around). Things have since gotten even more complicated. The popularity of SUVs and pickup trucks in the United States means there is "some evidence," as Ray Krammes told me, that cars are now getting *higher.* Is it time to lower the dead dog? See Ezra Hauer, "Safety in Geometric Design Standards," Toronto, Ontario, 1999. Retrieved from http://ca.geocities.com/hauer@rogers.com/Pubs/SafetyinGeometricDesign.pdf.

184 in the United States alone): Steve Moler, "Stop. You're Going the Wrong Way!" *Public Roads,* vol. 66, no. 2 (September–October 2002).

185 moving smoothly *triples:* The literature on weaving sections is surprisingly enormous, but for a good summary of weaving-section research and dynamics, see Richard Glad, John C. Milton, and David K. Olson, *Weave Analysis and Performance: The Washington State Case Study* (Olympia, Wash.: 2001).

185 safer and more efficient: See, for example, Richard W. Glad, Milton, and Olson, ibid.

185 be *less* safe: This information comes from an unpublished paper by Ezra Hauer, "Lane Width and Safety" (review of literature for the Interactive Highway Safety Design Model, 2000); accessed at http://ca.geocities.com/hauer@rogers.com/download.htm).

185 not statistically significant. See Karin M. Bauer, Douglas W. Harwood, Karen R. Richard, and Warren E. Hughes, "Safety Effects of Using Narrow Lanes and Shoulder-Use Lanes to Increase the Capacity of Urban Freeways," *Transportation Research Record: Journal of the Transportation Research Board,* vol. 1897 (2004). On a side note concerning the phrase "statistically significant," Ezra Hauer cautions that statisticians and policy makers often take the phrase "statistically not significant," when referring to a traffic-safety study, to mean there would be no cost or benefit to implementing or not implementing some policy or another. Hauer points as one example to a series of studies examining the adoption of "right turn on red" laws; all showed higher crash rates after right turn on red was adopted. None were "statistically significant," but all "pointed in the same direction": Allowing right turns on red led to more crashes. See Hauer, "The Harm Done by Tests of Significance," *Accident Analysis & Prevention,* vol. 36 (2004), pp. 495–500.

185 "to the road they see": See Hauer, "Lane Width and Safety," op cit.

186 already on the road: See Robert E. Dewar and Paul L. Olson, *Human Factors in Traffic Safety* (Tucson: Lawyers and Judges Publishing, 2002), p. 429. David Shinar writes of the "double jeopardy" of misidentified signing. "Misidentified signs compromise safety by taking more time from the driving task *and* leading drivers to make incorrect decisions. But signs that are interpreted as opposite of their intended meaning mislead the drivers who seem to respond to them as quickly as they do to signs that they identify correctly. Indicating that in these

infrequent cases the drivers are sure, but wrong." See Shinar, *Traffic Safety and Human Behavior* (Amsterdam: Elsevier, 2007), p. 168.

186 not put them up: See *Supplemental Advance Warning Devices: A Synthesis of Highway Practice,* National Cooperative Highway Research Program Synthesis 186 (Washington, D.C.: National Academy Press, 1993), p. 38.

186 fewer deer had crossed: See T. M. Pojar, D. F. Reed, and T. C. Reseigh, "Effectiveness of a Lighted, Animated Deer Crossing Sign," *Journal of Wildlife Management,* vol. 39, no. 1 (1975), pp. 87–91.

187 "deer-vehicle collision": See K. M. Gordon, S. H. Anderson, B. Gribble, and M. Johnson, "Evaluation of the FLASH (Flashing Light Animal Sensing Host) System in Nugget Canyon, Wyoming," Report No. FHWA-WY-01/03F, University of Wyoming, Wyoming Cooperative Fish and Wildlife Research Unit (Laramie, Wy.: July 2001).

187 MOOSE SIGNS AHEAD: The moose story comes from Robert Finch, "Moose Signs Ahead," *Orion,* July–August 2007, p. 7.

188 "they'll behave like that": Monderman's suspicion of traffic signs was not necessarily a radical stance. *The Manual on Uniform Traffic Control Devices,* the bible of American traffic engineers, itself has a warning about warning signs: "The use of warning signs," it notes, "should be kept to a minimum as the unnecessary use of warning signs tends to breed disrespect for all signs."

189 is cognitive dissonance: Whether a driver actually *gets* the ticket may depend on several factors, as a study by Thomas Stratmann and Michael Makowsky argued. "The farther the residence of a driver from the municipality where the ticket could be contested," they wrote, "the higher is the likelihood of a speeding fine, and the larger the amount of the fine. The probability of a fine issued by a local officer is higher in towns when constraints on increasing property taxes are binding, the property tax base is lower, and the town is more dependent on revenues from tourism." From Michael Makowsky and Thomas Stratmann, "Political Economy at Any Speed: What Determines Traffic Citations?," January 31, 2007; available at http://ssrn.com/abstract=961967.

190 dawn of the car itself: According to the research of one historian, the speed bump was first introduced in Chatham, New Jersey, on April 22, 1906. As reports noted, the paved stone in the road was meant to combat "automobile scorchers," as speeders were then known. See Peter Applebome, "Making a Molehill Out of a Bump," *New York Times,* April 19, 2006.

190 are to violate them: Drivers, it turns out, already tend to treat stop signs like "Slow" signs. A study by Michael DeCesare that looked at a sample of 2,390 vehicles at several intersections in the northeastern United States found that only 14 percent of the group came to a complete stop. Most drivers simply "paused," and those that did come to a complete stop often did so only because there were already other cars crossing through the intersection. Interestingly, no cars completely violated the stop signs, which implies that stop-sign visibility was not an issue. See "Behavior at Stop Sign Intersections: A Matter of Convenience and Threat of Danger," paper presented at the Annual Meeting of the Eastern Sociological Society, Boston, 1999.

190 to make up time: See, for example, Gerald L. Ullman, "Neighborhood Speed Control—U.S. Practices," *ITE Compendium of Technical Papers* (1996), pp. 111–15, and Richard F. Beaubein, "Controlling Speeds on Residential Streets," *ITE Journal*, April 1989, pp. 37–39.

190 time to speed: Reid Ewing, "U.S. Experience with Traffic Calming," *Institute of Transportation Engineers Journal*, August 1997, p. 30.

191 to these trips: Crysttal Atkins and Michael Coleman, "Influence of Traffic Calming on Emergency Response Times," *Institute of Transportation Engineers Journal*, August 1997.

191 "fatigue of getting upstairs": Charles Dickens, "Street Accidents," *All the Year Round*, vol. 8 (1892; repr.), p. 499.

192 often without supervision: For a good history of the *woonerven*, see Michael Southworth and Eran Ben-Joseph, *Streets and the Shaping of Towns and Cities* (New York: McGraw-Hill, 1996).

195 acting accordingly in the moment: Interestingly, this idea has had precedents here and there in the history of traffic engineering. In July 1927, the American magazine *Nation's Traffic* reported on a novel signal system at a four-way intersection that featured white lights instead of the traditional amber. When the lights in all four directions shone white, it signified that cars in all four directions could make left turns. Instead of mayhem during the evening rush hour, the writer reported, "We saw four streams of traffic making left turns at the same time . . . without the scraping of a fender." The local police chief made an interesting observation: "We have taught these people to sort of care for themselves." From Gordon Sessions, *Traffic Devices: Historical Aspects Thereof* (Washington, Institute of Traffic Engineers, 1971), p. 50.

195 the less we see: This is particularly true at roundabouts. An observational study in Finland found that drivers entering a roundabout were less likely to look to the right and more likely to violate the right-of-way of cyclists crossing to the right. See Heikki Summala and Mikko Rasanen, "Top-Down and Bottom-Up Processes in Driver Behavior at Roundabouts and Crossroads," *Transportation Human Factors*, vol. 2, no. 1 (2000), pp. 29–37.

195 around 20 miles per hour: When I presented this theory to Daniel Lieberman, a professor of biological anthropology at Harvard University's Skeletal Biology Lab, he answered, via e-mail: "I would agree with you that natural selection would have to have permitted the skeleton to survive falls from running and other such natural events, but we were never designed to be hit by 1-ton cars going at 60 MPH (a lot of momentum). But is running the highest natural force a body experiences? Not sure. We also got hit on the head, attacked by saber-tooths, etc. But it is clearly the case that running is a common way to injure ourselves since we are naturally awkward unstable creatures more likely to fall than quadrupeds, and more likely to get injured by a fall (farther to go). So you might indeed be right."

197 running a red light: In 2000, for example, more than one thousand people were killed in crashes caused by someone running a red light, according to the Federal Highway Administration. Figure retrieved from http://safety.fhwa.dot.gov/intersections/comm_rlrfaq.htm.

197 people on foot: This raises the question of what happens at intersections without "Walk"/"Don't Walk" signals. Picture the mayhem as ill-behaved pedestrians cross willy-nilly without being told when to do so. At the time of this writing, it was possible to see this in New York City (although plans were afoot to install pedestrian signals). Simply go to Park Avenue, anywhere from Forty-sixth Street to Fifty-sixth Street. There you will notice that not only are the traffic lights smaller but there are no pedestrian crossing signals (called "ped heads"). For unique structural reasons owing to a commuter train running underneath the street, traffic authorities for years were unable to install the necessary foundation for a standard signal. So are pedestrians hit by cars more frequently at these corners? A five-year "crash map" put together by the city DOT showed that there were no more pedestrians struck in that section of Park Avenue than in the areas immediately to the north and south that had ped heads. This suggests that pedestrians on those blocks were forced to more actively assess the danger posed by cars. The crash map was put together by the New York City Department of Transportation's Pedestrian Projects office and was supplied to me by Michael King. To fully assess the actual risk faced by pedestrians at those sections of Park Avenue versus other areas, and the reasons why, a comprehensive study would have to be undertaken to determine pedestrian volumes and analyze the causative factors of the crashes. If more pedestrians are struck at those corners, the reason might also have to do with the below-standard traffic signals for vehicles. As Michael Primeggia, the deputy commissioner at the city's Department of Transportation, noted to me in a conversation, *vehicle* crashes are higher at those corners; particularly "right-angle collisions," which are often attributed to a car's failure to stop at a red.

197 their own green light: This is why engineers often install the "leading pedestrian interval," or LPI, which gives an "exclusive phase" of a few seconds or so to the walker, to give him a head start and allow him to assert his authority in the crosswalk. This, of course, slows vehicular traffic flow. The most radical example of a pedestrian-only phase is the so-called Barnes dance, named after New York City's longtime traffic commissioner, in which pedestrians are given the "Walk" signal in both directions and cars in *all* directions must wait. The Barnes dance was not actually invented in New York City, as is often thought, but in Barnes's previous posting of Denver. After he unveiled an all-way pedestrian phase, a local scribe wrote, "Barnes has made the people so happy they're dancing in the streets"; hence the Barnes dance. See Henry Barnes, *The Man with Red and Green Eyes* (New York: Dutton, 1965), p. 116.

197 for the health of pedestrians: D. F. Preusser, W. A. Leaf, K. B. Debartla, and R. D. Blomberg, *The Effects of Right-Turn-on-Red on Pedestrians and Bicycle Accidents*, NHTSA-DOT/HS-806/182 (Dunlap and Associates, Darien, Conn.: October 1981).

197 law than while not: In a study that looked at a year's worth of pedestrian and bicycle fatalities (1997), drivers were found to be "at least partly culpable" in 71 percent of the cases. See Charles Komanoff, "Killed by Automobile: Death in the Streets in New York City, 1994–1997," March 1999. In 2004, nearly one-third of all pedestrians killed in New York City were killed in the crosswalk of

an intersection. Of all the pedestrian fatalities, the majority (114, or 67.5 percent) were not attributed to any action by the pedestrian, while the categories that reasonably indicate pedestrian blame ("darting, running, or stumbling into road," "improper crossing of roadway or intersection," "failure to obey traffic control devices, traffic officers, traffic laws, etc.," and "walking, playing, working in roadway") total 48 of 169 fatalities, or roughly 28 percent. See Claire E. McKnight, Kyriacos Mouskos, Camille Kamga, et al., *NYMTC Pedestrian Safety Study*, Institute for Transportation Systems, City University of New York; prepared for the New York Metropolitan Transportation Council, February 27, 2007.

198 must navigate several lanes: The undisputed king of marked crosswalk studies is Charles Zegeer, at the University of North Carolina. See Charles V. Zegeer, J. Stewart, and H. Huang, *Safety Effects of Marked Versus Unmarked Crosswalks at Uncontrolled Locations: Executive Summary and Recommended Guidelines, 1996–2001* (Washington, D.C.: Federal Highway Administraion, March 2002); available at http://www.walkinginfo.org/pdf/r&d/crosswalk_021302.pdf.

198 make things safer: See David R. Ragland and Meghan Fehlig Mitman, "Driver/Pedestrian Understanding and Behavior at Marked and Unmarked Crosswalks," U.C. Berkeley Traffic Safety Center, Paper UCB-TSC-RR-2007-4, July 1, 2007; http://repositories.cdlib.org/its/tsc/UCB-TSC-RR-2007-4. See also Meghan Fehlig Mitman and David R. Ragland, "What They Don't Know Can Kill Them," U.C. Berkeley Traffic Safety Center, Paper UCB-TSC-TR-2007-2, April 1, 2007; http://repositories.cdlib.org/its/tsc/UCB-TSC-TR-2007-2.

198 a good thing for pedestrians: Conversely, knowing traffic laws such as right-of-way can be dangerous. A study in Finland that looked at collisions between cars and bicycles found that while only 11 percent of cars reported seeing the bicyclist before the crash, some 68 percent of cyclists reported seeing the car—and 92 percent of those who noticed the car assumed it would yield the right-of-way. See Summala and Rasanen, "Top-Down and Bottom-Up Processes," op. cit.

198 of their own safety: One reason for this is the "multiple-threat collision," in which one driver stops but a driver in the next lane does not, most likely because his view of that pedestrian is blocked. This was described to me in a conversation with Charlie Zegeer at the University of North Carolina, a traffic-safety researcher who has spent more time than anyone studying the problems of getting pedestrians across the road safely. See also Zegeer, Stewart, and Huang, *Safety Effects*, op. cit.

199 in the face of oncoming traffic: M. Winnet, S. Farmer, J. Anderson, and R. Lockwood, "An Evaluation of the Effect of Removing White Centre Lines," report prepared for the Wiltshire County Council by CEEMA Ltd. and TRL Limited.

199 to drive faster: This is an old saw in traffic engineering. In the 1922 book *Good Roads*, for example, author James McConaghie notes that "it has been found that by placing a series of lines on the pavement, dividing the space up into its maximum number of traffic lanes, a greater speeding up of traffic has been the result." Quoted in Sessions, *Traffic Devices*, p. 104.

199 insufficiently wide bike lanes: See D. L. Harkey and J. R. Stewart, "Evaluation of Shared-Use Facilities for Bicycles and Motor Vehicles," Transportation Research Record 1578, Transportation Research Board, Washington, D.C., 1997. For a less scientifically rigorous but no less interesting report, see Pete Ownes, "The Effect of Cycle Lanes on Cyclists' Road Space," Warrington Cycle Campaign, October 2005. Other studies have made the point that bicycle lanes reduce the amount of vehicle "displacement"—that is, how much they veer toward the center line or even into the other lane—and that bicycles themselves stay on a straighter path in the presence of lanes. See Bonnie J. Kroll and Melvin R. Ramey, "Effects of Bike Lanes on Driver and Bicyclist Behavior," *Journal of Transportation Engineering*, vol. 103, no. 2 (March–April 1977), pp. 243–56, and S. R. McHenry and M. J. Wallace, *Evaluation of Wide Curb Lanes as Shared Lane Bicycle Facilities*, Report FHWA/MD-85/06, Maryland Department of Transportation, Baltimore, August 1985.

200 system was more dangerous!: The Laweiplein information comes from an unpublished study by Jeroen van Doome and Jelmer Herder of the Leeuwarden Technical College. The data is still preliminary and, as in all such studies, it can be difficult to immediately attribute reasons for increases or decreases in crashes. There may still be lingering "novelty effects" in the scheme, as well as a possibility of a "regression to the mean," whereby statistical entities such as crash statistics possess a natural tendency to fluctuate. More time will be needed to fully assess the scheme. The reader might well wonder whether the safety and traffic improvements made in Drachten could have been achieved by simply converting the space to a conventional roundabout. But the Leeuwarden report notes that the traffic improvements at Drachten outperform what would be expected using modeling for a "conventional roundabout." Hamilton-Baillie pointed out to me in an e-mail that the geometry of the scheme differs from that of a conventional roundabout: "By narrowing the entrances and exits—they are 6 meters wide—there's very little flaring. It doesn't seem to be a problem for traffic to just allow pedestrians and bicycles to just filter through." The Laweiplein design, he maintains, avoids some of the problems of how to accommodate pedestrians and bicycles, a common criticism of roundabout schemes. On the idea that users thought the system was more dangerous, when it was statistically not, there is evidence that this kind of distortion is not uncommon. In a study conducted on the University of North Carolina at Chapel Hill campus, a group of students were surveyed as to what they thought the most dangerous areas on campus were for pedestrians. Some locations that people thought were "safe" had actually had a number of crashes, even more so than areas they labeled "dangerous." See R. J. Schneider, R. M. Ryznar, and A. J. Khattak, "An Accident Waiting to Happen: A Spatial Approach to Proactive Pedestrian Planning," *Accident Analysis & Prevention*, vol. 36, no. 2 (March 2004), pp. 193–211.

201 made on bicycles: From "Cycling for Everyone: The Key to Political and Public Support," by John Pucher, Rutgers University. Document retrieved from www.policy.rutgers.edu/faculty/pucher/BikeSummit2007COMP_Mar25.pdf on April 8, 2007.

202 for minor injuries: The Kensington High Street statistics are found in Graeme Swinburne, "Report on Road Safety in Kensington High Street," Royal Borough of Kensington and Chelsea, London.

203 "for no inconsiderable time": Charles Dickens, *Sketches by Boz* (1835; repr. London: Penguin Classics, 1996), p. 92.

204 "you need freeways": Walter Kulash, of Glatting Jackson, described to me a similar tension in terms of traffic flow. "One thing we have learned," he said, "is that streets are always a bundle of competing interests. There is always going to be less of one thing if there is more of another thing. If there is more seclusion and streets are by their very layout incapable of carrying any through traffic . . . then a negative is going to pop up somewhere else. And that negative is unbearable arterial streets."

205 twelfth-deadliest road in America: Scott Powers, "Colonial One of Nation's Most Dangerous Roads," *Orlando Sentinel*, November 21, 2004. The U.S. 19 information is taken from a survey conducted by NBC's *Dateline*; see Josh Mankiewicz, "Dangerous Roads," *Dateline*, June 7, 2005.

206 would have deemed safer: For the details of Eric Dumbaugh's studies I have drawn on several sources. The first is his unpublished Ph.D. dissertation: "Safe Streets, Livable Streets: A Positive Approach to Urban Roadside Design" (Georgia Institute of Technology, August 2005). I also used a related article: Eric Dumbaugh, "Safe Streets, Livable Streets," *Journal of the American Planning Association*, vol. 71, no. 3 (Summer 2005), pp. 283–300.

207 26 to 30 miles per hour: National Highway Traffic Safety Administration, "Literature Review on Vehicle Travel Speeds and Pedestrian Injuries," DOT HS 809 021, October 1999.

207 by some 10 percent: See, for example, M. Martens, S. Comte, and N. Kaptein, "The Effects of Road Design on Speed Behavior: A Literature Review," Technical Research Centre of Finland VTT, Espoo, 1997. Moreover, a survey of street segments in Connecticut, Massachusetts, and Vermont revealed that on-street parking itself seems to have a safety benefit. The researchers write: "Our results suggest that on-street parking can also help to create a safer environment. While this statement seems to contradict most of the existing research, the reality is that lower speed roads (less than 35 mph) with on-street parking have far less severe and fatal crashes. In fact, lower speed streets without parking had a severe and fatal crash rate more than two times higher than the streets with parking. We also showed conclusively that drivers tended to travel slower in the presence of features such as on-street parking and small building setbacks. Slower vehicle speeds provide pedestrians, cyclists, and drivers more time to react, and when a crash does occur, the chance of it being life-threatening is greatly reduced." See Wesley Marshall, Norman Garrick, and Gilbert Hansen, "Reassessing On-Street Parking," paper presented at the Transportation Research Board meeting, January 2008, Washington, D.C.

207 "roadside conditions": Richard F. Weingroff, "President Dwight D. Eisenhower and the Federal Role in Highway Safety" (Washington, D. C.: Federal Highway Administration, 2003), retrieved at http://www.fhwa.dot.gov/infrastructure/safety.htm.

209 they felt it was safer: N. J. Ward and G. J. S. Wilde, "Driver Approach Behaviour at an Unprotected Railway Crossing Before and After Enhancement of Lateral Sight Distances: An Experimental Investigation of a Risk Perception and Behavioural Compensation Hypothesis," *Safety Science*, vol. 22 (1996), pp. 63–75.

209 raise property values: See, for example, S. E. Maco and E. G. McPherson, "A Practical Approach to Assessing Structure, Function, and Value of Street Tree Populations in Small Communities," *Journal of Arboriculture*, vol. 29, no. 2 (March 2003).

210 from roadsides for decades: In a 1941 Chicago planning study titled *Subdivision Regulation*, for example, the author, Harold Lautner, wrote: "While it has been customary in the past to plant street trees between the street curb and the pedestrian walk, an alternate procedure is now recommended as preferable in some cases. Trees planted along the street curb increase the severity of motor accidents and in turn are easily subjected to traffic injury . . . and except on very wide streets, curb planted trees crowd in upon the traveled way. To plant street trees on the property side of pedestrian walks, away from the pavement and traffic, seems more desirable, *particularly on residential streets*" (emphasis in original). This would, of course, not only increase the speed of passing traffic, posing more of a risk to pedestrians, but would also remove a potential barrier to a car striking a pedestrian. From Southworth and Ben-Joseph, *Streets and the Shaping of Towns and Cities*, op. cit., p. 88.

## Chapter Eight: How Traffic Explains the World

217 same space as New York City: This figure is taken from Richard L. Forstall, Richard P. Green, and James B. Pick, "Which Are the Largest: Why Published Populations for Major Urban Areas Vary So Greatly." Accessed from the University of Illinois–Chicago "City Futures" conference Web site, http://www.uic.edu/cuppa/cityfutures/.

218 same lane as the cyclists: Dinesh Mohan, *The Road Ahead: Traffic Injuries and Fatalities in India* (New Delhi: Transportation Research and Injury Prevention Programme, Indian Institute of Technology; 2004), pp. 1–30.

220 but *before*: Lu Huapu, Shi Qixin, and Masato Iwasaki, "A Study on Traffic Characteristics at Signalized Intersections in Beijing and Tokyo," Tsinghua University, *Proceedings of EASTS (The 2nd Conference of the Eastern Asia Society for Transportation Studies)*.

220 would mean "stop": This story is discussed in Keesing's Research Report, *The Cultural Revolution in China* (New York: Scribner, 1967), p. 18.

220 "can he actually *overtake*": Kenneth Tynan, *The Diaries of Kenneth Tynan* (New York: Bloomsbury, 2002), p. 101.

220 the entire street: The journalist Jan Wong, writing about Beijing in the 1980s, reported that "even state-owned cars were so rare that most Beijing intersections lacked traffic lights. Stop signs were non-existent. At night, cars were *required* to douse headlights to avoid blinding cyclists. With only a handful of

vehicles on the road, no one worried about one car smashing into another in the dark." See *Jan Wong's China* (Toronto: Doubleday Canada, 1999), p. 212.

221 as a social good?: For a good discussion of Mao's "lawlessness" concept, see Chapter 10 of Zhengyuan Fu, *Autocratic Tradition and Chinese Politics* (Cambridge: Cambridge University Press, 1993).

221 public morality and civic culture: See, for example, Wen-shun Chi, *Ideological Conflicts in Modern China: Democracy and Authoritarianism* (New York: Transaction Publishers, 1986), p. 56.

221 "superior to them": This quote comes from "Moral Embarrassment," *Shanghai Star*, August 11, 2001.

221 "rights by litigation": Albert H. Y. Chen, "Toward a Legal Enlightenment: Discussions in Contemporary China on the Rule of Law," *UCLA Pacific Basin Law Journal*, vol. 17 (2000).

223 drive on the right: The information about which side of the road different countries drive on was obtained from Peter Kincaid's exhaustive treatise *The Rule of the Road: An International Guide to History and Practice* (New York: Greenwood Press, 1986).

223 violation of the standard: The flashing of headlights in Europe also seems to be bluntly effective at getting people to move over. As a study of Austrian highway behavior showed, while demographic factors explained which drivers tended to drive faster and tailgate more aggressively (men driving expensive cars, as you might expect), there was also what the author called an "instrumental function"—the urge to "dominate" other drivers seemed to be the most effective way to encourage them to move over. "It was found that drivers who approach to under ten meters behind the camera car were more likely to displace the driver ahead," the authors wrote. "Furthermore, drivers who approached faster displaced others more effectively." Klaus Atzwanger and B. Ruso, in *Vision in Vehicles VI* (Amsterdam: Elsevier Science B.V., 1999), p. 197.

223 confusing array of laws: See, for example, the Web site maintained by John Carr, http://www.mit.edu/~jfc/right.html.

224 rights have been violated: George McDowell, an economist at Virginia Tech, has offered the fascinating theory that a country's traffic behavior is reflected in its economic system. In the United States, the supposed "free market" is, he argues, instead an "open market," in which "rules, both formal and informal, govern behavior. Opportunistic behavior is expected and even encouraged but within a strict set of parameters." In China, however, he argues, the system is better described as a "free market," where "the only rule is caveat emptor." The Chinese system of what he calls "advantage" means that horns are used less as a means to signal "road rage" but more to "notify other vehicles that you are there and will not give way." Advantage "is gained," he writes, "exploited by the person who gained it, and accepted by the person bested." In the United States this acceptance is less likely to occur. See George R. McDowell, "The Market as Traffic: An Economic Metaphor," *Journal of Economic Issues*, vol. 38 (2004), pp. 270–74.

224 acts more personally: American roads are also more crowded than the expensive Italian *autostrada*. This brings up the issue that it may be more difficult for drivers to "get over" and meet the demands of the driver to the rear; there is also the larger issue that giving up an entire lane to a few people wishing to go fast, with all the lane changing that entails, can be poor use of the traffic network.

224 fairness and equality: According to the political scientist Robert Putnam, this dynamic is more prevalent in the southern regions of Italy. These, he argued, have historically lacked a strong civic culture, being dominated instead by feudalistic patronage relationships and an "amoral familism"—worry about yourself and trust that everyone will look after themselves. Instead of "horizontal" networks of reciprocal relations and trust among the community, Putnam argues, the south has been dominated by more vertical, patron-client-style relationships. From Robert Putnam, *Making Democracy Work: Civic Traditions in Modern Italy* (Princeton: Princeton University Press, 1993).

224 *jaywalking*: The historian Peter Norton, in an exemplary article, traces the etymology of the word to at least 1909, well before the 1917 Boston usage registered by the second edition of the *Oxford English Dictionary* in 1989. Norton traces the rise of the word in the popular imagination as pedestrians saw gradually eroded their longstanding right to a shared use of city streets, in favor of a historically unprecedented edict, as described by one writer, upon the arrival of the automobile: "The streets are for vehicle traffic, the sidewalks for pedestrians." Jaywalking, in essence, marginalized and even criminalized what had been standard urban behavior. This was done ostensibly in the name of safety, but as Norton notes, its real aim was to clear urban streets for the increased circulation of vehicular traffic (other, potentially more effective, safety measures like speed "governors" for cars were overridden by motoring interests). Peter D. Norton, "Street Rivals: Jaywalking and the Invention of the Motor Age Street," *Technology and Culture*, vol. 48 (April 2007), pp. 331–59.

224 in which he was raised: Aksel Sandemose, *A Fugitive Crosses His Tracks* (New York: Alfred A. Knopf, 1936).

226 rules of grammar: Sanford W. Gregory Jr. compared traffic behavior in Egypt to a "verdant grammar," one not "yet ripened by centuries of social-interactive maturation." The arrival of mass driving in Egypt, he suggests, happened too quickly for Western traffic patterns to be institutionalized, so instead a kind of pidgin or creole language was formed, with distinct rules, as is often the case "when mature speakers of diverse dominant language groups meet." Without time to create a formal order of its own, Egypt's drivers invented a brutally effective slang of sorts. Gregory commented that this seemed based more on eye contact and informal signals than in the West. See Gregory, "Auto Traffic in Egypt as a Verdant Grammar," *Social Psychology Quarterly*, vol. 48, No. 4 (December 1985), pp. 337–48.

227 each side of the street: This story is mentioned in William Muray, *City of the Soul: A Walk in Rome* (New York: Crown, 2003), p. 26.

227 Mythological status: H. V. Morton, in his 1957 travelogue A *Traveler in Rome*, observed, while riding in a taxi: "The cars around us, which were traveling just

as fast as we were, swerved aside by one of those instinctive Italian motoring movements not unlike birds in formation who part and form again" (1957; repr., New York: Da Capo, 2002), p. 135.

227 one-fifth of the traffic: Michele Faberi, Marco Martuzzi, and Franco Pirrami, *Assessing the Health Impact and Social Costs of Mopeds: Feasibility Study in Rome* (Rome: World Health Organization, 2004), p. xvii.

228 fewer riders wear helmets: The helmet-use rates come from F. Servadei, C. Begliomini, E. Gardini, M. Giustini, F. Taggi, and J. Kraus, "Effect of Italy's Motorcycle Helmet Law on Traumatic Brain Injuries," *Injury Prevention*, vol. 9, no. 3 (2003), pp. 257–60.

228 collisions with cars: Giuseppe Latorre, Giuliano Bertazzoni, Donato Zotta, Edward Van Beeck, and Gualtiero Ricciardi, "Epidemiology of Accidents Among Users of Two-Wheeled Motor Vehicles: A Surveillance Study in Two Italian Cities," *European Journal of Public Health*, vol. 12, no. 2 (2002), pp. 99–103.

229 (and getting away with it): R. B. Cialdini, L. J. Demaine, B. J. Sagarin, D. W. Barrett, K. Rhoads, and P. L. Winter, "Managing Social Norms for Persuasive Impact," *Social Influence*, vol. 1 (2006), pp. 3–15.

229 behavior either way: There have been several studies of jaywalking and model behavior. See, for instance, Monroe Lefkowitz, Robert R. Blake, and Jane Srygley Mouton, "Status Factors in Pedestrian Violation of Traffic Signals," *Journal of Abnormal and Social Psychology*, vol. 51 (1955), pp. 704–06, and Brian Mullen, Carolyn Copper, and James E. Driskell, "Jaywalking as a Function of Model Behavior," *Personality and Social Psychology Bulletin*, vol. 16, no. 2 (1990), pp. 320–30.

230 are famously orderly: Joe Moran makes the point that people in England have been "complaining about the disintegration of queue discipline for almost as long as they have been lauding the queue as the essence of British decency—perhaps because this myth carries such symbolic weight that it cannot be sustained by the necessarily messier reality." From Joe Moran, *Queuing for Beginners* (London: Profile Books, 2007), p. 92.

230 more in theory than reality: Liu Shinan argues that Chinese do queue up when queuing itself is the norm: "We queue where we are accustomed to queue, for example, at a cinema booking office or at the cashier's counter in a supermarket. In many places where we are not accustomed to queue, however, we do not queue—for example, in front of an elevator or subway door." Liu Shinan, "Behavior of Tourists Has No Quick Fix," *China Daily*, November 10, 2006.

230 to be slight: One study found the correlation between "service quality" and tipping to be just 0.07 percent. See Michael Conlin, Ted O'Donohue, and Michael Lynn, "The Norm of Restaurant Tipping," *Journal of Economic Behavior and Organization*, vol. 52 (2003), pp. 297–321. For an excellent overview of the quite extensive academic literature on tipping, I recommend the work of Ofer Azar, an economist at Ben-Gurion University of the Negev in Israel, particularly "The Social Norm of Tipping: A Review," *Journal of Applied Social Psychology*, vol. 37, no. 2 (2007), pp. 380–402.

230 "obeying the law": See Amir Licht, "Social Norms and the Law: Why People Obey the Law," a working paper available at http://www.faculty.idc.ac.il/licht/papers.htm.

231 to nearly 84,000: Sheng-Yong Wang, Gui-Bo Chi, Chun-Xia Jing, Xiao-Mei Dong, Chi-Peng Wu, and Li-Ping Li, "Trends in Road Traffic Crashes and Associated Injury and Fatality in the People's Republic of China, 1951–1999," *Injury Control and Safety Promotion*, vol. 10, nos. 1–2 (2003), pp. 83–87.

231 roughly 49 million: *New York Times*, July 22, 1951.

231 Smeed's law: R. J. Smeed, "Some Statistical Aspects of Road Safety Research," *Journal of the Royal Statistical Society, Series A (General)*, vol. 112, no. 1 (1949), pp. 1–34.

232 as low as 10 percent: Vinand M. Nantulya and Michael R. Reich, "The Neglected Epidemic: Road Traffic Injuries in Developing Countries," *British Medical Journal*, May 2002, pp. 1139–41.

232 a staggering 80 percent: Mohan, *The Road Ahead*, op. cit. pp. 1–30.

232 onto the same thoroughfare: In a discussion paper for the World Bank, Christopher Willoughby notes that "the current problems of motorization seem not generally to result from its occurring at lower per capita income levels, or more rapidly, than in the countries which coped with it reasonably satisfactorily in earlier years; it also grew very fast there for prolonged periods, especially in France (and Germany). The problems tend to be connected rather with the higher concentration of national population, economic activity and motorization itself in one or a very few major cities, at times when those cities are also increasing in size and population much more rapidly than was the case in Europe or Japan." From Christopher Willoughby, "Managing Motorization," Discussion Paper TWU-42, World Bank; available at: http://www.world-bank.org/transport/publicat/twu_42.pdf.

233 nearly 100 percent: For a fascinating discussion of history of automobile insurance in China and recent reforms, see J. Tim Query and Daqing Huang, "Designing a New Automobile Insurance Pricing System in China: Actuarial and Social Considerations," *Casualty Actuarial Society Forum*, Winter 2007.

233 to West Germany's 130: Flaura K. Winston, Craig Rineer, Rajiv Menon, and Susan P. Baker, "The Carnage Wrought by Major Economic Change: Ecological Study of Traffic Related Mortality and the Reunification of Germany," *British Medical Journal*, vol. 318 (June 19, 1999), pp. 1647–50.

233 begin to accelerate: See Richard Dahl, "Heavy Traffic Ahead: Car Culture Accelerates," *Environmental Health Perspectives*, April 2005.

233 Maureen Cropper shows: Elizabeth Kopits and Maureen Cropper, "Traffic Fatalities and Economic Growth," *Accident Analysis & Prevention*, vol. 37 (2005), pp. 169–78.

234 terms of traffic safety: Based on statistics from the International Traffic Safety Data and Analysis Group; retrieved on January 13, 2007, from http://cemt.org/IRTAD/IRTADPUBLIC/we2.html.

234 some 160 deaths per 10,000 vehicles: *World Report on Road Traffic Injury Prevention* (Geneva: World Health Organization and World Bank, April 4, 2004).

234 "to use the buses": BBC, February 28, 2001. Accessed from: http://news.bbc.co.uk/2/hi/africa/1186572.stm.

234 slightly higher in Belgium): *Pocket World in Figures 2007* (London: Economist, 2007).

234 risk of traffic fatalities: See Theodore E. Keeler, "Highway Safety, Economic Behavior, and Driving Environment," *American Economic Review*, vol. 84, no. 3 (1994), pp. 684–93, and Reid Ewing, Richard A. Schieber, and Charles V. Zegeer, "Urban Sprawl as a Risk Factor in Motor Vehicle Occupant and Pedestrian Fatalities," *American Journal of Public Health*, vol. 93, no. 9 (2003), pp. 1541–45.

235 Belgium had 522: *World Report on Road Traffic Injury Prevention*, op. cit., p. 198.

235 fairness of the process: Tom R. Tyler, *Why People Obey the Law* (Princeton, N.J.: Princeton University Press, 2006).

236 The information on Belgium's traffic enforcement comes from Lode Vereeck and Lieber Deben, "An International Comparison of the Effectiveness of Traffic Safety Enforcement Policies," unpublished paper, Limburg University, Belgium, 2003.

236 lowest crash rates in the world: Retrieved from the International Road Traffic and Accident Database (IRTAD), at http://cemt.org/IRTAD/IRTADPUBLIC/we2.html.

236 after-tax income: Before 1999, fines were based on *pre*tax income, says Heikki Summala of the Traffic Research Unit at the University of Helsinki. This means fines have dropped between 20 and 60 percent, but at the same time minimum fines were raised, so revenue has in fact increased. E-mail correspondence with Heikki Summala, November 9, 2007.

236 Jaakko Rytsölä: The Finnish speeding ticket information comes from Steve Stecklow, "Finnish Drivers Don't Mind Sliding Scale, but Instant Calculation Gets Low Marks," *Wall Street Journal*, January 2, 2001.

237 return to shortly: A Finnish public-opinion poll in 2001 found that 66 percent of male drivers and 73 percent of male nondrivers felt the fine system was fair, while 77 percent of female drivers and 78 percent of female nondrivers thought it fair. The data comes from a study (in Finnish): T. Lappi-Seppälä, "Public Opinion and the 1999 Reform of the Day-Fine System," National Research Institute of Legal Policy, Publication No. 195, Helsinki, 2002. Thanks to Heikki Summala for providing the numbers.

237 rather stagnant: In 2003, for example, according to Eurostat, it grew just .50 percent. Data obtained from http://epp.eurostat.ec.europa.eu. Had the GDP risen, there may have been an increase in fatalities, reflecting the higher amounts of driving due to economic vitality—but it certainly would not have been by enough to offset the huge reductions in fatalities.

237 been in a crash): E. Lagarde, M. Chiron, and S. Lafont. "Traffic Ticket Fixing and Driving Behaviours in a Large French Working Population," *Journal of Epidemiology and Community Health*, vol. 58 (2004), pp. 562–68.

237 hundreds of traffic fatalities: Alexandre Dorozynski, "French Elections Can Kill," *British Medical Journal*, November 3, 2001, p. 1021.

237 The lesson is: At least one analysis posits that income equality is related in a linear fashion to traffic fatalities; e.g., in both poor, and, to a lesser extent, wealthy countries, the traffic fatality rate may be affected by the level of income equality. Perhaps not surprisingly, the Scandinavian countries, among the leaders in income equality, also rank near the top in traffic safety. See Nejat Anbarci, Monica Escaleras, and Charles Register, "Income, Income Inequality and the 'Hidden Epidemic' of Traffic Fatalities," No. 5002, Working Papers from Department of Economics, College of Business, Florida Atlantic University. Retrieved from http://econpapers.repec.org/paper/falwpaper/05002.htm.

237 *and* traffic fatalities: This relationship is argued in, among other sources, D. Treisman, "The Causes of Corruption: A Cross-National Study," *Journal of Public Economics*, no. 76 (June 2000), pp. 399–457.

237 income and traffic fatalities: See Nejat Anbarci, Monica Escaleras, Monica Register, and Charles A. Register, "Traffic Fatalities and Public Sector Corruption," *Kyklos*, vol. 59, no. 3 (August 2006), pp. 327–44; available at http://ssrn.com/abstract=914243.

238 of Europe's road fatalities: See "Fools and Bad Roads," *Economist*, March 22, 2007.

238 rewards inefficient firms: For a good review of the various debates over corruption and growth, see P. Bardhan, "Corruption and Development: A Review of Issues," *Journal of Economic Literature*, vol. 35 (September 1997), pp. 1320–46.

238 beneath the acceptable "minimum": See Daniel Kaufmann, "Corruption: The Facts," *Foreign Policy*, no. 107 (Summer 1997), pp. 114–31.

239 *because* of corruption: The most extreme case of this may be Lagos, the largest city in Nigeria and predicted to be among the world's largest cities in the next decade. The average commuter in Lagos is said to face myriad challenges. These begin with the crumbling roads and infrastructure, which have scarcely been repaired since being erected in the oil boom of the 1970s; they themselves are a kind of symbol of the endemic corruption of Nigeria, where close to $400 billion in oil revenues were sequestered out of the country in a forty-year period. Other challenges include arbitrary fees charged at will by bus drivers and their quasi-official associates, the *agberos*, not to mention the numerous unofficial roadblocks, manned by gangs of unemployed "area boys," that drivers must navigate. The multiple levels of corruption present in—and contributing to—Lagos's epic "go-slows" were demonstrated in an astonishing story told by the journalist George Packer. While riding on the streets of Lagos, Packer's driver was stopped by an *agbero*, who demanded money to help the driver negotiate *another* bribe, with the official traffic police. The traffic cop intervened, if only to collect the bribe—not doing so, it seemed, would actually make the police officer look as if he were derelict in his duty. See George Packer, "The Megacity: Decoding the Chaos of Lagos," *New Yorker*, November 26, 2006. See also Adewale Ajayi, *Nigerian Tribune*, March 2, 2007; and Osise Dibosa, "Olubunmi Peters and Ferma," *This Day*, June 12, 2007.

239 take their place: Benjamin A. Olken and Patrick Barron, "The Simple Economics of Extortion: Evidence from Trucking in Aceh," NBER Working Paper No. 13145, National Bureau of Economic Research, June 2007.

240 "work repairing potholes": Robert Guest, "The Road to Hell Is Unpaved," *Economist*, December 19, 2002.

240 "actual driving skill": The Delhi driving-license experiment is detailed in Marianne Bertrand, Simeon Djankov, Rema Hanna, and Sendhil Mullainathan, "Does Corruption Produce Unsafe Drivers?" NBER Working Paper No. 12274, National Bureau of Economic Research, June 2006.

240 "clarity of purpose": This line comes from Pavan K. Varma, *Being Indian* (London: Penguin Books, 2005), p. 79.

241 some 150,000 tickets: Raymond J. Fisman and Edward Miguel, "Cultures of Corruption: Evidence from Diplomatic Parking Tickets," NBER Working Paper No. W12312 (June 2006). Retrieved at http://ssrn.com/abstract=910844.

241 the city of London: Retrieved from *Channel Four News Online*, http://www.channel4.com/news/articles/society/environment/diplomatic+ccharge+bill+tops+45m/569892.

241 pays the charge: Nicola Woolcock, "Nations Unite to Join a Boycott of Congestion Charge," *Times* (London), February 21, 2007.

241 norms regarding them: This is why we can often see compliance with traffic laws differing even *within* a country. In Italy, corruption is more endemic in the south than the north, for reasons, as mentioned in an earlier note, having to do with varying degrees of civic culture. And so as the state seems to gradually wither away the farther south you go, so too does the traffic behavior come to have less to do with the law. In 2000, a national helmet law was passed for motorcyclists of any age. Afterward, usage rates in the north were reported as high as 95 percent. In the south, however, they were only as high as 70 percent, and as low as 50 percent. For corruption levels, see Alfredo del Monte and Erasmo Papagni, "The Determinants of Corruption in Italy: Regional Panel Data Analysis," *European Journal of Political Economy*, vol. 23 (June 2007), pp. 379–96. For helmet-use rates, see F. Servadei, C. Begliomini, E. Gardini, M. Giustini, F. Taggi, and J. Kraus, "Effect of Italy's Motorcycle Helmet Law on Traumatic Brain Injuries," *Injury Prevention*, vol. 9, no. 3 (2003), pp. 257–60.

242 casualties there will be: See D. Parker, J. T. Reason, A. S. R. Manstead, and S. G. Stradling, "Driving Errors, Driving Violations and Accident Involvement," *Ergonomics*, vol. 38 (1995), pp. 1036–48.

243 more women in government: Anand Swamy, Stephen Knack, Young Lee, and Omar Azfar, "Gender and Corruption," Center for Development Economics, Department of Economics, Williams College, 2000.

## Chapter Nine: Why You Shouldn't Drive with a Beer-Drinking Lawyer

244 our brains as we drive: Research has shown that the various aspects of driving, everything from following a traffic rule (e.g., specifying a one-way street) to navigating a set of directions to anticipating the actions of other drivers, seem to trigger discrete activity in a variety of brain regions and networks. Researchers

at University College London, for example, have monitored drivers as they "drove" the detailed recreation of London found in the popular video game *The Getaway.* See H. J. Spiers and E. A. Maguire, "Neural Substrates of Driving Behaviour," *NeuroImage,* vol. 36 (2007), pp. 245–55.

245 fifty thousand times a year: P. G. Martin and A. L. Burgett, "Rear-End Collision Events: Characterization of Impending Crashes," *Proceedings of the First Human-Centered Transportation Simulation Conference* (Iowa City: University of Iowa, 2000).

246 walks away alive: See Jack Stuster, "The Unsafe Driving Acts of Motorists in the Vicinity of Large Trucks," U.S. Department of Transportation, Federal Highway Administration, Office of Motor Carriers and Highway Safety, February 1999.

246 should probably fear: See L. J. Armony, D. Servan-Schreiber, J. D. Cohen, and J. E. LeDoux, "An Anatomically-Constrained Neural Network Model of Fear Conditioning," *Behavioral Neurocience,* vol. 109 (1995), pp. 246–56.

246 dangerous nature of trucks: Opinion surveys of car drivers tend to find mostly negative opinions of truck drivers' behavior. See, for example, Robert S. Moore, Stephen LeMay, Melissa L. Moore, Pearson Lidell, Brian Kinard, and David McMillen, "An Investigation of Motorists' Perceptions of Trucks on the Highways," *Transportation Journal,* January 5, 2001.

247 responsibility in the crash: Daniel Blower, "The Relative Contribution of Truck Drivers and Passenger Vehicles to Truck-Passenger Vehicle Traffic Crashes," report prepared for the U.S. Department of Transportation, Federal Highway Administration, Office of Motor Carriers, June 1998.

248 is actually the case): This may be the "availability heuristic" at work again. Large trucks, in part because they are driven longer distances and tend to be on the road at the same time as most motorists, seem to be more prevalent than they really are. A Canadian study found that while motorists believed that the number of trucks on the roads was rising, the number actually *dropped* during the period in question (while the number of cars grew). See Gordon G. Baldwin, "Too Many Trucks on the Road?" Transportation Division, Statistics Canada, Ottawa.

248 "risk as analysis": Paul Slovic, Melissa L. Finucane, Ellen Peters, and Donald G. MacGregor, "Risk as Analysis and Risk as Feelings: Some Thoughts About Affect, Reason, Risk, and Rationality," *Risk Analysis,* vol. 24, no. 2 (2004), pp. 311–23.

249 50 years of driving: Data retrieved on May 5, 2007, from http://hazmat.dot.gov/riskrngmt/riskcompare.htm.

249 the lifetime probability: P. Slovic, B. Fischhoff, and S. Lichtenstein, "Accident Probabilities and Seat Belt Usage: A Psychological Perspective," *Accident Analysis & Prevention,* vol. 13 (1978), pp. 281–85.

249 "the danger of leaving home": William H. Lucy, "Mortality Risk Associated with Leaving Home: Recognizing the Relevance of the Built Environment," *American Journal of Public Health,* vol. 93, no. 9 (September 2003), pp. 1564–69.

250 eleven times that: This figure was provided to me by Per Garder, a professor of civil and environmental engineering at the University of Maine. Using the required risk exposure levels as quoted by the Occupational Safety and Health Administration (in "Occupational Exposure to Asbestos," Federal Register 59:40964-41161, 1994, and OSHA Preambles, "Blood Borne Pathogens," 29 CFR 1910.1030, Federal Register 56:64004, 1991: 29206), Garder notes that the risk of dying over a lifetime in manufacturing and service employment, respectively, "must be less than 1.8 and 1.0 deaths per 1,000 employees." By those standards, Garder extrapolates if 1 person in a 1,000 were "allowed" to die in traffic over an average of 77 years of life, 1 person in 77,000 would thus be allowed to die in America this year in a traffic accident. Using America's population of 300 million, 1 in 77,000 would be 3,896 people. But the fatality figure was over 11 times that. In other words, if traffic were an industry—whether heavy manufacturing or service—it would have been shut down a long time ago.

250 every thirty-two minutes: Fatality statistics were taken from *Traffic Safety Facts 2004* (Washington, D.C.: National Highway Traffic Safety Administration, 2005).

250 3 out of every 1,000: Clifford Winston, Vikram Maheshri, and Fred Mannering, "An Exploration of the Offset Hypothesis Using Disaggregate Data: The Case of Airbags and Antilock Brakes," *Journal of Risk Uncertainty*, vol. 32 (2006), pp. 83–99.

251 raises the crash risk: M. G. Lenné, T. J. Triggs, and J. R. Redman, "Time of Day Variations in Driving Performance," *Accident Analysis & Prevention*, vol. 29, no. 4 (1997), pp. 431–37, and G. Maycock, "Sleepiness and Driving: The Experience of U.K. Car Drivers," *Accident Analysis & Prevention*, vol. 29, no. 4 (1997), pp. 453–62.

251 day to be on the road: As David Klein and Julian Waller noted, the posting of holiday traffic fatalities presents several problems. "Although absolute numbers may serve a purpose in indicating the raw impact of highway crashes on the nation or on a community," they write, "their use provides only a partial indication of magnitude and often a misleading indication of trends. First, fatality figures ignore the 1.5 to 3 million annual non-fatal injuries—which may represent a social cost far higher than the 56,000 fatalities. Second, the 'holiday death toll' may give drivers an unjustified feeling of anxiety on holiday weekends and a false sense of security on weekdays if it persuades them that the holiday incidence is substantially higher than on weekdays." From Klein and Waller, "Causation, Culpability and Deterrence in Highway Crashes," prepared for the Department of Transportation, July 1970, p. 27.

251 week before or after: C. M. Farmer and A. F. Williams. "Temporal Factors in Motor Vehicle Crash Deaths," *Injury Prevention*, vol. 2 (2005), pp. 18–23.

251 should be about $8,000: Steven D. Levitt and Jack Porter, "How Dangerous Are Drinking Drivers?," *Journal of Political Economy*, vol. 109, no. 6 (2001), pp. 1198–1237. The authors rely on a clever statistical trick that does not require knowing the actual number of drinking and sober drivers on the road (a

number that would be extremely hard to come by in any case) but, rather, uses an extrapolation taken from the relative proportion of sober and drunk drivers involved in two-car crashes. Levitt and Porter generate their relative risk numbers by looking at two-car crashes and "the relative frequency of accidents involving two drinking drivers, two sober drivers, or one of each." This information, they argue, "is sufficient to separately identify both the relative likelihood of causing a fatal crash on the part of drunk and sober drivers and the fraction of drivers on the road who have been drinking."

252 doubling of the speed: H. C. Joksch, "Velocity Change and Fatality Risk in a Crash: A Rule of Thumb," *Accident Analysis & Prevention*, vol. 25, no. 1 (1993), pp. 103–04.

252 doing 30 miles per hour: Allan F. Williams, Sergey Y. Krychenko, and Richard A. Retting, "Characteristics of Speeders," *Journal of Safety Research*, vol. 37 (2006), pp. 227–32.

252 get into more crashes: See, for example, Williams, Kyrychenko, and Retting. "Characteristics of Speeders," ibid.

252 additional 5 kilometers per hour: See C. N. Kloeden, A. J. McLean, V. M. Moore, and G. Ponte, "Travelling Speed and the Risk of Crash Involvement," NHMRC Road Accident Research Unit, University of Adelaide, November 1997.

252 "relatively high speed drivers": David Solomon, *Accidents on Main Rural Highways Related to Speed, Driver, and Vehicle* (Washington, D.C.: U.S. Department of Commerce, Bureau of Public Roads, 1964).

253 flow in smooth harmony: The speed-variance argument was most famously taken up by Charles Lave, "Speeding, Coordination, and the 55 MPH Limit," *American Economic Review*, vol. 75, no. 5 (December 1985), pp. 1159–64. Interestingly, in a point that has not been emphasized by those later citing Lave, he writes: "Although I have found no statistically discernible effect from speed, per se, this does not necessarily imply that it is safe to raise the speed limit, for we do not know what effect a higher limit would have on the speed variance." If the speed limit is 65 miles per hour but many people are driving 75, it does not necessarily follow that raising it to 75 miles per hour will reduce speed variance or make things safer. Do we want the drivers who feel comfortable at a lower level forced to go faster? Do we *want* Grandma and Grandpa driving 75 miles per hour?

253 held by young males: T. Horberry, L. Hartley, K. Gobetti, F. Walker, B. Johnson, S. Gersbach, and J. Ludlow, "Speed Choice by Drivers: The Issue of Driving Too Slowly," *Ergonomics*, vol. 47, no. 14 (November 2004), pp. 1561–70.

253 at low speeds: For elaboration on this point, see Kloeden, McLean, Moore, and Ponte, "Travelling Speed," op. cit.

253 involved a stopped vehicle: Ronald K. Knipling, "IVHS Technologies Applied to Collision Avoidance: Perspectives on Six Target Crash Types and Countermeasures," technical paper presented at the Safety and Human Factors session of 1993 IVHS America Annual Meeting, Washington, D.C., April 14–17, 1993.

253 not hold for individuals: Gary A. Davis, "Is the Claim That 'Variance Kills' an Ecological Fallacy?," *Accident Analysis & Prevention*, vol. 34 (2002), pp. 343–46. With the Solomon curve, Davis argues that one cannot determine the individual driver's crash risk by looking at the whole. Solomon's curve, maintains Davis, is a purely mathematical effect that says little about how the world works, "like saying an object is heavy because it weighs more." Another problem with the Solomon curve is that it does not explain causes. If twenty cars slowing for traffic congestion—and thus going below the median speed—were struck by ten cars traveling at the median and ten cars traveling above the median, the resulting "curve" would indeed suggest that slower drivers were the most at risk of being in a crash. But looking at each crash individually, one would conclude that the faster-moving cars had actually been the source of the risk for the slower-moving cars. As an example of a ecological fallacy, the statistician David Freedman has compared the income levels of U.S. states against the percentage of foreign-born residents in each. Doing this, one could make a statistically robust "correlation" that says foreign-born residents of the United States earn more than native-born residents, when actually the *opposite* is true. See David A. Freedman, "Ecological Inference and the Ecological Fallacy," in *International Encyclopedia of the Social & Behavioral Sciences*, vol. 6, ed. N. J. Smelser and Paul B. Baltes (New York: Pergamom, 2001), pp. 4027–30.

253 in the same direction: E. C. Cerrelli, "1996 Traffic Crashes, Injuries, and Fatalities—Preliminary Report," Report No. DOT HS 808 543, National Highway Traffic Safety Administration, March 1997. I was alerted to this finding by an excellent report summarizing the various speed issues. See Jack Stuster and Zail Coffman (1998), *Synthesis of Safety Research Related to Speed and Speed Limits*, FHWA-RD-98-154 (Washington, D.C.: Federal Highway Administration, 1998).

254 whose teams had lost: D. A. Redelmeier and C. L. Stewart, "Do Fatal Crashes Increase Following a Super Bowl Telecast?" *Chance*, vol. 18, no. 1 (2005), pp. 19–24.

254 have been drinking: R. G. Smart, "Behavioral and Social Consequences Related to the Consumption of Different Beverage Types," *Journal of Studies on Alcohol*, vol. 57 (1996), pp. 77–84.

254 at .08 to .1 percent: R. P. Compton, R. D. Blomberg, H. Moskowitz, M. Burns, R. C. Peck, and D. Fiorentino, "Crash Risk of Alcohol Impaired Driving," *Proceedings of the 16th International Conference on Alcohol, Drugs and Traffic Safety*, CD-ROM (Montreal, Société de l'Assurance Automobile du Québec, 2002).

254 BAC of zero: R. F. Borkenstein, R. F. Crowther, R. P. Shumate, W. B. Ziel, and R. Zylman, "The Role of the Drinking Driver in Traffic Accidents," Bloomington, Indiana, Department of Police Administration and Indiana University, 1964.

254 "handling" a small intake: See, for example, Leonard Evans, *Traffic Safety* (Bloomfield Hills: Science Serving Society, 2004), p. 246.

254 shown up in other studies: P. L. Zador, S. A. Krawchuk, and R. B. Voas, *Relative Risk of Fatal and Crash Involvement by BAC, Age and Gender* (Rockville, Md.: Westat, April 2000).

254 statistically less safe: Paul M. Hurst, David Harte, and William Frith, "The Grand Rapids Dip Revisited," *Accident Analysis & Prevention*, vol. 26, No. 5 (1994), pp. 647–54.

255 ratio is even higher: Evans, *Traffic Safety*, op. cit., p. 44.

255 the rate is .36: David Gerard, Paul S. Fischbeck, Barbara Gengler, and Randy S. Weinberg, "An Interactive Tool to Compare and Communicate Traffic Safety Risks: Traffic STATS," Center for the Study and Improvement of Regulation, Carnegie Mellon University, Transportation Research Board 07-1332, November 2006.

255 to prove that they are: They also kill others more often. A study in the United Kingdom found, for example, that pedestrians were roughly 1.5 times more likely to die when they were hit by a male driver than a female driver. *Car Make and Model: The Risk of Driver Injury and Car Accident Rates in Great Britain: 1994*, Transport Statistics Report (London: HMSO, 1995).

256 more likely to drink: National Institute on Alcohol Abuse and Alcoholism. "Drinking in the United States: Main Findings from the 1992 National Longitudinal Alcohol Epidemiologic Survey (NLAES)," *U.S. Alcohol Epidemiologic Data Reference Manual*, vol. 6 (Bethesda, Md.: National Institute of Health, 1998).

256 less likely to wear helmets: C. Peek-Asa and J. F. Kraus, "Alcohol Use, Driver, and Crash Characteristics Among Injured Motorcycle Drivers, *Journal of Trauma*, vol. 41 (1996), pp. 989–93.

256 those who are sober: See, for example, R. D. Foss, D. J. Beirness, and K. Sprattler, "Seat Belt Use Among Drinking Drivers in Minnesota," *American Journal of Public Health*, vol. 84, no. 11 (1994), pp. 1732–37.

256 attributed to the driver: Emmanuel Lagarde, Jean-François Chastang, Alice Gueguen, Mireille Coeuret-Pellicer, Mireille Chirion, and Sylviane Lafont, "Emotional Stress and Traffic Accidents: The Impact of Separation and Divorce," *Epidemiology*, vol. 15, no. 6 (November 2006).

256 and gender differences): G. Whitlock, R. Norton, T. Clark, R. Jackson, and S. MacMahon, "Motor Vehicle Driver Injury and Marital Status: A Cohort Study with Prospective and Retrospective Driver Injuries," *Injury Prevention*, vol. 10 (2004), pp. 33–36.

256 Spain to California: See, for example, T. Reuda-Domingo and P. Lardelli-Claret, "The Influence of Passengers on the Risk of the Driver Causing a Car Collision in Spain: Analysis of Collisions from 1990 to 1999," *Accident Analysis & Prevention*, vol. 36 (2004), pp. 481–89, and Judy A. Geyer and David R. Ragland, "Vehicle Occupancy and Crash Risk," UCB-TSC-RR-2004-16, Berkeley, Institute of Transportation Studies, 2004; paper accessed at http://repositories.cdlib.org/its/tsc/UCB-TSC-RR2004-16.

256 if there's a passenger: Actually, if one *is* involved in a crash, a passenger is still a good bet. The added mass, it has been suggested, could reduce a driver's fatality risk in a frontal collision by 7.5 percent. See Leonard Evans, "Causal Influence of Car Mass and Size on Driver Fatality Risk," *American Journal of Public Health*, vol. 91, no. 7 (July 2001), pp. 1076–81.

256 passengers in the car: Geyer and Ragland, "Vehicle Occupancy," op. cit.

256 with passengers onboard: Li-Hui Chen, Susan P. Baker, Elisa R. Braver, and Guohua Li, "Carrying Passengers as a Risk Factor for Crashes Fatal to 16- and 17-Year-Old Drivers," *Journal of the American Medical Association*, vol. 283 (2000), pp. 1578–82.

257 held for female drivers): B. G. Simons-Morton, N. Lerner, and J. Singer, "The Observed Effects of Teenage Passengers on Risky Driving Behavior of Teenage Drivers," *Accident Analysis & Prevention*, vol. 37 (2005), pp. 973–82.

257 their male comrades: Ronald Kotulak, "Increase in Women Doctors Changing the Face of Medicine," *Jerusalem Post*, August 2, 2007.

257 alcohol-related fatal crash: Information on crashes in Montana and New Jersey is drawn from Rajesh Subramanian, "Alcohol-Related Fatalities and Fatality Rates by State, 2004–2005," DOT HS 810 686, National Highway Traffic Safety Administration, December 2006: available at http://www.nhtsa.dot.gov.

257 found on rural roads: *Growing Traffic in Rural America: Safety, Mobility and Economic Challenges in America's Heartland* (Washington, D.C.: Road Information Program, March 2005).

258 any other road: ibid.

258 (nearly 75 percent in 2005): *Chicago Tribune*, January 12, 2005.

258 or even while driving: Laura K. Barger, Brian E. Cade, Najib F. Aya, et al., "Extended Work Shifts and the Risk of Motor Vehicle Crashes Among Interns," *New England Journal of Medicine*, vol. 352, no. 2 (January 13, 2005).

258 vehicle on the road: This does not have to do entirely with the vehicle, of course. As Charles Kahane of the National Highway Traffic Safety Administration points out, pickup trucks, at least historically, have tended to be driven more often in rural environments and more often by men—two risk-inflating variables. See Charles J. Kahane, "Vehicle Weight, Fatality Risk and Crash Compatibility of Model Year 1991–99 Passenger Cars and Light Trucks," National Highway Traffic Safety Administration Report DOT HS 809 662, October 2003.

258 other kind of vehicle: See, for example, Gerard, Fischbeck, Gengler, and Weinberg, "An Interactive Tool," op. cit.

258 pickups also impose: Several hundred people per year in the United States are also killed riding in the unprotected cargo beds of pickup trucks. See C. L. Anderson, P. F. Agran, D. G. Winn, and S. Greenland, "Fatalities to Occupants of Cargo Areas of Pickup Trucks," *Accident Analysis & Prevention*, vol. 32, no. 4 (2000), pp. 533–40.

258 on drivers of other vehicles: See Marc Ross and Tom Wenzel, "The Effects of Vehicle Model and Driver Behavior on Risk," *Accident Analysis & Prevention*, vol. 37 (2005), pp. 479–94.

259 more energy in a crash: See Marc Ross, Denna Patel, and Tom Wenzel, "Vehicle Design and the Physics of Traffic Safety," *Physics Today*, January 2006, pp. 49–54.

259 drivers of smaller cars: Leonard Evans, "Mass Ratio and Relative Driver Fatality Risk in Two-Vehicle Crashes," *Accident Analysis & Prevention*, vol. 25 (1993), pp. 609–16.

259 "was maintained very well": Thanks to Gabriel Bridger for pointing this out. See http://www.iihs.org for results.

259 in the *New Yorker*: Malcolm Gladwell, "Big and Bad," *New Yorker*, January 12, 2004.

260 Wenzel have pointed out: Tom Wenzel and Marc Ross, "Are SUVs Really Safer Than Cars? An Analysis of Risk by Vehicle Type and Model," Lawrence Berkeley National Laboratory Seminar, July 30, 2002, Washington, D.C. Similarly, the Chevrolet Camaro (or Pontiac Firebird) and Chevy Corvette are equally risky to their own drivers, but the Corvette poses less risk to others. The researchers suspect it may be because of the Corvette's fiberglass body and lower profile, both of which might cause less damage to others.

261 the statistically safest demographic?: Sometimes the statistics confound expectations. Take the Volvo V70 station wagon and the two-door BMW 3 Series. The first car conjures visions of staid Scandinavian safety and innocuous suburban commuting, while the image of the latter is of a small sports car piloted by the typically aggressive "Beemer" driver. Yet according to the Insurance Institute for Highway Safety, from 2002 through 2005, the U.S. fatality rate (per million registered vehicle years) for both cars was identical. I have no way of qualifying the difference, and there are a raft of potential statistical problems, but this leads to all kinds of speculation: Did the BMW have better crash protection? Was the safer driving of the Volvo owner offset by inferior handling? Perhaps Volvo wagons carried more passengers or logged more miles? Are BMW drivers better drivers? Or was it just a statistical fluke? As Marc Ross remarked to me in an e-mail correspondence, the relatively small number of fatalities in either car means that any variations in how the data is handled can easily throw off the results. There are a host of small factors that can corrupt the data, he explained: "For example, how long was the model in question on the road in the first year. If the model came out early, then the 'exposure' to crashes was relatively long in that first year. If the model came out late, the exposure was short in that first year. A different complication is that models that don't sell so well tend to stay on the dealer's lot, but some of them get registered [for tax reasons] by the dealer but aren't being driven while they stay on the lot." Therefore they have less exposure to traffic risk than might appear.

261 more than women: See, for example, Pew Research Center, "As the Price of Gas Goes Up, the Nation's Odometer Slows Down," August 8, 2006; available at http://pewresearch.org.

261 wear seat belts less often: V. Vasudevan and S. Nambisan, *Safety Belt Usage Surveys: Final Project Report* (Las Vegas: Transportation Research Center, University of Nevada, Las Vegas, 2006).

261 trucks without seat belts: Jeremy Diener and Lilliard E. Richardson, "Seat Belt Use Among Rural and Urban Pickup Truck Drivers," Report 4–2007, Institute of Public Policy, University of Missouri, July 2007.

261 involved in a fatal crash: See National Highway Traffic Safety Administration, "Alcohol Involvement in Fatal Motor Vehicle Traffic Crashes, 2003," DOT HS 809 822, March 2005.

261 versus white ones?: S. Newstead and A. D'Elia, "An Investigation into the Relationship Between Vehicle Colour and Crash Risk," Monash University Accident Research Centre, Report 263, 2007.

261 rental cars: In a conversation, Sheila "Charlie" Klauer noted that in the VTTI's aforementioned 100-car naturalistic study, both younger and older drivers of leased cars were involved in more risky driving events than the owners of private vehicles. "The leased vehicle drivers were involved in just slightly more events than were the private vehicle drivers. It was consistent," she said. "It's kind of a rental car phenomenon, that's what we're hypothesizing. I think we are all a little bit more reckless when we're in a rental car than in our own car." I was unable to find any study in the U.S. that had tackled this question head-on, although the multiplicity of drivers any rental car has and the varieties of exposure would make it difficult to gauge risk. A study in Jordan did report a higher crash rate among rental cars, though this was complicated by the fact that younger drivers (a riskier group to begin with) seemed to be over-represented among car renters. See Adli H. Al-Balbissi, "Rental Cars Unique Accident Trends," *Journal of Transportation Engineering*, vol. 127, no. 2 (March–April 2001), pp. 175–77.

261 (less regard for life?): Guy Stecklov and Joshua R. Goldstein, "Terror Attacks Influence Driving Behavior in Israel," *Proceedings of the National Academy of Science*, vol. 101, no. 40 (2004), pp. 14551–56.

262 than in the front: Evans, *Traffic Safety*, op. cit., p. 56.

262 some 28,500 lives: C. Hunter Sheldon, *Journal of the American Medical Association*, November 5, 1955.

262 from seat belts: John Adams notes that from 1970 to 1978, in a sample of major Western countries that adopted seat-belt laws during the period, "the group of countries that had not passed seat-belt laws experienced a greater decrease [in fatalities] than the group that had passed laws." In the United Kingdom, he writes, the drop in fatalities in 1983, the first full year after the belt law was passed, was "nothing remotely approaching" the predicted decline of one thousand deaths a year. The only segment of fatalities that dropped dramatically, he notes, was fatalities during the "drink-drive hours" of early Saturday and Sunday mornings—in response, he argues, to a stepped-up campaign against drunk driving. The drop in fatalities at other times, he suggests, was no higher than the annual 3 percent decrease already taking place. "No studies have been done to explain why," he writes, "after the seat-belt law came into effect in Britain, seat belts were so extraordinarily selective in saving the lives of those who are over the alcohol limit and driving between 10 at night and 4 in the morning." See John Adams, "Britain's Seat-Belt Law Should Be Repealed," draft of a paper for publication in *Significance*, March 2007.

263 colors would make more sense): R. G. Mortimer, "A Decade of Research in Rear Lighting: What Have We Learned?," in *Proceedings of the Twenty-first Conference of the American Association for Automotive Medicine* (Morton Grove, Ill.: AAAM, 1977), pp. 101–22.

263 improved reaction times: See, for example, J. Crosley and M. J. Allen, "Automobile Brake Light Effectiveness: An Evaluation of High Placement and

Accelerator Switching," *American Journal of Optometry and Archives of American Academy of Optometry*, vol. 43 (1966), pp. 299–304. For a good history of brake lights and the various issues involved, see D. W. Moore and K. Rumar, "Historical Development and Current Effectiveness of Rear Lighting Systems," Report No. UMTRI-99-31, 1999, University of Michigan Transportation Research Institute, Ann Arbor.

263 cut by 50 percent: The trial is described in John Voevodsky, "Evaluation of a Deceleration Warning Light for Reducing Rear-End Automobile Collisions," *Journal of Applied Psychology*, vol. 59 (1974), pp. 270–73.

263 to around 15 percent: Charles Farmer, "Effectiveness Estimates for Center High Mounted Stop Lamps: A Six-Year Study," *Accident Analysis & Prevention*, vol. 28, no. 2 (1996), pp. 201–08.

263 crashes by 4.3 percent: See Suzanne E. Lee, Walter W. Wierwille, and Sheila G. Klauer, "Enhanced Rear Lighting and Signaling Systems: Literature Review and Analyses of Alternative System Concepts," DOT HS 809 425, National Highway Traffic Safety Administration, March 2002.

263 inventors had hoped: Critics of the chimsil have attributed its underwhelming impact in part to the idea that drivers do not necessarily brake when they see brake lights illuminated. The chimsil, this critique goes, offers more information, but more of the *same* information. It says nothing, for example, about how quickly a car is decelerating or whether it has, in fact, stopped—a key consideration given the majority of rear-end collisions involving stopped cars. The work of R. G. Mortimer has provided the most thoroughgoing critique of the chimsil. See, for example, R. G. Mortimer, "The High-Mounted Brake Light: The 4% Solution," Society of Automotive Engineers Technical Paper 1999-01-0089, 1999.

263 *by* someone else: L. Evans and P. Gerrish, "Anti-lock Brakes and Risk of Front and Rear Impact in Two-Vehicle Crashes," *Accident Analysis & Prevention*, vol. 28 (1996), pp. 315–23.

263 non-ABS drivers did: Elizabeth Mazzae, Frank S. Barickman, and Garrick J. Forkenbrock, "Driver Crash Avoidance Behavior with ABS in an Intersection Incursion Scenario on Dry Versus Wet Pavement," Society of Automotive Engineers Technical Paper, 1999-01-1288, 1999.

263 braking the wrong way: A. F. Williams. and J. K. Wells, "Driver Experience with Antilock Brake Systems," *Accident Analysis & Prevention*, vol. 26 (1994), pp. 807–11.

263 "close to zero": Charles J. Kahane, "Preliminary Evaluation of the Effectiveness of Antilock Brake Systems for Passenger Cars," NHTSA Report No. DOT HS 808 206, December 1994.

263 "has never been explained"): Insurance Institute for Highway Safety, *Status Report*, vol. 35, no. 4 (April 15, 2000).

264 guide, it will not: Nick Bunkley, "Electronic Stability Control Could Cut Fatal Highway Crashes by 10,000," *New York Times*, April 6, 2007. One key difference to note with ESC versus ABS is that ESC functions on its own—it does not need to be used "correctly," as in the case of ABS.

264 railroad safety improvements: Charles Francis Adams, in his 1879 book *Notes*

*on Railroad Accidents*, wrote: "It is a favorite argument with those who oppose the introduction of some of these improvements, or who make excuses for want of them, that their servants are apt to become more careless from the use of them, in consequence of the extra security which they are believed to afford; and it is desirable to consider how much truth there is in this assertion." As it happens, Adams did not subscribe to this early offset hypothesis: "The risk is proved by experience to be very much greater without them than with them; and, in fact, the negligence and mistakes of servants are found to occur most frequently, and generally with the most serious results, not when the men are over-confident in their appliances or apparatus, but when, in the absence of them, they are habituated to risk in the conduct of the traffic." Interestingly, though, in a passage that still applies today, he noted that accidents at grade crossings, then as now, seemed to happen under what would be presumed to be the least likely, or "safest," of conditions: "The full average of accidents of the worst description appear to have occurred under the most ordinary conditions of weather, and usually in the most unanticipated way. This is peculiarly true of accidents at highway grade crossings. These commonly occur when the conditions are such as to cause the highway travelers to suppose that, if any danger existed, they could not but be aware of it." From Charles Francis Adams, *Notes on Railroad Accidents* (New York: G. P. Putnam's Sons, 1879).

264 "the highway death rate": Sam Peltzman, "The Effects of Automobile Safety Regulation," *Journal of Political Economy*, vol. 83, no. 4 (August 1976), pp. 677–726.

264 reason to feel less safe: Decades later, people are still sifting through the data, trying to refute or defend Peltzman's hypothesis. He has been questioned for, among other things, including motorcyclists in his count of nonoccupant fatalities—that is, along with pedestrians and cyclists—as if they were a similar beast. (Annual motorcycle registrations were also growing, it has been argued, and many motorcyclists, in any case, die in single-vehicle crashes, which are presumably not the result of car drivers acting more aggressively.) See, for example, Leon S. Robertson, "A Critical Analysis of Peltzman's 'The Effects of Automobile Safety Regulation,' " *Journal of Economic Issues*, vol. 2, no. 3 (September 1977), pp. 587–600. Others have suggested that people may not have been driving more aggressively but simply *more*—driving more in the newer cars because they felt safer (arguably a form of behavioral adaptation itself). See Robert B. Noland, "Traffic Fatalities and Injuries: Are Reductions the Result of 'Improvements' in Highway Design Standards?," paper submitted to Annual Meeting of the Transportation Research Board, November 10, 2000. A study by a pair of Harvard economists that paid specific attention to how many people were actually using seat belts (again, something that can only be guessed at) found no evidence for a Peltzman effect. The authors did, however, conclude that fatalities had not dropped by nearly as much as government regulators had predicted. See Alma Cohen and Liran Einav, "The Effects of Mandatory Seat Belt Laws on Driving Behavior and Traffic Fatalities," Discussion Paper No. 341, Harvard Law School, November 2001; downloaded on

February 12, 2007, from http://www.law.harvard.edu/programs/olin_center/. Peltzman was also criticized for not separating, or "disaggregating," the regulated vehicles from the nonregulated vehicles (to see, for example, if the cars with safety upgrades were overrepresented in fatal pedestrian crashes). This argument was made by Leon Robertson and Barry Pless, "Does Risk Homeostasis Theory Have Implications for Road Safety," *British Medical Journal*, vol. 324 (May 11, 2002), pp. 1151–52.

265 to be riskier drivers: This point is made explicit in the discussion of the fictional Fred earlier in the chapter, but see, too, P. A. Koushki, S. Y. Ali, and O. Al-Saleh, "Road Traffic Violations and Seat Belt Use in Kuwait: Study of Driver Behavior in Motion," *Transportation Research Record*, vol. 1640 (1998), pp. 17–22; see also T. B. Dinh-Zarr, D. A. Sleet, R. A. Shults, S. Zaza, R. W. Elder, J. L. Nichols, R. S. Thompson, and D. M. Sosin, "Reviews of Evidence Regarding Interventions to Increase the Use of Safety Belts," *American Journal of Preventive Medicine*, vol. 21, no. 4, Supp. 1 (2001), pp. 48–65, and D. F. Preusser, A. F. Williams, and A. K. Lund, "The Effect of New York's Seat Belt Use Law on Teenage Drivers," *Accident Analysis & Prevention*, vol. 19 (1987), pp. 73–80.

265 not wearing their belts: Evans, *Traffic Safety*, op. cit., p. 89.

265 "frequently get into accidents": Russell S. Sobel and Todd M. Nesbit, "Automobile Safety Regulation and the Incentive to Drive Recklessly: Evidence from NASCAR," *Southern Economic Journal*, vol. 74, no.1 (2007).

265 suits and helmets: This point is made by Stephen J. Dubner and Steven D. Levitt in "How Many Lives Did Dale Earnhardt Save?" *New York Times*, February 19, 2006. They note that if NASCAR drivers had died at the same rate as American drivers in general in a five-year period, fifteen drivers should have died—instead, none did. This raises the interesting point that Earnhardt's death became something of a spur for greater safety on NASCAR racetracks, where no single death in the general population of drivers seems capable of prompting a similar response.

266 fatalities by some 25 percent: A. J. McLean, B. N. Fildes, C. J. Kloeden, K. H. Digges, R. W. G. Anderson, V. M. Moore, and D. A. Simpson, "Prevention of Head Injuries to Car Occupants: An Investigation of Interior Padding Options," Federal Office of Road Safety, Report CR 160, NHMRC Road Accident Research Unit, University of Adelaide and Monash University Accident Research Centre.

266 seat belts and air bags: Sam Peltzman, "Regulation and the Natural Progress of Opulence," lecture presented at the American Enterprise Institute, September 8, 2004, AEI-Brookings Joint Center for Regulatory Studies, Washington, D.C.

266 Simpson has suggested: Joe Simpson, writing about "super-share ice screws" and other technological innovations, notes that "one would have thought these welcome developments would have made the sport considerably safer. Unfortunately climbers now throw themselves onto ice climbs that would have been unheard-of only a decade ago." He then draws a comparison to his car, a "rust bucket of a Mini" that "left you with no illusions as to what a small cube of

twisted metal it could instantly become if you hit anything." As a result, he writes, "I drove with a modicum of caution." From *The Beckoning Silence* (Seattle: Mountaineers Books, 2006), p. 105.

266 for more "safety": The Mount McKinley information comes from a fascinating study by R. Clark and Dwight R. Lee, "Too Safe to Be Safe: Some Implications of Short- and Long-Run Rescue Laffer Curves," *Eastern Economic Journal*, vol. 23, no. 2 (Spring 1997), pp. 127–37. It is true that many more people were climbing the mountain by the century's end, but it is also true that many more climbers were needing to be rescued. In 1976 alone, the study notes, there were thirty-three rescues, one out of every eighteen climbs—almost as many as the total number prior to 1970.

267 no-pull fatality: Vic Napier, Donald Self, and Carolyn Findlay, "Risk Homeostasis: A Case Study of the Adoption of a Safety Innovation on the Level of Perceived Risk," paper submitted to the American Society of Business and Behavioral Sciences meeting, Las Vegas, February 22, 2007.

267 our willingness for risk: O. Adebisi and G. N. Sama, "Influence of Stopped Delay on Driver Gap Acceptance Behavior," *Journal of Transportation Engineering*, vol. 3, no. 115 (1989), pp. 305–15.

267 fatal crashes goes *down:* Daniel Eisenberg and Kenneth E. Warner, "Effects of Snowfalls on Motor Vehicle Collisions, Injuries, and Fatalities," *American Journal of Public Health*, vol. 95, no.1 (January 2005), pp. 120–24.

268 perfect risk "temperature": Robertson and Pless, "Does Risk Homeostasis Theory Have Implications for Road Safety," op. cit.

268 the lookout for cars: For a Palo Alto report, see Alan Wachtel and Diana Lewiston, "Risk Factors for Bicycle-Motor Vehicle Collisions at Intersections," *ITE Journal*, September 1994. See also L. Aultmann-Hall and M. F. Adams. "Sidewalk Bicycling Safety Issues," *Transportation Research Record*, no. 1636, 1998, pp. 71–76. For a fascinating and in-depth discussion of bicycle risk and safety issues, see Jeffrey A. Hiles, "Listening to Bike Lanes," September 1996; retrieved on November 14, 2006, at http://www.wright.edu/~jeffrey.hiles/essays/listening/ contents.html.

268 than those without them): See, for example, Lasse Fridstrom, "The Safety Effect of Studded Tyres in Norwegian Cities," *Nordic Road and Transport Research*, no. 1 (2001), as well as Veli-Pekka Kallberg, H. Kanner, T. Makinen, and M. Roine, "Estimation of Effects of Reduced Salting and Decreased Use of Studded Tires on Road Accidents in Winter," *Transportation Research Record*, vol. 1533 (1995).

269 drivers of larger cars: Paul Wasielewski and Leonard Evans, "Do Drivers of Small Cars Take Less Risk in Everyday Driving?," *Risk Analysis*, vol. 5, no. 1 (1985), pp. 25–32.

269 higher speeds and more lanes: D. Walton and J. A. Thomas, "Naturalistic Observations of Driver Hand Positions," *Transportation Research Part F: Traffic Psychology and Behavior*, vol. 8 (2005), pp. 229–38.

269 lower feelings of risk: D. Walton and A. Thomas, "Measuring Perceived Risk: Self-reported and Actual Hand Positions of SUV and Car Drivers," *Transpor-*

*tation Research Part F: Traffic Psychology and Behaviour*, vol. 10, issue 3 (May 2007).

269 talking on a cell phone: Lesley Walker, Jonathan Williams, and Konrad Jamrozik, "Unsafe Driving Behaviour and Four Wheel Drive Vehicles: Observational Study," *British Medical Journal*, vol. 333, issue 17558 (July 8, 2006), p. 71.

270 "car in front": Sten Fossser and Peter Christensen, "Car Age and the Risk of Accidents," TOI Report 386, Institute of Transport Economics, Norway, 1998.

270 more than old cars: This was suggested to me in a conversation with Kim Hazelbaker, senior vice president of the Insurance Institute for Highway Safety, May 19, 2007.

270 drive it more often: When I asked Leonard Evans, one of the leading authorities on traffic safety in the United States, what kind of car he drove, his answer made an impression on me. "In terms of a certain mind-set I drive a very unsafe car," he said. "It's about the least expensive, lightest car that my former employer manufactured: the Pontiac Sunfire." It is well over a decade old.

270 *and* killed in war: Shaoni Bhattacharya, "Global Suicide Toll Exceeds War and Murder," *New Scientist*, September 8, 2004.

271 struck by lightning: John Mueller, "A False Sense of Insecurity," *Regulation*, vol. 27, no. 3 (Fall 2004), pp. 42–46.

271 stricter cell phone laws): Frank McKenna, a professor of psychology at the University of Reading, points out that people have commonly resisted previous traffic and other health safety measures, ranging from wearing seat belts to restricting workplace smoking, on the grounds that they impinge upon "freedoms." There is also, in public policy, a tendency to avoid legislating behaviors that do not violate John Stuart Mill's "harm principle"—that is, this thinking maintains that laws should be passed only to "prevent harm to others," not for the "physical or moral" good of any individual. As McKenna argues, even though drunk driving and not wearing seat belts were once considered legitimate behavior, the social costs of these behaviors, as with workplace smoking, were eventually recognized. This raises the question, however, of why speeding, which can cause "harm to others," is so widely tolerated. It may be that, as has been argued in this book, people are often simply not aware of their speed, or of the potential risks they are assuming in driving at a high speed. This may help contribute to a perceived lack of "legitimacy" on the part of authorities in trying to mount stricter enforcement campaigns. Police are faced with a well-known quandary: Be too lenient in enforcing strict speed limits, and drivers' speeds will creep up; be too strict, and "they risk strain on public acceptability." McKenna concludes that the current public acceptance of regularly driving above speed limits may at some point look as retrograde as workplace smoking: "It is noted that the perceived legitimacy of action can change considerably over time and interventions that would not be perceived as legitimate at one point in time may be considered uncontroversial at a later point in time." See Frank P. McKenna, "The Perceived Legitimacy of Intervention: A Key Feature for Road Safety," AAA Foundation for Traffic Safety, 2007.

271 drive rather than fly: "Consequences for Road Traffic Fatalities of the Reduction in Flying Following September 11, 2001," Michael Sivak and Michael Flannagan, *Transportation Research Part F: Traffic Psychology and Behavior*, vol. 7, nos. 4–5 (July–September 2004), pp. 301–05.

271 assigned to counterterrorism: Carl Ingram, "CHP May Get to Hire 270 Officers," *Los Angeles Times*, June 2, 2004, p. B1. In the article, one police officer points out that Timothy McVeigh was caught on a "routine traffic stop." Eerily enough, Mohammed Atta, the ringleader of the September 11 participants, was ticketed once for speeding and once for driving without a license; the license he finally got was suspended when he failed to appear in court.

271 raising speed limits: Elihu D. Richter, Lee S. Friedman, Tamar Berman, and Avraham Rivkind, "Death and Injury from Motor Vehicle Crashes: A Tale of Two Countries," *American Journal of Preventative Medicine*, vol. 29, no. 5 (2005), pp. 440–50. The authors implicate several other differences, including the steep rise in ownership of SUVs and other light trucks in the United States in the 1990s, as well as higher rates of driving under the influence of alcohol.

271 would have been killed: This point was raised in a letter by Leonard Evans in response to the previous article. *American Journal of Preventative Medicine*, vol. 30, no. 6 (2006), p. 532.

272 "psychophysical numbing": D. Fetherstonhaugh, P. Slovic, S. Johnson, and J. Friedrich, "Insensitivity to the Value of Human Life: A Study of Psychophysical Numbing," *Journal of Risk and Uncertainty*, vol. 14, no. 3 (1997), pp. 282–300.

272 of a terrible disease: Karen E. Jenni and George Lowenstein, "Explaining the 'Identifiable Victim Effect,' " *Journal of Risk Uncertainty*, vol. 14 (1997), pp. 235–37.

272 only *one* more child: Paul Slovic, "If I Look at the Mass I Will Never Act: Psychic Numbing and Genocide," *Judgement and Decision Making*, vol. 2, no. 2 (April 2007), pp. 1–17.

272 all who died: One exception to this is found at streetsblog.org, which has tracked fatalities and crashes in the New York metropolitan area.

272 "dread" and "novelty": B. Fischhoff, P. Slovic, S. Lichtenstein, S. Read, and B. Combs, "How Safe Is Safe Enough? A Psychometric Study of Attitudes Towards Technological Risks and Benefits," *Policy Sciences*, vol. 9 (1978), pp. 127–52.

272 (like nuclear power): In New York City, an undercurrent of public opinion says that bicycles are "dangerous." Neighborhoods have fought against the addition of bike lanes for this very reason. Yet one could count the number of people killed by bicycles in New York City each year on one hand, with a few fingers left over, while many times that number of people are killed or severely injured by cars. When I met with Ryan Russo, an engineer with the New York City Department of Transportation, I could not help but hear the echo of several of the reasons why we misperceive risk. "It's silent and it's rare," he told me, when I asked about New Yorkers' antipathy toward cyclists. "As opposed to cars, which make noise and are prevalent. You don't see it because it's smaller, you don't hear it approach because it's silent, and you don't expect it because it's

not prevalent." A close call with a cyclist, no matter how less dangerous statistically, stands out as the greater risk than a close call with a car, even though—or in fact precisely *because*—pedestrians are constantly having near-hazardous encounters with turning cars in crosswalks.

272 seem to be misperceived: A classic case, pointed out by Leonard Evans, is the specter of "vehicle recalls." Every month or so, the news announces that some particular model of car has a potential defect. These recalls haunt us, raising our hackles with a constant stream of exploding tires and potentially faulty brakes. The cumulative result of this, Evans suggested, is that we may come to feel that the greatest threat to a driver's safety is the improper functioning of his or her vehicle. "They will say on the news there are 'no injuries reported,'" Evans said. We may feel relieved; the system works. "But the previous night there might have been a thousand people injured in crashes. And we're told it's the *recall* that is important."

272 those killed by lightning): An analysis by AAA found 10,037 incidents of "violent and aggressive driving" between January 1, 1990, and August 31, 1996, that led to the deaths of 218 people. An estimated 37 percent of those cases involved a firearm. Cited by David K. Willis of AAA in *Road Rage: Causes and Dangers of Aggressive Driving; Hearings Before the Subcommittee on Surface Interpretation of the House Committee on Transportation and Infrastructure*, 105th Congress, 1st Session, 1997. As Michael Fumento has pointed out, in the same time span that these 218 "aggressive driving" deaths were registered, some 290,000 people were killed on the road. See Fumento, "'Road Rage' vs. Reality," *Atlantic Monthly*, August 1998.

273 than pistol-packing drivers: Traces of the sleeping pill Ambien, not taken as prescribed, have been showing up in the bloodstreams of drivers involved in crashes. See Stephanie Saul, "Some Sleeping Pill Users Range Far Beyond Bed," *New York Times*, March 8, 2006. But many other drugs of the kind that typically warn users not to "operate heavy machinery" while taking them also show up in the bodies of drivers (who apparently forget that cars are heavy machinery); for example, dextromethorphan, a synthetic analogue of codeine that appears frequently in over-the-counter medicines. See Amy Cochems, Patrick Harding, and Laura Liddicoat, "Dextromethorphan in Wisconsin Drivers," *Journal of Analytical Toxicology*, vol. 31, no. 4 (May 2007), pp. 227–32.

273 if they pick the numbers: This phenomenon was described by psychologist Ellen Langer, who called it the "illusion of control." See E. J. Langer, "The Illusion of Control," *Journal of Personality and Social Psychology*, vol. 32, no. 2 (1975), pp. 311–28.

273 real dangers cars present: Consider, for example, the fact that, in the United States at least, hardly any children walk to school anymore—the figure has dropped from 48 percent in 1969 to under 15 percent in 2001. One perceived reason is "stranger danger." But abductions, by strangers or family members, the U.S. Department of Justice has noted, make up only 2 percent of violent crimes against juveniles. Riding in the family car, and not "stranger danger," is the greatest risk to people aged four to thirty-seven in the United States (and

many other places). The car is actually a risk before it even leaves the driveway. In 2007, more than two hundred children were killed in the United States in "nontraffic fatalities," a grim category that includes everything from "back-over" incidents (typically in "safe" SUVs) to the hyperthermia of children unintentionally left in cars. For abduction statistics, see D. Finklehor and R. Ormrod, "Kidnapping of Juveniles: Patterns from NIBRS," *Juvenile Justice Bulletin*, June 2000. Children's walk-to-school rates come from Reid Ewing, Christopher V. Forinash, and William Schroeer, "Neighborhood Schools and Sidewalk Connections: What Are the Impacts on Travel Mode Choice and Vehicle Emissions?," *TR News*, vol. 237 (March–April 2005). School bus fatality risks are taken from Ann M. Dellinger and Laurie Beck, "How Risky Is the Commute to School," *TR News*, vol. 237 (March–April 2005).

273 more dangerous it is: This information comes from a study by William Lucy, a University of Virginia professor of urban planning. His findings are based on two key mortality indices: chance of being killed by a stranger and risk of being killed in traffic. See Lucy, "Mortality Risk Associated with Leaving Home: Recognizing the Relevance of the Built Environment," *American Journal of Public Health*, vol. 93, no. 9 (September 2003), pp. 1564–69.

274 (roughly 22 miles per hour): In 2006, there were 14 traffic fatalities recorded in Bermuda, though that number was set to rise to 20 in 2007. See Tim Smith, "Call for Greater Police Presence to Tackle Road Deaths 'Epidemic,' " *Royal Gazette*, November 24, 2007. This is actually quite a high number for a country with a population of some 66,000 (not including the many tourists who visit). Typically, however, 80 percent of these fatalities involve the riders or passengers of motorbikes, and a high percentage of those involve tourists who are either unfamiliar with the roads (or the bikes) or presumably have been drinking. Tourists in Bermuda are estimated to be almost six times at risk for being injured on a motorbike than are local residents. See M. Carey, M. Aitken, "Motorbike Injuries in Bermuda: A Risk for Tourists," *Annals of Emergency Medicine*, vol. 28, Issue 4, pp. 424–29. Other studies have shown tourists to be overrepresented in car crashes. See C. Sanford, "Urban Medicine: Threats to Health of Travelers to Developing World Cities," *Journal of Travel Medicine*, vol. 11, no. 5 (2004), pp. 313–27. John Adams brought up the Bermuda example in his book *Risk and Freedom: The Record of Road Safety* (Cardiff: Transport Publishing Projects, 1985), p. 2. He quotes, in turn, Herman Kahn, *The Next 200 Years* (New York: William Morrow, 1976), p. 168.

274 cars and cyclists: Based on a conversation with city manager Judie Zimomra and police department records specialist Bob Conklin. Zimomra noted that there were traffic fatalities in the 1990s, but subsequent enforcement and engineering efforts have proven successful. The lesson: Speed is important, but hardly the only issue.

274 lowers crash risks: C. N. Kloeden, A. J. McClean, and G. Glonek, "Reanalysis of Travelling Speed and Risk of Crash Involvement in Adelaide, South Australia," Australian Transport Safety Bureau Report CR 207, April 2002.

274 Adams calls "hypermobility": See John Adams, "Hypermobility: Too Much of a

Good Thing?," Royal Society for the Arts Lecture, November 21, 2001. Retrieved at http://www.geog.ucl.ac.uk/~jadams/publish.htm.

274 roughly half the crashes: See Cherian Varghese and Umesh Shankar, "Restraint Use Patterns Among Fatally Injured Passenger Vehicle Occupants," DOT HS 810 595, National Highway Traffic Safety Administration, May 2006.

274 slow level of 35 miles per hour: From a report prepared by Michael Paine, based on data taken from the U.S. National Highway Traffic Safety Administration from 1993 to 1997; retrieved from http://users.tpg.au/users/mpaine/speed/html.

275 (among other things): An observational study of a random sample of drivers in New York City found that those talking on a hands-free device were more likely to engage in other distracting activities (e.g., smoking, eating, grooming) than those speaking on a handheld cell phone. As the researchers observed, the drivers "may be trading one automobile-related risk for another." See "Driving Distractions in New York City," Hunter College, November 2007.

## Epilogue: Driving Lessons

278 to pass the front: For an excellent discussion of the physics of oversteering and understeering, as well as driving in general, see Barry Parker, *The Isaac Newton School of Driving: Physics and Your Car* (Baltimore: Johns Hopkins University Press, 2003).

279 to maintain our course?: W. O. Readinger, A. Chatziastros, D. W. Cunningham, J. E. Cutting, and H. H. Bülthoff, "Gaze-Direction Effects on Drivers' Abilities to Steer a Straight Course," *TWK Beiträge zur 4. Tübinger Wahrnehmungskonferenz*, ed. H. H. Bülthoff, K. R. Gegenfurtner, H. A. Mallot, R. Ulrich. Knirsch, Kirchentellinsfurt, 149 (2001). Available at http://www.kyb.mpg.de/publication.html?publ=67

279 "doing so at all": See W. O. Readinger, A. Chatziastros, D. W. Cunningham, H. H. Bülthoff, and J. E. Cutting, "Gaze-Eccentricity Effects on Road Position and Steering," *Journal of Experimental Psychology: Applied*, vol. 8, no. 4 (2002), pp. 247–58.

280 "might be your English teacher": Actually, the traditional model of high school driver's ed—usually classroom instruction plus on-road time—has been largely discredited. The reasons have less to do with the worth or validity of learning the rules of the road than with the fact that such programs, rather than helping to produce safer drivers, just seem to put more unsafe drivers on the road at a younger age. A number of studies have come to this conclusion, but see, in particular, J. Vernick, G. Li, S. Ogaitis, E. MacKenzie, S. Baker, and A. Gielen, "Effects of High School Driver Education on Motor Vehicle Crashes, Violations, and Licensure," *American Journal of Preventive Medicine*, vol. 16, no. 1 (1999), pp. 40–46; M. F. Smith, "Research Agenda for an Improved Novice Driver Education Program: Report to Congress, May 31, 1994," DOT HS 808 161, National Highway Traffic Safety Administration, retrieved from www.nhtsa

.dot.gov/people/injury/research/pub/drive-ed.pdf; and I. Roberts and L. Kwan, "School Based Driver Education for the Prevention of Traffic Crashes," *Cochrane Database of Systematic Reviews*, no. 2 (2006).

280 skills needed to drive: Thanks to Leonard Evans for this reference.

280 stock-car drivers: A. F. Williams and B. O'Neill, "On-the-Road Driving Records of Licensed Race Drivers," *Accident Analysis & Prevention*, vol. 6 (1974), pp. 263–70.

281 "not going fast enough": Thanks to Leonard Evans for Andretti Quote.

281 to go next: Vision researchers studied the eye and head movements of Formula 3 racer Tomas Scheckter as he drove on the Mallory Park circuit in Leceistershire, England. They suggested that Scheckter, because he had learned the layout of the track, actually moved his head in the direction in which he wanted to go before he adjusted his steering. See Michael F. Land and Benjamin W. Tatler, "Steering with the Head: The Visual Strategy of a Racing Driver," *Current Biology*, vol. 11 (2001), pp. 1215–20.

282 to avoid a crash: For an excellent summary of the research, see Lisa D. Adams, "Review of the Literature on Obstacle Avoidance Maneuvers: Braking Versus steering," Report No. UMTRI-94-19, University of Michigan Transportation Research Institute, Ann Arbor, August 1994.

282 the only thing to do: Jeffrey Muttart raises the idea of "operant conditioning" in "Factors That Influence Drivers' Response Choice Decisions in Video Recorded Crashes," *Society of Automotive Engineers Journal*, 2005.

282 to their full power: See Rodger J. Koppa and Gordon G. Hayes, "Driver Inputs During Emergency or Extreme Vehicle Maneuvers," *Human Factors*, vol. 18, no. 4 (1976), pp. 361–70.

282 the obstacle is moving: D. Fleury, F. Fernandez, C. Lepesant, and D. Lechner, "Analyse typologique des manoeuvres d'urgence en intersection," *Rapport de recherche INRETS*, no. 62 (1988), quoted in Lisa D. Adams, 1994.

282 to the point where we do nothing: Michael A. Dilich, Dror Kopernik, and John M. Goebelbecker, "Evaluating Driver Response to a Sudden Emergency: Issues of Expectancy, Emotional Arousal, and Uncertainty," *Safety Brief*, vol. 20, no. 4 (June 2002). A frequent occurrence in driving simulator studies that seek to evaluate how drivers respond to unexpected obstacles or hazards is that a small number of subjects often have "no response." A French study, for example, in which drivers on a test track had to react to an inflatable "dummy car," found that 4 percent of subjects did nothing, simply "freezing." See Christian Collett, Claire Petit, Alain Priez, and Andre Dittmar, "Stroop Color-Word Test, Arousal, Electrodermal Activity and Performance in a Critical Driving Situation," *Biological Psychology*, vol. 69 (2005), pp. 195–203.

283 car was going to do: D. Lechner and G. Maleterre, "Emergency Maneuver Experimentation Using a Driving Simulator," Society of Automotive Engineers Technical Paper No. 910016, 1991; referenced in Dilich, Kopernik and Goebelbecker, op. cit.

283 "living room on wheels": Micheline Maynard, "At Chrysler, Home Depot Still Lingers," *New York Times*, October 30, 2007.

285 warnings he or she might disregard: See, for example, M. P. Manser, N. J. Ward, N. Kuge, and E. R. Boer, "Influence of a Driver Support System on Situation Awareness and Information Processing in Response to Lead Vehicle Braking," *Proceedings of the Human Factors and Ergonomics Society Forty-eighth Annual Meeting* (New Orleans, Human Factors and Ergonomics Society, 2004), pp. 2359–63, and "Crash Warning System Interfaces," DOT HS 810 697, January 2007.

285 be able to react accordingly: This is one of the problems that plague automation. Barry Kantowitz at the University of Michigan notes that automation "works fine up to a certain point, and then it fails utterly and completely." He uses the example of a plane crash in which the autopilot, in attempting to correct for an imbalance in fuel, tipped the plane to the point where the autopilot couldn't control it any longer. "So essentially it did the equivalent of ringing a bell and telling the pilot, 'Okay, you take over now,' " he says. "You have a pilot who's unaware there's a problem. He's 'out of the loop.' He has to very quickly figure out what the hell happened." But when people fail, they have what he calls a "graceful degradation. They fail slowly instead of abruptly. They can cope with it a little better." Design theorist Donald Norman gives a driving example in his book *The Design of Future Things*: A friend was driving with adaptive cruise control. This is the device that measures the distance away in time, in speed, of the vehicle in front, and keeps the car automatically at a safe distance. But, Norman notes, his friend suddenly moved to exit the freeway, forgetting the ACC was on. The car, thinking it suddenly had clear road ahead, chose to accelerate at the very moment it should have been decelerating. Automation is supposed to relieve the driver of having to pay attention, but in this case, if the driver hadn't been paying attention there would have likely been a severe crash. Norman argues that while full automation would be safer than human manual driving, the "difficulty lies in the transition towards full automation, when different vehicles will have different capabilities, when only some things will be automated, and when even the automation that is installed will be limited in capability." See Donald Norman, *The Design of Future Things* (New York: Basic Books, 2007), p. 116.

285 memory playing tricks): A group of psychologists at the University of Nottingham showed subjects a series of eight-second film clips of "dangerous" and "safer" situations that had been digitally manipulated to play at a range of faster or slower speeds (but always for eight seconds). Subjects were more likely to have judged the "dangerous" films as having been sped up. "If real dangerous events are remembered as if time slowed down," the authors write, "this will create an expectation that videos of such events should run slowly. . . . Because the actual speed of the video does not slow down, viewers will judge films of dangerous events as having been sped up." See Peter Chapman, Georgina Cox, and Clara Kirwan, "Distortion of Drivers' Speed and Time Estimates in Dangerous Situations," in *Behavioral Research in Road Safety* (London: Transport for London, 2005), pp. 164–74.

# Index

# PENGUIN PSYCHOLOGY

**BLINK: THE POWER OF THINKING WITHOUT THINKING**
MALCOLM GLADWELL

'Astonishing … *Blink* really does make you rethink the way you think' *Daily Mail*

An art expert sees a ten-million-dollar sculpture and instantly spots it's a fake. A marriage analyst knows within minutes whether a couple will stay together. A firefighter suddenly senses he has to get out of a blazing building …

This book is all about those moments when we 'know' something without knowing why. Malcolm Gladwell explores the phenomenon of 'blink', showing how a snap judgement can be far more effective than a cautious decision. By trusting your instincts, he reveals, you'll never think about thinking in the same way again …

**OUTLIERS: THE STORY OF SUCCESS**
MALCOLM GLADWELL

'Makes geniuses look a bit less special, and the rest of us a bit more so' *Time*

Why do some people achieve so much more than others? Can they lie so far out of the ordinary? What makes them exceptional?

In this provocative and inspiring book, Malcolm Gladwell looks at everyone from rock stars to professional athletes, software billionaires to scientific geniuses, to show that the story of success is far more surprising, and more fascinating, than we could ever have imagined. Gladwell reveals that success is as much about where we're from and what we do, as who we are – and that no one, not even a genius, ever makes it alone.

read more 

# PENGUIN BUSINESS/POLITICS

**NUDGE:**
**IMPROVING DECISIONS ABOUT HEALTH, WEALTH AND HAPPINESS**
RICHARD H. THALER & CASS R. SUNSTEIN

*Nudge* is the book that changes the way we think about choice, showing how we can influence people, improving decisions about health, wealth and happiness.

Using eye-opening real-life examples, Richard H. Thaler and Cass R. Sunstein show that no choice is ever presented in a neutral way.

The question is: when do we need a nudge in the right direction?

'I love this book. It is one of the few books I've read recently that fundamentally changes the way I think about the world' Steven D. Levitt, co-author of *Freakonomics*

'Hot stuff … an idea whose time seems to have come' *Sunday Times*

'Hugely influential … choice architects are everywhere' *Guardian*

'All the rage … the issue is not "to nudge or not to nudge"; it is how to nudge well' *Daily Telegraph*

read more 

# PENGUIN SCIENCE

---

**THE DRUNKARD'S WALK:**
**HOW RANDOMNESS RULES OUR LIVES**
LEONARD MLODINOW

Randomness and uncertainty surround everything we do.
So why are we so bad at understanding them?

Leonard Mlodinow exposes the truth about the success of sporting heroes and film stars, reveals the psychological illusions that prevent us under-standing everything from stock-picking to wine-tasting, winning the lottery to road safety and even explains how to make sense of a blood test.

*The Drunkard's Walk* is an exhilarating, eye-opening guide to understanding our random world – read it, so you won't be left a victim of chance.

'A wonderfully readable guide to how the mathematical laws of randomness affect our lives' Stephen Hawking, author of *A Brief History of Time*

'Delightful … Our lives may be shaped by chance, but they are enriched by awareness – just the sort of awareness that this fascinating book will give you' *Guardian*

'Mlodinow writes in a breezy style, interspersing probabilistic mind-benders with portraits of theorists … The result is a readable crash course in randomness' *The New York Times*

'Please read *The Drunkard's Walk* by Leonard Mlodinow, a history, explanation, and exaltation of probability theory … The results are mind-bending' *Fortune*

---

read more

# NASSIM NICHOLAS TALEB

**THE BLACK SWAN**
THE IMPACT OF THE HIGHLY IMPROBABLE

What have the invention of the wheel, Pompeii, the Wall Street Crash, Harry Potter and the internet got in common? Why should you never run for a train or read a newspaper? What can Catherine the Great's lovers tell us about probability?

This book is all about Black Swans: the random events that underlie our lives, from bestsellers to world disasters. Their impact is huge; they're nearly impossible to predict; yet after they happen we always try to rationalize them. A rallying cry to ignore the 'experts', *The Black Swan* shows us how to stop trying to predict everything and take advantage of uncertainty.

'An idiosyncratically brilliant new book' Niall Ferguson

'Great fun … brash, stubborn, entertaining, opinionated, curious, cajoling' Stephen J. Dubner, author of *Freakonomics*

**FOOLED BY RANDOMNESS**
THE HIDDEN ROLE OF CHANCE IN LIFE AND IN THE MARKETS

Everyone wants to succeed in life. But what causes some of us to be more successful than others? Is it really down to skill and strategy – or something altogether more unpredictable? This book is the word-of-mouth sensation that will change the way you think about business and the world. It is all about luck: more precisely, how we perceive luck in our personal and professional experiences. Nowhere is this more obvious than in the markets – we hear an entrepreneur has 'vision' or a trader is 'talented', but all too often their performance is down to chance rather than skill. It is only because we fail to understand probability that we continue to believe events are non-random, finding reasons where none exist. This irreverent bestseller has shattered the illusions of people around the world by teaching them how to recognize randomness. Now it can do the same for you.

'Brilliant' John Kay

# *He just wanted a decent book to read ...*

Not too much to ask, is it? It was in 1935 when Allen Lane, Managing Director of Bodley Head Publishers, stood on a platform at Exeter railway station looking for something good to read on his journey back to London. His choice was limited to popular magazines and poor-quality paperbacks – the same choice faced every day by the vast majority of readers, few of whom could afford hardbacks. Lane's disappointment and subsequent anger at the range of books generally available led him to found a company – and change the world.

*'We believed in the existence in this country of a vast reading public for intelligent books at a low price, and staked everything on it'*
***Sir Allen Lane, 1902–1970, founder of Penguin Books***

The quality paperback had arrived – and not just in bookshops. Lane was adamant that his Penguins should appear in chain stores and tobacconists, and should cost no more than a packet of cigarettes.

Reading habits (and cigarette prices) have changed since 1935, but Penguin still believes in publishing the best books for everybody to enjoy. We still believe that good design costs no more than bad design, and we still believe that quality books published passionately and responsibly make the world a better place.

So wherever you see the little bird – whether it's on a piece of prize-winning literary fiction or a celebrity autobiography, political tour de force or historical masterpiece, a serial-killer thriller, reference book, world classic or a piece of pure escapism – you can bet that it represents the very best that the genre has to offer.

**Whatever you like to read – trust Penguin.**